Crops: Growth, Yield and Quality

Crops: Growth, Yield and Quality

Edited by Stanley Hill

SYRAWOOD
PUBLISHING HOUSE
New York

Published by Syrawood Publishing House,
750 Third Avenue, 9th Floor,
New York, NY 10017, USA
www.syrawoodpublishinghouse.com

Crops: Growth, Yield and Quality
Edited by Stanley Hill

International Standard Book Number: 978-1-64740-063-7 (Hardback)

Cataloging-in-Publication Data

Crops : growth, yield and quality / edited by Stanley Hill.
p. cm.
Includes bibliographical references and index.
ISBN 978-1-64740-063-7
1. Crops. 2. Crops--Growth. 3. Crop yields. 4. Crops--Quality. 5. Crop science. 6. Agriculture. I. Hill, Stanley.
SB91 .C76 2022
630--dc23

TABLE OF CONTENTS

PREFACE

Crops are grown and harvested in agriculture for subsistence and profit. Some of the crops that contribute to human food supply are rice, wheat, sugarcane, corn, etc. The yield of a crop is measured per unit area of the land cultivated, and the seed generation of the crop. To improve yield, measures like the use of hybrid seeds, agrichemicals, genetically engineered crops and high mechanization are adopted. Intensive crop farming or the industrial production of crops involves innovation in agricultural machinery, genetic engineering technology, modern farming methods, etc. Food crops are also being designed with better nutrient profiles, quality and resistance to diseases, pests, spoilage, environmental conditions and chemical treatments with the aid of genetic engineering. Such genetically modified crops are being widely used by farmers, due to their high yields and reduction in the use of pesticides. This book discusses the fundamentals as well as modern approaches of crop production. It strives to provide a fair idea about crop growth, yield and quality, and help develop a better understanding of the latest advances within this field. For all readers who are interested in agriculture, the case studies included in this book will serve as an excellent guide to develop a comprehensive understanding.

After months of intensive research and writing, this book is the end result of all who devoted their time and efforts in the initiation and progress of this book. It will surely be a source of reference in enhancing the required knowledge of the new developments in the area. During the course of developing this book, certain measures such as accuracy, authenticity and research focused analytical studies were given preference in order to produce a comprehensive book in the area of study.

This book would not have been possible without the efforts of the authors and the publisher. I extend my sincere thanks to them. Secondly, I express my gratitude to my family and well-wishers. And most importantly, I thank my students for constantly expressing their willingness and curiosity in enhancing their knowledge in the field, which encourages me to take up further research projects for the advancement of the area.

Editor

1

Comparison between grafting and cutting as vegetative propagation methods for conilon coffee plants

Saul de Andrade Júnior[1*], Rodrigo Sobreira Alexandre[1], Edilson Romais Schmildt[1], Fábio Luiz Partelli[1], Maria Amélia Gava Ferrão[2] and Aldo Luiz Mauri[3]

*[1]Programa de Pós-graduação em Agricultura Tropical, Departamento de Ciências Agrárias e Biológicas, Centro Universitário Norte do Espírito Santo, Universidade Federal do Espírito Santo, Rod. BR 101 Norte, Km 60, 29932-540, Bairro Litorâneo, São Mateus, Espírito Santo, Brazil. [2]Empresa Brasileira de Pesquisa Agropecuária, Vitória, Espírito Santo, Brazil. [3]Instituto Capixaba de Pesquisa, Assistência Técnica e Extensão Rural, Fazenda Experimental de Marilândia, Marilândia, Espírito Santo, Brazil. *Author for correspondence. E-mail: saul_agronomo@hotmail.com*

ABSTRACT. The purpose of this study was to assess the growth of conilon coffee tree plantlets that were propagated by grafting and cutting. The experiment was conducted at the plantlet production site of Incaper's Experimental Farm in the city of Marilândia, Espírito Santo State. For grafting, plantlets derived from the seed propagation of *Coffea canephora* cv. Robusta Tropical (ENCAPER 8151) were used as rootstocks, and six clones of cv. Conilon Vitória (INCAPER 8142) were used as the grafts. The cutting was performed with six clones that were used for grafting. The experimental design consisted of randomized blocks of twelve treatments with five repetitions composed of twelve plantlets. On the hundred and fifth day, the averages of the variables were assessed and compared by the Scheffé test at a probability of 5%. The grafted plantlets were superior for almost all of the characteristics assessed, which suggests that it is possible to propagate conilon coffee trees.

Keywords: *Coffea canephora*, coffee, growth, cloning.

Introduction

Approximately 100 species of the genus *Coffea* have been described (DAVIS et al., 2006), of which only two, *C. arabica* (arabian coffee) and *C. canephora* (Robusta coffee), are economically exploited. In Brazil, almost all of the varieties grown are *C. canephora* cv. Conilon, which is better known as Robusta coffee (BRAGANÇA et al., 2001). Espírito Santo is the largest national producer of this variety (CONAB, 2012).

The 'Conilon' coffee tree is diploid with 2n = 22 chromosomes and self-incompatible, reproducing by crossed fertilization. This incompatibility is gametophytic (CONAGIN; MENDES, 1961; BERTHAUD, 1980). Because it is allogamous, it promotes differentiations in plant height, vegetative vigor, size, color and fruit shape (FERREIRA et al., 2005). Thus, when selecting progeny of the 'Conilon' coffee tree, genetic divergence studies, which take into account various characteristics including productivity, must be performed (FERREIRA et al., 2005; CECON et al., 2008; FERRÃO et al., 2008; NIKHILA et al., 2008; KITILA et al., 2011).

To ensure the permanence of specific characteristics of each genotype of the *C. canephora* coffee tree, cutting is used as a means of propagation (WEIGEL; JURGENS, 2002; PARTELLI et al., 2006b), creating populations of elite clones with desirable agronomic features including precocity, productive stability, architecture uniformity, higher productivity, fruit quality, and escalated cropping. The possibility of independent multiplication regardless of the time of the year was one of the fundamental factors to consolidate this technology; in addition, plants propagated through cutting exhibit higher productivity than plants propagated by seeds (JUNQUEIRA et al., 2006; PARTELLI et al., 2006b).

Although there are gains obtained through the planting of clones, problems that arise from the genetic narrowing of the species and shallower radicular systems can be observed in the field. A few clones in the field have shown allometric growth, exhibiting uneven relationships between their aerial parts and radicular systems and in many cases resulting in the death of the plants. Bragança et al. (2010) observed sigmoidal growth for conilon coffee where there is no balance between the foliar area and the total biomass of the plant.

Within the context of aiming to balance plant growth and ensure the permanence of the features obtained by vegetative propagation, grafting is a precise tool for adding a radicular system that has originated through sexual reproduction. Grafting effectively enables a new balancing of the plant's biomass with quick results in arabian coffee beans (OLIVEIRA et al., 2004), contributing to increased productivity (FAHL et al., 1998).

Thus, the purpose of the present study was to assess the growth of conilon coffee plantlets produced by grafting and cutting.

Material and methods

This experiment was conducted at the plantlet production site of the Incaper Experimental Farm in the city of Marilândia, Espírito Santo State, which is located at an altitude of 102 m, a latitude of 19°24'25" South and a longitude of 40°32'20" West. The climate classification proposed by Köppen is tropical moist with an annual average temperature of 24°C (FEITOSA et al., 1999).

Plantlets derived from the seed propagation of *C. canephora* cv. Robusta Tropical (ENCAPER 8151) were used as rootstocks and 13 clones of cv. Conilon Vitória (INCAPER 8142) were used for grafting (FERRÃO et al., 2007). Grafting was performed after selecting six diverging precocious (8 and 12v), average (2 and 7v) and late (5 and 13v) clones.

Seeding was performed directly into plastic bags. The plantlets were separated by two-centimeter bamboo rulers to increase the collection diameter more quickly. The planting mixture that was employed was composed of 70% underground earth and 30% coffee straw; for each cubic meter, 2 kg of dolomitic limestone, 5 kg of simple super phosphate and 1 kg of potassium chloride were added (PREZOTTI et al., 2007). This standard mixture is still widely used in the production of coffee plantlets due to its chemical composition and easily purchasable components (DIAS; MELO, 2009).

An analysis of the mixture was performed after it was evenly combined, and it was verified to be at pH 6.0. It contained 1.1 dag kg^{-1} of organic matter and 550, 31.4, 56, 40, 1.7 and 49 mg dm^{-3} of K, Zn, Fe, Mn, Cu and B, respectively. The concentrations of Ca, Mg, Al, H+Al and CTC (t) were 5.4, 1.1, 0.0, 1.4, 7.9 and 7.9 $cmol_c\ dm^{-3}$, respectively, and 85% of bases were saturated.

The forking, filled gap mode method of grafting was employed and was performed 90 days after the seeding of 'Robusta Tropical' when the plantlets already presented a diameter that was compatible with the grafts. Concomitant with the grafting process, the planting of the rootstocks of the precocious (8 and 12v), average (2 and 7v) and late (5 and 13v) maturation 'Conilon Vitória' clones was performed.

The plantlets, after being grafted to the ones derived from cutting, were stored under a 50% shading roof with an intermittent nebulization irrigation system to ensure proper conditions of light and moisture for their growth.

When the plantlets had two pairs of leaves, they were individually identified and grouped in blocks under a palm tree cover; the cover was later removed slowly until full sunlight was achieved to ensure that they were ready for definitive planting 150 days after grafting.

When the plantlets were under controlled climate conditions, their assessments were performed. Plant height (cm) was measured using a millimeter-graded ruler; stem diameter was measured with the aid of a digital caliper; foliar area (cm^2) was measured with the (Licor Inc., Lincon, Nebrasca, US), model LI-COR LI-3000 the radicular system volume (mL) was measured with a graduated cylinder; and masses of the fresh matter of the leaves (g), stems (g), stem of the primary graft (g), roots (g) and final roots (g) and the total fresh matter (g) were quantified using an Ohaus Adventurer^{TM} model ARD 110 digital scale with a precision of two decimal places.

The relationship between the mass of the dry matter of the aerial parts and the dry matter of the root, the RPAR, and the relationship between the height of the aerial parts and their diameter, the

RAD, were collected. The Dickson quality index (DQI), obtained by the equation [total dry matter mass/(RAD+RPAR)] (DICKSON et al., 1960), was verified. To determine the mass of the dry matter of the leaves (g), stems (g), primary graft stems (g) and roots (g), these structures were collected individually dully individualized, placed in labeled paper bags and subsequently conditioned in a forced circulation chamber at 70°C for 72 hours until they achieved a constant weight.

The experimental design consisted of randomized blocks of twelve treatments and five plots, each composed of twelve plantlets. The averages of the treatments were submitted to an analysis of variance. If significant differences were detected between the treatments, the averages were compared using the Scheffé test at a probability of 5% (STEEL et al., 1997), which compared the grafting and grafted treatment groups. In an attempt to investigate which propagated material was superior in growth or which material would exhibit unsatisfactory growth, thus indicating a potential incompatibility between the graft and the rootstock, the Scheffé test was performed within the cloned and grafted groups and also by a joint comparison of the precocious (8 and 12v), average (2 and 7v) and late (5 and 13v) maturation degrees.

All of the analyses were performed with the assistance of the GENES (CRUZ, 2013) computer program.

Results and discussion

Thirty days after grafting, all of the plantlets exhibited welding in the grafted area, and when a randomly chosen plantlet was lifted using the graft, it was able to support its own weight showing a strong connection with the rootstock. Such observations agree with Dickson (2000), who stated that welding occurs within a few days; in this initial stage, there is a formation of parenchymal cells at the graft's interface, which fills the gap at the grafting spot through hardening and promotes the junction of the graft with the rootstock.

Grafting resulted in junctions 91.5% of the time and demonstrated promise as a method for the propagation of conilon coffee plants, whereas the traditional cutting system resulted in junctions 96.8% of the time.

The grafted plantlet was determined to be superior if it was comparable to the cutting for the following characteristics: foliar area; leaf, stem, primary graft stem, aerial parts, root and total fresh matter masses; root volume; the number of leaves; plant height; the relationship between plant height and stem diameter; and the Dickson quality index. For the stem diameter, no significant difference was detected between the cloned and grafted groups. For the relationship between the dry matter mass of the aerial parts and the dry matter mass of the root, the averages of the cloned group were greater than the averages of the grafted group as shown in Table 1.

Table 1. Averages of the foliar area variables (FA), fresh leaf matter mass (FLMM), dry leaf matter mass (DLMM), fresh stem matter mass (FSMM), dry stem matter mass (DSMM), primary graft stem fresh matter mass (PGSFMM), primary graft stem dry matter mass (PGSDMM), aerial parts fresh matter mass (APFMM), aerial parts dry matter mass (APDMM), fresh root matter mass (FRMM), dry root matter mass (DRMM), thin root dry matter mass (TRDMM), root volume (RV), total fresh matter mass (TFMM), total dry matter mass (TDMM), the number of leaves (NL), plant height (PH), stem diameter (SD), the relationship between plant height and stem diameter (HDR), the relationship between the aerial parts dry matter mass and the root dry matter mass (APRR) and the Dickson quality index (DQI) for the grafting and cutting propagation methods(1).

Variables	Propagation methods		$\lvert \hat{C}_i \rvert$	S the 5%
	Grafting	Cutting		
FA (cm^2)	187.400	298.700	-111.4	26.680*
FLMM (g)	4.864	7.439	-2.575	0.725*
DLMM (g)	1.230	1.988	-0.758	0.225*
FSMM (g)	3.583	6.238	-2.655	0.427*
DSMM (g)	1.037	2.096	-1.059	0.165*
PGSFMM (g)	2.154	3.800	-1.646	0.241*
PGSDMM (g)	0.705	1.490	-0.786	0.103*
APFMM (g)	8.448	13.670	-5.230	1.076*
APDMM (g)	2.268	4.085	-1.817	0.369*
FRMM (g)	3.466	7.604	-4.138	1.487*
DRMM (g)	0.822	2.149	-1.328	0.220*
TRDMM (g)	0.339	0.870	-0.530	0.135*
RV (cm^3)	3.550	7.248	-3.698	1.444*
TFMM (g)	11.910	21.283	-9.369	2.031*
TDMM (g)	3.089	6.234	-3.145	0.564*
NL	9.481	10.940	-1.457	1.064*
PH (cm)	6.710	9.475	-2.765	1.145*
SD (mm)	3.171	3.324	-0.153	0.196^{ns}
HDR	2.122	2.859	-0.737	0.378*
APRR	2.774	1.913	0.861	0.195*
DQI	0.633	1.323	-0.689	0.148*

(1) ^{ns}indicates non-significant and * indicates significant differences when the grafting and cutting propagation methods are compared with the Scheffé test at a probability of 5%.

The superior results of grafting propagation when compared with cloned propagation can be explained by the following fact: at the moment of grafting on day 90, there is already a rootstock radicular system that has exhibited satisfactory growth. Thus, it is possible that the entire reserve of the graft was directed to produce biomass for the aerial parts, ensuring their expressive growth against the same clone materials. Other authors (FAHL et al., 1998; TOMAZ et al., 2006, 2008) that have studied the effects of the *C. canephora* rootstock on the growth of grafts of *C. arabica* observed an increased growth of the aerial parts, longer radicular lengths and larger volumes even when Catuaí Vermelho IAC 15 was combined with rootstocks 'ES26' and 'ES23', which were effective and produced dry matter when magnesium was applied.

The influence of the radicular system on the plant's growth and tolerance to environmental stress has become evident with the use of the reciprocal grafting technique (HARTMANN et al., 2002). For example, it was verified that grafting dry-sensitive materials with rootstocks that are tolerant to dry environments results in an increased tolerance to dryness and the effective use of water by sensitive plants (SILVA et al., 2010).

The grafted plantlets exhibited better results for all of the radicular system variables when compared with the same genotypes propagated by cutting (Table 1). The rhizogenic process resulted in a greater thin root dry matter mass in grafted plantlets (0.87 g) when compared with the plantlets propagated by cutting (0.44 g). This feature is fundamental for the initial growth of the plantlets because it ensures a higher efficiency of water and nutrient absorption (HARMAND et al., 2004), and according to Laclau et al. (2001), the spatial distribution of some roots determines their high capacity for nutrient absorption. However, robust plantlets that exhibit a higher percentage of root emission are better able to withstand environmental stress conditions, ensuring higher post-planting survival rates (FREITAS et al., 2005).

The root volume varied between the grafted materials (7.24 cm^3) and the plantlets propagated by cutting (3.55 cm^3), which can be partially explained by the characteristic of the roots derived from seed propagation. It may also be an indication of higher exploitation of the radicular system. The main physiological mechanisms of differential tolerance to dryness among the 'Conilon' genotypes are already known; these mechanisms are governed by the effective extraction of water from the ground and by the water usage rates of the plants (DAMATTA et al., 2003; PINHEIRO et al., 2005; DAMATTA; RAMALHO, 2006). Thus, grafted plants can be more adaptable to adverse field conditions after they have developed a larger radicular system.

To perform agronomic and physiological studies involving vegetal growth, it is necessary to know the characteristics of the aerial parts of the plant because the leaf is the location of the synthesis of photo-assimilated elements that are important for root growth (BLANCO; FOLEGATTI, 2003); the foliar area is one of the most important of these characteristics (FAVARIN et al., 2002; PARTELLI et al., 2006a).

Table 2 shows the characteristics of the aerial parts, such as foliar area, the number of leaves, and the dry matter masses of the leaf and aerial parts that were presented for grafting. They are related to a higher luminous absorption surface, which results in a carbon fixation gain by the plant; that is, a higher foliar area gives the plant a greater capacity to produce and store photo-assimilated elements. These observations support results obtained by Pauletto et al. (2001), Hartmann et al. (2002), who stated that factors which govern the absorption and translocation of water, nutrients and endogenous substances favor the growth of the aerial parts.

Understanding growth characteristics and their interactions will determine how close are to the ideal plantlet production. In this sense, stem diameter, plant height and the relationship between stem diameter and plant height varied between the grafting and cutting propagation systems, from 3.17 to 3.32 mm, 6.7 to 9.5 cm and 2.1 to 2.8, respectively. Stem diameter was the only variable that did not exhibit a difference as shown in Table 1. Similar values were obtained by Marana et al. (2008) for *C. arabica*. One of the important characteristics that determines the quality of the plantlet is the collection diameter, which represents a 10.4% contribution, and has the advantage of not being a destructive method (GOMES et al., 2002). According to Zonta et al. (2009), lower collection diameters are associated with longer watering shifts. The shoot height and diameter are easy to measure, non-destructive and technically accepted as a good measure of the performance potential of seedlings. The resulting index of the relationship between shoot height and stem diameter expresses the balance of growth, also called the robustness quotient, and is considered to be one of the most accurate indexes because it provides information on how thin the seedlings are.

The Dickson quality index is a well-established morphological measure. It is a good standard measure of plantlet quality because its calculation takes the robustness and distribution balance of the biomass into account, and it also assesses various important morphological characteristics (DICKSON et al., 1960). The present study found values of 0.63 for cutting propagations and 1.32 for grafting propagations. According to Dickson et al. (1960), the higher the value, the better the quality of the plantlets. In this case, quality indicates a better balance between the shoot and root systems, resulting in more seedlings adapted to the harsh conditions of the field, which ensures lower rates of death after planting.

Non-significant interactions were observed for both grafting and cutting among the precocious, average and later maturation season groups for all characteristics assessed on Tables 2 and 3, except for the foliar area in Table 2. The significant interactions between the precocious and average groups within the cutting treatment may have occurred because materials 8v and 12v were more vigorous, which was highlighted in the plantlet production. It implies vigorous seedlings that in the field adapted better to the adverse conditions of climate and soil.

Table 2. Averages of the foliar area variables (FA), fresh leaf matter mass (FLMM), dry leaf matter mass (DLMM), fresh stem matter mass (FSMM), dry stem matter mass (DSMM), primary graft stem fresh matter mass (PGSFMM), primary graft stem dry matter mass (PGSDMM), aerial parts fresh matter mass (APFMM), aerial parts dry matter mass (APDMM), fresh root matter mass (FRMM), dry root matter mass (DRMM), thin root dry matter mass (TRDMM), root volume (RV), total fresh matter mass (TFMM), total dry matter mass (TDMM), the number of leaves (NL), plant height (PH), stem diameter (SD), the relationship between plant height and stem diameter (HDR), the relationship between the aerial parts dry matter mass and the root dry matter mass (APRR) and the Dickson quality index (DQI) for plants propagated by the cutting method. A joint comparison was performed among the precocious (8 and 12v), average (2 and 7v) and late (5 and 13v) maturation groups(1).

Variables	Cutting			$\lvert \hat{C}_i \rvert$	S the 5%
	Maturation season				
	Precocious	Average	Late		
FA (cm^2)	205.546	158.021	-	47.519	46.216*
FLMM (g)	5.222	4.099	-	1.122	1.256^{ns}
DLMM (g)	1.346	1.039	-	0.307	0.390^{ns}
FSMM (g)	3.830	3.281	-	0.549	0.740^{ns}
DSMM (g)	1.109	0.939	-	0.170	0.285^{ns}
PGSFMM (g)	2.355	2.047	-	0.307	0.419^{ns}
PGSDMM (g)	0.755	0.661	-	0.095	0.179^{ns}
APFMM (g)	9.052	7.381	-	1.671	1.864^{ns}
APDMM (g)	2.455	1.978	-	0.476	0.639^{ns}
FRMM (g)	3.683	3.082	-	0.601	2.576^{ns}
DRMM (g)	0.861	0.769	-	0.092	0.381^{ns}
TRDMM (g)	0.371	0.312	-	0.059	0.235^{ns}
RV (cm^3)	3.666	3.017	-	0.649	2.501^{ns}
TFMM (g)	12.735	10.463	-	2.272	3.518^{ns}
TDMM (g)	3.316	2.747	-	0.569	0.977^{ns}
NL	10.150	8.650	-	1.500	1.843^{ns}
PH (cm)	6.918	6.137	-	0.781	1.983^{ns}
SD (mm)	3.304	3.110	-	0.194	0.339^{ns}
HDR	2.095	1.986	-	0.109	0.654^{ns}
APRR	2.845	2.639	-	0.206	0.338^{ns}
DQI	0.667	0.605	-	0.062	0.257^{ns}
FA (cm^2)	205.546	-	198.549	6.996	46.216^{ns}
FLMM (g)	5.222	-	5.273	-0.051	1.257^{ns}
DLMM (g)	1.346	-	1.306	0.039	0.391^{ns}
FSMM (g)	3.830	-	3.639	0.192	0.740^{ns}
DSMM (g)	1.109	-	1.063	0.046	0.285^{ns}
PGSFMM (g)	2.355	-	2.063	0.292	0.419^{ns}
PGSDMM (g)	0.756	-	0.698	0.057	0.179^{ns}
APFMM (g)	9.052	-	8.912	0.140	1.864^{ns}
APDMM (g)	2.455	-	2.369	0.085	0.639^{ns}
FRMM (g)	3.683	-	3.634	0.049	2.576^{ns}
DRMM (g)	0.861	-	0.835	0.026	0.381^{ns}
TRDMM (g)	0.371	-	0.336	0.035	0.235^{ns}
RV (cm^3)	3.666	-	3.967	-0.301	2.501^{ns}
TFMM (g)	12.735	-	12.545	0.190	3.518^{ns}
TDMM (g)	3.316	-	3.205	0.111	0.977^{ns}
NL	10.150	-	9.644	0.506	1.843^{ns}
PH (cm)	6.918	-	7.077	-0.159	1.983^{ns}
SD (mm)	3.304	-	3.098	0.206	0.339^{ns}
HDR	2.095	-	2.285	-0.190	0.654^{ns}
APRR	2.845	-	2.838	0.007	0.338^{ns}
DQI	0.667	-	0.628	0.039	0.257^{ns}
FA (cm^2)	-	158.026	198.549	-40.523	46.216^{ns}
FLMM (g)	-	4.099	5.273	-1.173	1.257^{ns}
DLMM (g)	-	1.039	1.306	-0.267	0.391^{ns}
FSMM (g)	-	3.281	3.638	-0.357	0.740^{ns}
DSMM (g)	-	0.939	1.063	-0.124	0.285ns

Continue...

...Continuation

Variables	Cutting Maturation season Precocious	 Average	 Late	$\lvert\hat{C}_i\rvert$	S the 5%
PGSFMM (g)	-	2.047	2.063	-0.015	0.419^{ns}
PGSDMM (g)	-	0.661	0.698	-0.037	0.179^{ns}
APFMM (g)	-	7.381	8.912	-1.531	1.864^{ns}
APDMM (g)	-	1.978	2.369	-0.391	0.639^{ns}
FRMM (g)	-	3.082	3.634	-0.552	2.576^{ns}
DRMM (g)	-	0.769	0.835	-0.066	0.381^{ns}
TRDMM (g)	-	0.312	0.336	-0.024	0.235^{ns}
RV (cm^3)	-	3.017	3.967	-0.950	2.501^{ns}
TFMM (g)	-	10.460	12.545	-2.082	3.518^{ns}
TDMM (g)	-	2.747	3.205	-0.458	0.978^{ns}
NL	-	8.650	9.644	-0.994	1.843^{ns}
PH (cm)	-	6.137	7.077	-0.940	1.983^{ns}
SD (mm)	-	3.110	3.098	0.012	0.339^{ns}
HDR	-	1.986	2.285	-0.299	0.654^{ns}
APRR	-	2.639	2.838	-0.199	0.338^{ns}
DQI	-	0.605	0.628	-0.023	0.257^{ns}

(1)^{ns} indicates non-significant and * indicates significant differences when comparisons are made within the cutting propagation method by the Scheffé test at a probability of 5%; the precocious, average and late maturation times were taken into account.

Table 3. Averages of the foliar area variables (FA), fresh leaf matter mass (FLMM), dry leaf matter mass (DLMM), fresh stem matter mass (FSMM), dry stem matter mass (DSMM), primary graft stem fresh matter mass (PGSFMM), primary graft stem dry matter mass (PGSDMM), aerial parts fresh matter mass (APFMM), aerial parts dry matter mass (APDMM), fresh root matter mass (FRMM), dry root matter mass (DRMM), thin root dry matter mass (TRDMM), root volume (RV), total fresh matter mass (TFMM), total dry matter mass (TDMM), the number of leaves (NL), plant height (PH), stem diameter (SD), the relationship between plant height and stem diameter (HDR), the relationship between the aerial parts dry matter mass and the root dry matter mass (APRR) and the Dickson quality index (DQI) for plants that were propagated by the grafting method. A joint comparison was performed among the precocious (8 and 12v), average (2 and 7v) and late (5 and 13v) maturation groups$_{(1)}$.

Variables	Grafting Maturation season Precocious	 Average	 Late	$\lvert\hat{C}_i\rvert$	S the 5%
FA (cm^2)	292.817	305.369	-	-12.552	46.216^{ns}
FLMM (g)	7.569	7.324	-	0.245	1.257^{ns}
DLMM (g)	2.025	1.936	-	0.089	0.391^{ns}
FSMM (g)	6.242	6.159	-	0.083	0.740^{ns}
DSMM (g)	2.110	2.058	-	0.052	0.285^{ns}
PGSFMM (g)	3.865	3.618	-	0.247	0.419^{ns}
PGSDMM (g)	1.515	1.437	-	0.078	0.179^{ns}
APFMM (g)	13.812	13.483	-	0.329	1.864^{ns}
APDMM (g)	4.135	3.994	-	0.141	0.639^{ns}
FRMM (g)	7.601	7.723	-	-0.122	2.576^{ns}
DRMM (g)	2.253	2.001	-	0.252	0.381^{ns}
TRDMM (g)	0.923	0.832	-	0.091	0.235^{ns}
RV (cm^3)	7.326	7.300	-	0.026	2.501^{ns}
TFMM (g)	21.413	21.205	-	0.208	3.518^{ns}
TDMM (g)	6.388	5.995	-	0.393	0.978^{ns}
NL	11.610	9.988	-	1.622	1.843^{ns}
PH (cm)	9.028	10.289	-	-1.261	1.983^{ns}
SD (mm)	3.380	3.323	-	0.057	0.339^{ns}
HDR	2.676	3.108	-	-0.432	0.654^{ns}
APRR	1.839	2.007	-	-0.168	0.338^{ns}
DQI	1.42	1.183	-	0.237	0.257^{ns}
FA (cm^2)	292.817	-	298.041	-5.224	46.216^{ns}
FLMM (g)	7.569	-	7.425	0.144	1.257^{ns}
DLMM (g)	2.025	-	2.004	0.021	0.391^{ns}
FSMM (g)	6.242	-	6.316	-0.073	0.740^{ns}
DSMM (g)	2.110	-	2.122	-0.011	0.285^{ns}
PGSFMM (g)	3.865	-	3.919	-0.054	0.419^{ns}
PGSDMM (g)	1.515	-	1.519	-0.003	0.179^{ns}
APFMM (g)	13.812	-	13.741	0.071	1.864^{ns}
APDMM (g)	4.135	-	4.125	0.010	0.639^{ns}
FRMM (g)	7.601	-	7.489	0.112	2.576^{ns}
DRMM (g)	2.253	-	2.195	0.058	0.381^{ns}
TRDMM (g)	0.923	-	0.855	0.068	0.235^{ns}
RV (cm^3)	7.326	-	7.117	0.209	2.501^{ns}
TFMM (g)	21.413	-	21.231	0.182	3.518^{ns}
TDMM (g)	6.388	-	6.321	0.067	0.978^{ns}
NL	11.610	-	11.216	0.394	1.843^{ns}
PH (cm)	9.028	-	9.109	-0.081	1.983^{ns}
SD (mm)	3.380	-	3.268	0.112	0.339^{ns}
HDR	2.676	-	2.793	-0.117	0.654^{ns}
APRR	1.839	-	1.892	-0.053	0.338^{ns}

Continue...

...Continuation

Variables	Grafting			$\lvert \hat{C}_i \rvert$	S the 5%
	Maturation season				
	Precocious	Average	Late		
DQI	1.420	-	1.366	0.054	0.257[ns]
FA (cm²)	-	305.369	298.041	7.328	46.216[ns]
FLMM (g)	-	7.324	7.425	-0.101	1.257[ns]
DLMM (g)	-	1.936	2.004	-0.068	0.391[ns]
FSMM (g)	-	6.159	6.316	-0.157	0.740[ns]
DSMM (g)	-	2.058	2.122	-0.063	0.285[ns]
PGSFMM (g)	-	3.618	3.919	-0.301	0.419[ns]
PGSDMM (g)	-	1.437	1.519	-0.081	0.179[ns]
APFMM (g)	-	13.483	13.741	-0.258	1.864[ns]
APDMM (g)	-	3.994	4.125	-0.131	0.639[ns]
FRMM (g)	-	7.723	7.489	0.234	2.576[ns]
DRMM (g)	-	2.001	2.195	-0.194	0.381[ns]
TRDMM (g)	-	0.832	0.855	-0.023	0.235[ns]
RV (cm³)	-	7.300	7.117	0.183	2.501[ns]
TFMM (g)	-	21.205	21.231	-0.026	3.518[ns]
TDMM (g)	-	5.995	6.321	-0.326	0.978[ns]
NL	-	9.988	11.216	-1.228	1.843[ns]
PH (cm)	-	10.289	9.109	1.180	1.983[ns]
SD (mm)	-	3.323	3.268	0.055	0.339[ns]
HDR	-	3.108	2.793	0.315	0.654[ns]
APRR	-	2.007	1.892	0.115	0.338[ns]
DQI	-	1.183	1.366	-0.183	0.257[ns]

(1)[ns] indicates non-significant and *indicates significant differences when comparisons are made ,within the grafting propagation method with the Scheffé test at a probability of 5%; the precocious, average and late maturation times were taken into account.

Although there was divergence between the genotypes with regard to height and plant architecture, color, leaf and fruit size and even in productivity, they behaved homogeneously with regard to growth as adults independent of their propagation method and degree of maturation as shown in Tables 2 and 3. For the clones of conilon coffee studied here, the adult plants have many different characteristics but the seedlings do not because they proved very homogeneous in their growth independent of the propagation system.

The cultivation of coffee, which has been commercially ongoing and exploited for many years, justifies the care taken during the production of the plantlets. Plantlets of good origins and quality are vital to the achievement of a long-lasting, productive crop and to enable the amortization of implementation costs with greater efficiency and sustainability. Therefore, the technology proposed in the present study, which consists of grafting segments of mature *C. canephora* plants onto *C. canephora* rootstocks that were propagated by seed, produces seedlings with root systems of greater mass and volume. Thus, the seedlings exhibit vigorous growth, which is certainly reflected in the number of field plants that can tolerate unfavorable abiotic conditions.

Conclusion

Based on the growth data, the method of grafting that changes the grafted cleft at the top results in perfect compatibility between the aerial parts and the rootstock.

The forting, full gap mode grafting method was found to be more effective in the production of plantlets when compared with propagation by cutting.

References

BERTHAUD, J. L. Incompatibilité chez *Coffea canephora* méthode de test et déterminisme génétique. **Café, Cacao, Thé**, v. 24, n. 1, p. 167-174, 1980.

BLANCO, F. F.; FOLEGATTI, M. V. A new method for estimating the leaf area index of cucumber and tomato plants. **Horticultura Brasileira**, v. 21, n. 4, p. 666-669, 2003.

BRAGANÇA, S. M.; MARTINES, H. E. P.; LEITE, H. G.; SANTOS, L. P.; LANI, J. A.; SEDIYAMA, C. S.; ALVAREZ, V. H. Acumulação de matéria seca pelo cafeeiro conilon. **Revista Ceres**, v. 57, n. 1, p. 48-52, 2010.

BRAGANÇA, S. M.; CARVALHO, C. H. S.; FONSECA, A. F. A.; FERRÃO, R. G. Variedades clonais de café conilon para o Estado do Espírito Santo. **Pesquisa Agropecuária Brasileira**, v. 36, n. 5, p. 765-770, 2001.

CECON, P. R.; SILVA, F. F.; FERREIRA, A.; FERRÃO, R. G.; POLICARPO, A. S. C.; DETMANN, E.; FARIA, P. N.; MORAIS, T. S. S. Análise de medidas repetidas na avaliação de clones de café 'Conilon'. **Pesquisa Agropecuária Brasileira**, v. 43, n. 9, p. 61-69, 2008.

CONAB-Companhia Nacional de Abastecimento. **1ª estimativa de safra/2012**. Brasília, 2012. Available from: <http://www.conab.gov.br/conteudos.php?a=1253&t=>. Access on: Mar. 13, 2012.

CONAGIN, C. H. T. M.; MENDES, A. J. T. Pesquisas citológicas e genéticas em três espécies de *Coffea*: autoincompatibilidade em *Coffea canephora*. **Bragantia**, v. 20, n. 34, p. 787-804, 1961.

CRUZ, C. D. Genes: a software package for analysis in experimental statistics and quantitative genetics. **Acta Scientiarum. Agronomy**, v. 35, n. 3, p. 271-276, 2013.

DAMATTA, F. M.; RAMALHO, J. D. C. Impacts of drought and temperature stress on coffee physiology and production: a review. **Brazilian Journal of Plant Physiology**, v. 18, n. 1, p. 55-81, 2006.

DAMATTA, F. M.; CHAVES, A. R. M.; PINHEIRO, H. A.; DUCATTI, C.; LOUREIRO, M. E. Drought tolerance of two field-grown clones of *Coffea canephora*. **Plant Science**, v. 164, n. 1, p. 111-117, 2003.

DIAS, R.; MELO. B. Proporção de material orgânico no substrato artificial para a produção de mudas de cafeeiro em tubetes. **Ciência e Agrotecnologia**, v. 33, n. 1, p. 44-152, 2009.

DICKSON, W. C. **Integrative plant anatomy**. Hartcourt: Academic Press, 2000.

DICKSON, A.; LEAF, A. L.; HOSNER, J. F. Quality appraisal of white spruce and white pine seedling stock in nurseries. **Forest Chronicle**, v. 36, n. 1, p. 10-13, 1960.

DAVIS, A. P.; GOVAERTS, R.; BRIDSON, D. M.; STOFFELEN, P. An annotated taxonomic conspectus of the genus *Coffea* (Rubiaceae). **Botanical Journal of the Linnean Society**, v. 152, n. 4, p. 465-512, 2006.

FAHL, J. I.; CARELLI, M. L. C.; GALLO, P. B.; COSTA, W. M.; NOVO, M. C. S. S. Enxertia de *Coffea arabica* sobre progênies de *C. canephora* e de *C. congensis* no crescimento, nutrição mineral e produção. **Bragantia**, v. 57, n. 2, p. 297-312, 1998.

FAVARIN, J. L.; DOURADO, D. N.; GARCÍA, A.; NOVA, N. A. V.; FAVARIN, M.; GRAÇA G. V. Equações para a estimativa do índice de área foliar do cafeeiro. **Pesquisa Agropecuária Brasileira**, v. 37, n. 6, p. 769-773, 2002.

FEITOSA, L. R.; CASTRO, L. L. F.; RESENDE, M.; ZANGRANDE, M. B.; STOCKING, M. S.; BOREL, R. M. A.; FULLIN, E. A.; CERQUEIRA, A. F.; SALGADO, J. S.; FEITOZA, H. N.; STOCK, L. A.; MANK, A. M.; DESSAUNE FILHO, N.; MANK, A. M.; FERINGA, W.; MARTINEZ, J. A. **Mapa das Unidades Naturais do Espírito Santo**. Vitória: Emcapa, 1999. (Mapa na escala 1:400 000, Colorido).

FERRÃO, R. G.; CRUZ, C. D.; FERREIRA, A.; CECON, P. R.; FERRÃO, M. A. G.; FONSECA, A. F. A.; CARNEIRO, P. C. S.; SILVA, M. F. Parâmetros genéticos em café conilon. **Pesquisa Agropecuária Brasileira**, v. 43, n. 1, p. 61-69, 2008.

FERRÃO, R. G.; FONSECA, A. F. A.; FERRÃO, M. A. G.; BRAGANÇA, S. M.; VERDIN FILHO, A. C.; VOLPI, P. S. Culivares de café conilon. In: FERRÃO, R. G.; FONSECA, A. F. A.; BRAGANÇA, S. M.; FERRÃO, M. A. G.; DEMUNER, L. H. (Ed.). **Café Conilon**. Vitória: Incaper, 2007. p. 203-225.

FERREIRA, A.; CECON, P. R.; CRUZ, C. D.; FERRÃO, R. G.; SILVA, M. F.; FONSECA, A. F. A.; FERRÃO, M. A. G. Seleção simultânea de *Coffea canephora* por meio da combinação de análise de fatores e índices de seleção. **Pesquisa Agropecuária Brasileira**, v. 40, n. 12, p. 1189-1195, 2005.

FREITAS, T. A. S.; BARROSO, G. D.; CARNEIRO, A. G. J; PENCHEL, M. R.; LAMÔNICA, R. K.; FERREIRA A. D. Desempenho radicular de mudas de eucalipto produzidas em diferentes recipientes e substratos. **Revista Árvore**, v. 29, n. 6, p. 853-861, 2005.

GOMES, J. M.; COUTO, L.; LEITE, H. G.; XAVIER, A.; GARCIA, S. L. R. Parâmetros morfológicos na avaliação da qualidade de mudas de *Eucalyptus grandis*. **Revista Árvore**, v. 26, n. 6, p. 655-664, 2002.

HARMAND, J. M.; NJITI, C. F.; REVERSAT, F. B.; PUING, H. Aboveground and belowground biomass, productivity and nutrient accumulation in tree improved fallows in the dry tropics of Cameroon. **Forest Ecology and Management**, v. 188, n. 1, p. 249-265, 2004.

HARTMANN, H. T.; KESTER, D. E.; DAVIES, F. T.; GENEVE, R. L. **Plant propagation**: principles and practices. 7th ed. Upper Sanddle River: Prentice Hall, 2002.

JUNQUEIRA, N. T. V.; LAGE, D. A. C.; BRAGA, M. F.; PEIXOTO, J. R.; SILVA, D. M.; BORGES, T. A.; ANDRADE, S. R. M. Reação a doença e produtividade de um clone de maracujazeiro-azedo propagado por estaquia e enxertia em estacas herbáceas de *Passiflora* silvestre. **Revista Brasileira de Fruticultura**, v. 28, n. 1, p. 97-100, 2006.

KITILA, O.; ALAMEREW, S.; KUFA. T.; GAREDEW, W. Variability of quantitative traits in limmu coffee (*Coffea arabica* L.) in Ethiopia. **International Journal of Agricultural Research**, v. 6, n. 6, p. 482-493, 2011.

LACLAU, J. P.; ARNAUD, M.; BOUILLET, J. P.; RANGER, J. Spatial distribution of *Eucapyptus* roots in a deep sandy soil in the Congo: relationships with the ability of the stand to take up water and nutrients. **Tree Physiology**, v. 21, n. 1, p. 129-136, 2001.

MARANA, J. P.; MIGLIORANZA, E.; FONSECA, E. P.; KAINUMA, R. H. Índices de qualidade e crescimento de mudas de café produzidas em tubetes. **Ciência Rural**, v. 38, n. 1, p. 39-45, 2008.

NIKHILA, K. R.; SURESHKUMAR, V. B.; MOHANAN, K. V.; SANTARAM, A. Association of agronomic characters in robusta coffee (*Coffea canephora* Pierre ex Froehner). **International Journal of Plant Breeding and Genetics**, v. 2, n. 1, p. 47-50, 2008.

OLIVEIRA, A. L.; GUIMARÃES, R. J.; SOUZA, C. A. S.; CARVALHO, J. A.; MENDES, A. N. G.; GUIMARÃES, R. S. Desenvolvimento de cafeeiros (*Coffea arabica* L.) enxertados submetidos a diferentes níveis de reposição de água. **Ciência e Agrotecnologia**, v. 28, n. 6, p. 1291-1298, 2004.

PAULETTO, D.; MOURÃO, F.; ALVES, F. A.; KLUGE, R. A.; SCARPARE, F.; ALEXIO, J. Produção e vigor da videira 'Niágara Rosada' relacionados com o porta-enxerto. **Pesquisa Agropecuária Brasileira**, v. 36, n. 1, p. 115-121, 2001.

PARTELLI, F. L.; VIEIRA, H. D.; DETMANN, E.; CAMPOSTRINI, E. Estimativa da área foliar do cafeeiro conilon a partir do comprimento da folha. **Revista Ceres**, v. 53, n. 306, p. 204-210, 2006a.

PARTELLI, F. L.; VIEIRA, H. D.; SANTIAGO, A. R.; BARROSO, D. G. Produção e desenvolvimento radicular de plantas de café 'Conilon' propagadas por sementes e por estacas. **Pesquisa Agropecuária Brasileira**, v. 41, n. 6, p. 949-954, 2006b.

PINHEIRO, H. A.; DAMATTA, F. M.; CHAVES, A. R. M.; LOUREIRO, M. E.; DUCATTI, C. Drought tolerance is associated with rooting depth and stomatal control of water use in clones of *Coffea canephora*. **Annals of Botany**, v. 96, n. 1, p. 101-108, 2005.

PREZOTTI, L. C.; GOMES, J. A.; DADALTO, G. G.; OLIVEIRA, J. A. **Manual de Recomendação de calagem e adubação para o Estado do Espírito Santo 5ª aproximação**. Vitória: SEEA/Incaper/Cedagro, 2007.

SILVA, V. A.; ANTUNES, W. C.; GUIMARÃES, B. L. S.; PAIVA, R. M. C.; SILVA, V. F.; FERRÃO, M. A. G.; DAMATTA, F. M.; LOUREIRO, M. E. Resposta fisiológica de clone de café conilon sensível à deficiência hídrica enxertado em porta-enxerto tolerante. **Pesquisa Agropecuária Brasileira**, v. 45, n. 5, p. 457-464, 2010.

STEEL, R. G. D.; TORRIE, J. H.; DICKEY, D. A. **Principles and procedures of statistics**: a biometrical approach. 3rd ed. New York: MacGraw-Hill Book Companies, 1997.

TOMAZ, M. A.; SAKIYAMA, N. S.; DAMATTA, F. M.; MARTINEZ, H. E. P.; CRUZ, C. D.; PEREIRA, A. A. Efeito do porta-enxerto nas trocas gasosas, á rea foliar e superfície de raiz de mudas de *Coffea arabica* L. **Revista Ceres**, v. 53, n. 306, p. 237-242, 2006.

TOMAZ, M. A.; MARTINEZ, H. E. P.; CRUZ, C. D.; FERRARI, R. B.; ZAMBOLIM, L.; SAKIYAMA, N. S. Diferenças genéticas na eficiência de absorção, na translocação e na utilização de K, Ca e Mg em mudas enxertadas de cafeeiro. **Ciência Rural**, v. 38, n. 6, p. 1540-1546, 2008.

WEIGEL, D.; JURGENS, G. Stem cells that make stems. **Nature**, v. 415, n. 1, p. 751-754, 2002.

ZONTA, J. H.; BRAUN, H.; REIS, E. F.; PAULUCIO, D.; ZONTA, J. B. Influência de diferentes turnos de rega e doses de hidroabsorvente no desenvolvimento inicial da cultura do café conillon (*Coffea canephora* Pierre). **Idesia**, v. 27, n. 3, p. 29-34, 2009.

2

Stevia rebaudiana (Bert) Bertoni: influence of osmotic stress and seed priming on seed germination under laboratory conditions

Keila Regina Hossa[1], José Walter Pedroza Carneiro[2*], Terezinha Aparecida Guedes[3] and Alessandro Lucca Braccini[2]

*[1]Programa de Pós-graduação em Agronomia, Universidade Estadual de Maringá, Av. Colombo, 5790, Bloco J-45, 87020-900, Maringá, Paraná, Brazil. [2]Departamento de Agronomia, Universidade Estadual de Maringá, Maringá, Paraná, Brazil. [3]Departamento de Estatística, Universidade Estadual de Maringá, Maringá, Paraná, Brazil. *Author for correspondence. E-mail: jwpcarneiro@uem.br*

ABSTRACT. The foremost factor necessary for plant growers cultivating large acreages of *Stevia rebaudiana* (Bert) Bertoni is the production of qualitative bedding plants. The objective of this study was to evaluate the influence of osmotic-priming on the uniformity of seed germination. First, we evaluated the percentage of normal seedlings from two seed samples harvested in 2011 and 2012. The seeds harvested in 2012 produced 71.4% normal seedlings and thus they were used in the next experiments. The seeds were subjected to osmotic stress using five concentrations of polyethylene glycol (PEG-6000) at -0.2, -0.4, -0.6, -0.8, and -1.0 MPa in contrast with distilled water. Based on these first results, only -0.8 and -1.0 MPa were evaluated in the third experiment. The seeds were immersed in both concentrations of polyethylene glycol (PEG-6000) for imbibing at 20°C for four, five, six, and seven days. Thereafter, we evaluated the time to the first normal seedling (Ti), time to the last normal seedling (Tf), percentage normal seedlings at the initial time (Pi) and percentage of normal seedlings at the end of every treatment (Pf). Osmotic priming increased the percentage of normal seedlings of the *Stevia rebaudiana* and reduced the time to the first and last germination events.

Keywords: osmotic priming, radicle protrusion, time to seed germination, sweeteners.

Introduction

Stevia rebaudiana (Bert) Bertoni is a short-day plant that produces many foliar steviol glycosides sweeter than sucrose (Yadav, Singh, Dhyani, & Ahuja, 2011; Brandle, Starratt, & Gijzen, 1998; Carneiro, 1990). These non-caloric glycosides have been recommended for diabetic, overweight and obese patients requiring natural alternatives to artificial sweeteners, such as the industrial aspartame and saccharin (Moraes, Donega, Cantrell, Mello, & McChesney, 2013; Lemus-Mondaca, Vega-Galvez, Zura-Bravo, & Ah-Hen, 2012; Geuns, 2003). Furthermore, the industrial use of Stevia goes beyond the presence of sweeteners because glycosides, such as rubusoside, which is also partially responsible for the sweet taste, have remarkable solubilizing properties that can be used by the pharmaceutical industry (Liu et al., 2011). The leaf composition indicates financial possibilities in the food and pet industries, reflecting the richness of these plants in dietary iron, calcium, and potassium, in addition to low levels of fat. The agro-industry of

Stevia rebaudiana also offers opportunities in the cosmetic industry because of the large number of essential oils. Recently, significant quantities of inulin were detected in the stems and roots of Stevia plants. Inulin is a reserve oligosaccharide with pre-biotic effects in humans, and currently, this compound is under high demand by the food industry (Oliveira et al., 2011). These recent chemical approaches prompt a new economic discussion for cultivating *Stevia rebaudiana* because the high crop yield of the stem by-products (Moraes et al., 2013) can be a complementary and safe source of net income for Stevia growers.

Although *Stevia rebaudiana* acreage has been expanding worldwide to supply the needs of various agro-industrial programs (Carneiro, 1990; Brandle et al., 1998; Ramesh, Singh, & Megeji, 2006), high and cost-effective crop yields depend on the field establishment of vigorous bedding plants. Furthermore, these bedding plants require seed production with high physiological quality evaluated by the percentage of seed germination over a period of ten days, seed vigor tests, or both.

The high-quality production of uniform bedding plants depends on high and uniform percentage of seed germination, and high seedling resistance to stressful weather conditions (Carneiro, Muniz, & Guedes, 1997; Heydecker, Higgins, & Turner, 1975). Seed imbibition using priming technology is one organic strategy in which the hydric conditions initiate the mechanisms essential to seed germination, but it does not induce the protrusion of radicles. Hydric stress, otherwise, reduces both the percentage and rate of germination events (Bewley & Black, 1985). Polyethylene glycol (PEG-6000) has been used in the osmotic priming technology (Carneiro & Guedes, 1992b) because it is chemically inert and does not has toxic effects on the seedlings. Michel and Kauffmann (1973) initially proposed the use of osmotic potential to evaluate germination responses under hydric stress. Currently, this method is broadly applied to the commercial seeds of numerous species.

The objective of this study was to evaluate the influence of osmotic priming on seed germination of *Stevia rebaudiana*. Specifically, the objectives were to select the better seeds for further experimental evaluation, assess the seed germination under hydric stress conditions, and evaluate the concentration and optimal period for osmotic priming treatments.

Material and methods

Three experiments were conducted in the Seed Science and Technology Laboratory, Iguatemi Research Station 20 km apart from the central campus at 23° 21' SL, 52° 04' WL and 542 m of altitude, and in the NAPD (Núcleo de Estudos Avançados em Ciência de Plantas Daninhas) at 20°20'SL, 52°04'WL and altitude of 506 m in the central campus of the State University at Maringá, Maringá County, North Paraná State, in Brazil. In 2011 and 2012, achene fruit-like seeds were harvested under glasshouse conditions from plants cultivated in 20-L black plastic pots. The seed samples were visually hand cleaned and graded as light and heavy based on the seed length and discriminated using a stereomicroscope (model SMZ-140 N2GG, Motic, from Asia). We eliminated from the samples the presence of immature, empty, or both achene fruit-like seeds. Subsequently, the hairy pappus was removed using hand friction on the surface of germination test papers. This method improves the contact of the seed with the germination medium (Carneiro & Guedes, 1992a; Carneiro, 1996). The samples were stored at 5± 1°C and 20% relative humidity before every experiment. Thereafter, the seeds were germinated at 25°C using germi-test paper (Cel 065, JProlab, Brazil) in germination boxes (11 x 11 x 5 cm, JProlab, Brazil).

First experiment: The germination responses from two seed lots harvested in 2011 and 2012 were evaluated to discriminate the samples with better percentages of seed germination. The experimental design was completely randomized with six replications, and every germination box was an experimental unit. Three germination test papers (Cel 065, JProlab, Brazil) were wetted two and half times their weight in grams with distilled, deionized water (ISTA, 2014; Brasil, 2009). The germination temperature was 25±1°C under daylight lamps for seven days, and subsequently we counted data from normal and abnormal seedlings, and ungerminated seeds (Carneiro & Guedes, 1992a).

Second experiment: The conditions of the first experiment were improved using eight replications and modified through the application of distilled, deionized water as the control (T_1). In contrast we prepared five concentrations of polyethylene glycol (PEG-6000, Labsynth, Brazil) according to Villela, Doni Filho, and Siqueira (1991): -0.2 MPa (T_2) with 119.571 g L^{-1}, -0.4 MPa (T_3) with 178.343 g L^{-1}, -0.6 MPa (T_4) with 223.664 g L^{-1}, -0.8 MPa (T_5) with 261.948 g L^{-1} and -1.0 MPa (T_6) with 295.713 g L^{-1} of PEG-6000. These concentrations were estimated by the following equation of Michel and Kaufmann (1973): $\Psi_{os} = (1.18 \times 10^{-2}) * C - (1.18 \times 10^{-4}) * C^2 + (2.67 \times 10^{-4}) * C * T + (8.39 \times 10^{-7}) * C^2 * T$, in which: Ψ_{os} = osmotic potential (bar); C = concentration (g L^{-1}) of polyethylene glycol, and T = temperature (°C). All samples were treated with Thiram 200 SC® at 0.2% (p/v).

The seeds evaluated in this experiment had the best germination responses in the first evaluation when the percentage of normal seedlings was 71.4%. The frequencies of seed germination were verified at eight-hour intervals when every seedling had roots measuring 2 mm in length. The test was finished after 384 hrs when no further germination events were observed. The abnormal seedlings were detected by the presence of one black spot in the differentiating region of the root system, and cotyledons indicating the presence of phytophagous insects. Furthermore, no germination events were observed at -0.8 and -1.0 MPa, and both concentrations were used in the third experiment in which we applied the osmotic-priming treatment.

Third experiment: The seeds were imbibed in a solution of polyethylene glycol 6000 at 20°C for four, five, six, and seven days in contrast with the control treated with distilled, deionized water. The germination paper was wetted with 20 mL of the PEG-6000 solution using a hypodermic syringe, and the germination box was sealed using polypropylene tape. After the priming period, we washed all the seeds with tap water for 1 min., followed by germination at 25°C for 196 hrs. The following response variables were recorded: time to the first normal seedling (Ti), time to the last normal seedling (Tf), percentage of normal seedlings (Pi) at the initial time (Ti), and percentage of total normal seedlings (Pf).

Data analyses

The germination data from these three experiments were reported by exploratory data analysis using box-and-whiskers plots according to Banzatto and Kronka (2008). Box-and-whiskers plots are based on sample statistics. This method uses communicative figures to show sample data when the variability is more prominent than the means (Krzywinski & Altman, 2014) as typically observed for *Stevia rebaudiana* seeds. The graphic ranges from the first (Q1) to the third quartile (Q3) of the distribution and represents the inter-quartile range (IQR). The line across the box indicates the median, and the whiskers are lines extending from Q1 and Q3 to their end points, typically defined as the most extreme data within Q1 – 1.5 × IQR and Q3 + 1.5 × IQR, respectively (Streit & Gehlenborg, 2014). Each outlier is represented as an individual mark. Alternatively, the minimum and maximum values from every data set are used as the end points for the whiskers. One benefit of the box-and-whiskers plots is the requirement of only five units for analysis, and these data are readily comparable across three or more samples. For n < 5, however, the best decision is to show the individual data points (Krzywinski & Altman, 2014). The control of variability during the germination and bedding plant development of *Stevia rebaudiana* produces uniform plants under crop field conditions.

Results and discussion

On average, the highest percentage (71.4%) of germination was obtained from seeds harvested in 2012 in contrast to the normal seedlings (67.0%) from seeds harvested in 2011, and both samples were stored at 5°C under 20% relative humidity. Stevia seeds are sensitive to storage conditions likely reflecting the quantitative oil contents evaluated by reddish embryos after immersion in Sudan IV for 20 minutes (Carneiro, 1990).

The average percentage of normal seedlings (71.4%) reflected the number of non-germinated seeds at 25°C, instead the level of the abnormal seedlings in the sample, despite their presence in the test. The elimination of the hairy pappus has also been motive of seedlings abnormalities. The variability was higher than in the seed sample harvested in 2011, when the reduced percentage of normal seedlings (67.0%) reflected the presence of abnormal seedlings. Based on the variability indicated by the inter-quartile ranges (IQR), both samples maintained the development stage $V_{1.7}$ (Carneiro, 2007). The selection of newly harvested seeds for use in the following experiments was based on a whisker fence value higher than 80.0% and asymmetry on the right (Figure 1a), suggesting room for DNA repair, if any, and a higher percentage of normal seedlings after the osmotic priming. This variability is typical in germination tests when the seed quality is not uniform. This non-uniformity requires further investigation to understand how to reduce variations in bedding plant production under nursery conditions.

In the second experiment, the time to first germination (T_i) was recorded at 64 hrs for seeds under control conditions (1). Along these levels of water stress, the seeds delayed a time to the first event of more than 150 hrs when evaluated under -0.6 MPa, except for two outliers. This finding might reflect the diversity in the investigated seed population in this study (Figure 2a). We did not find any radicle protrusion under osmotic stress conditions of -0.8 and -1.0 MPa (Figure 2b and c). The percentage of normal seedlings from the control was higher than the initial seed germination of 71.4%, suggesting significant physiological events the seed sample. Again the superior fence of the whisker was higher than 80% of normal seedlings. The graphical responses in Figure 2 indicate asymmetry in the data distribution from the treatment with -0.6 MPa, where the highest variability in time to germination was observed. In the control, the highest variability was observed in the time to first germination and initial percentage of normal seedlings (Figure 2a and c). These responses require more investigation concerning the best methodology for applying priming methods in Stevia seeds. In summary, concentrations of -0.8 and -1.0 MPa were the most favorable to study the osmotic priming of *Stevia rebaudiana* seeds

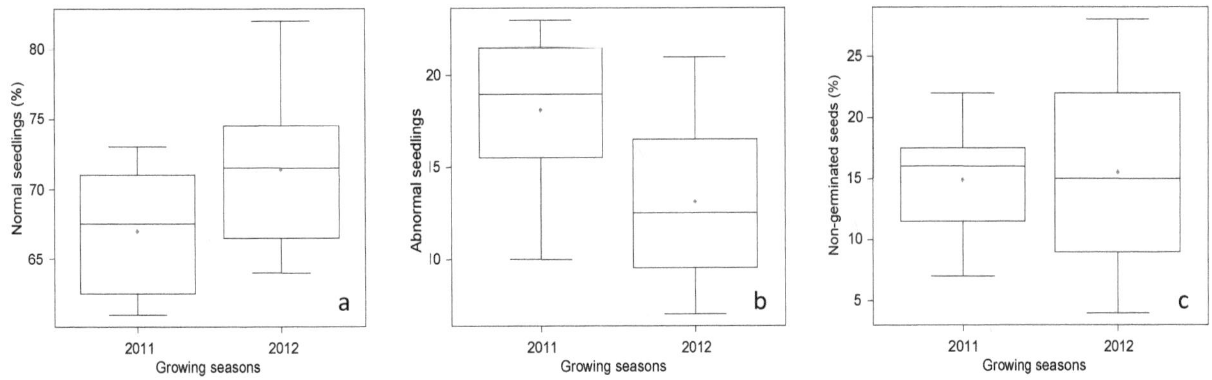

Figure 1. Box- and-whiskers plots of components from the germination test using *Stevia rebaudiana* (Bert) Bertoni seeds harvested in two growing seasons in the Iguatemi Research Station, Maringá County, Paraná State, Brazil.

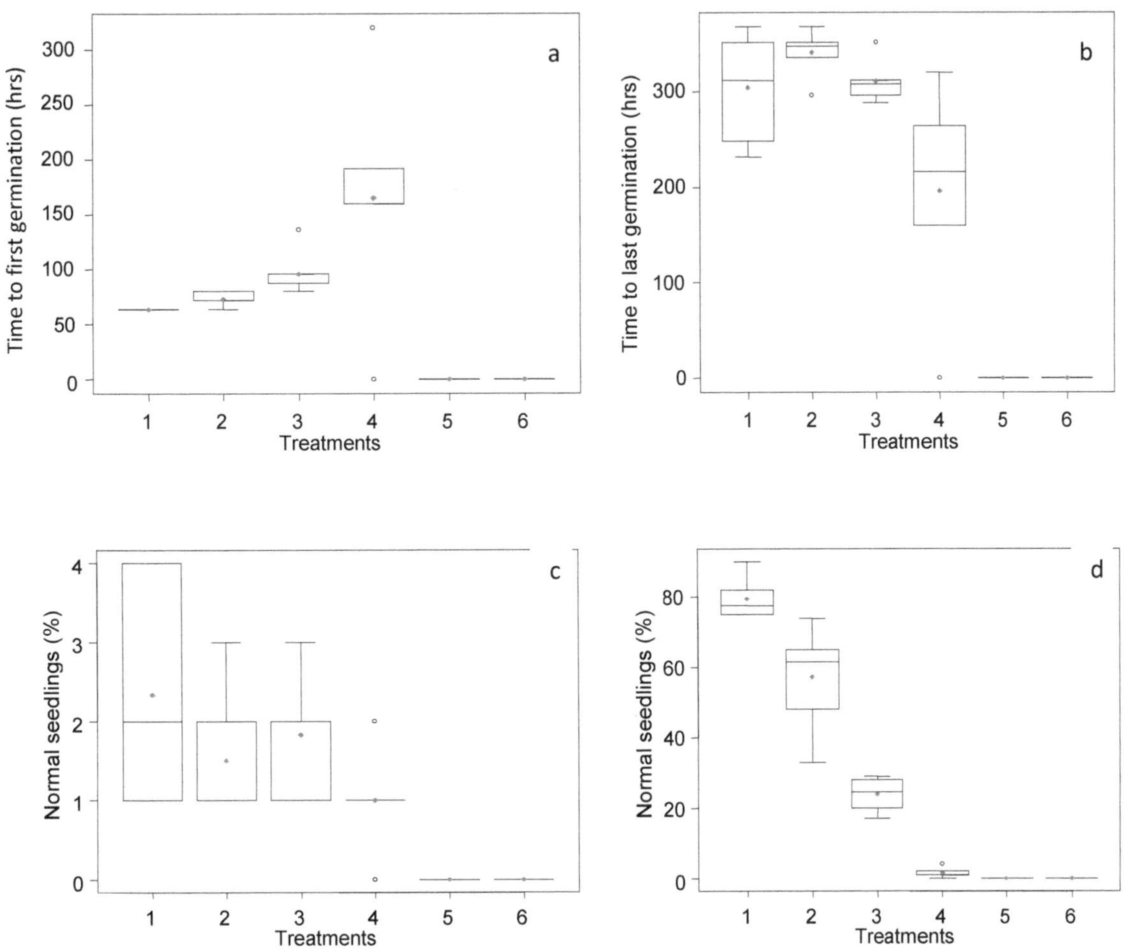

Figure 2. Performance of *Stevia rebaudiana* (Bert) Bertoni seeds under the control (1) and five osmotic priming treatments (2 = -0.2 MPa, 3 = -0.4 MPa, 4 = -0.6 MPa, 5 = -0.8 MPa, and 6 = -1.0 MPa).

In the third treatment, all the primed seeds experienced the first protrusion within 12 hrs after the beginning of the germination test (first records, data not shown), despite the number of days of priming. The average end of every treatment was different for six and seven days (Figure 3a), but the variability of these treatments was higher than expected, despite the similarity of the median and mean. The whiskers were larger on the right than in seeds primed for four and five days. Similar responses were reported for temperate carrot seeds treated with PEG-6000 at -1.0 MPa, when cumulative germination data were fit to the parameters of the Weibull distribution (Carneiro & Guedes, 1992b). The lower fence of the box-and-whiskers plots for seeds immersed in the PEG-6000 for six days was approximately 40 hrs, and the treatment for seven days delayed the last germination event to more than 160 hrs. This response contrasts with the six days of treatment, when the last observation was approximately 140 hrs (Figure 3a). In the first records, the percentage of normal seedlings reached values close to 80.0% (Figure 3b), but high variability was detected by the parameters of the box-and-whiskers plots, where five days had data skewed to the left. One explanation for skewness is the time required for all seeds to germinate.

Osmotic priming with PEG-6000 at -1.0 MPa for seven days showed a final percentage higher than 75%, and a small IQR range for 50% of normal seedlings. These higher responses suggest DNA repair by this osmotic treatment but with data still skewed to the left. Another explanation for these results is the non-uniform dimension of these seeds, particularly the seed length. Thus, the next step in seed technology research is to improve the seed grade using classification equipment based on the length or density of seeds. Osmotic treatment at -1.0 MPa increased the germination stage from $V_{1.7}$ (Figure 1a) to $V_{1.8}$ (Figure 3f), as verified using other priming treatments. However, the IQR was reduced compared with the control and treatment at -0.8 MPa. These results were similar to those of Carneiro and Guedes (1992b), when they improved the responses of primed carrot seeds at -1.0 MPa using PEG-6000. In summary, two foremost responses were observed: the reduction of IQR and the length of the whiskers, despite the stretching of the first quartile (Figures 1a and 3f). These responses also suggest the presence of cohorts during the germination period and these effects suggest further investigations.

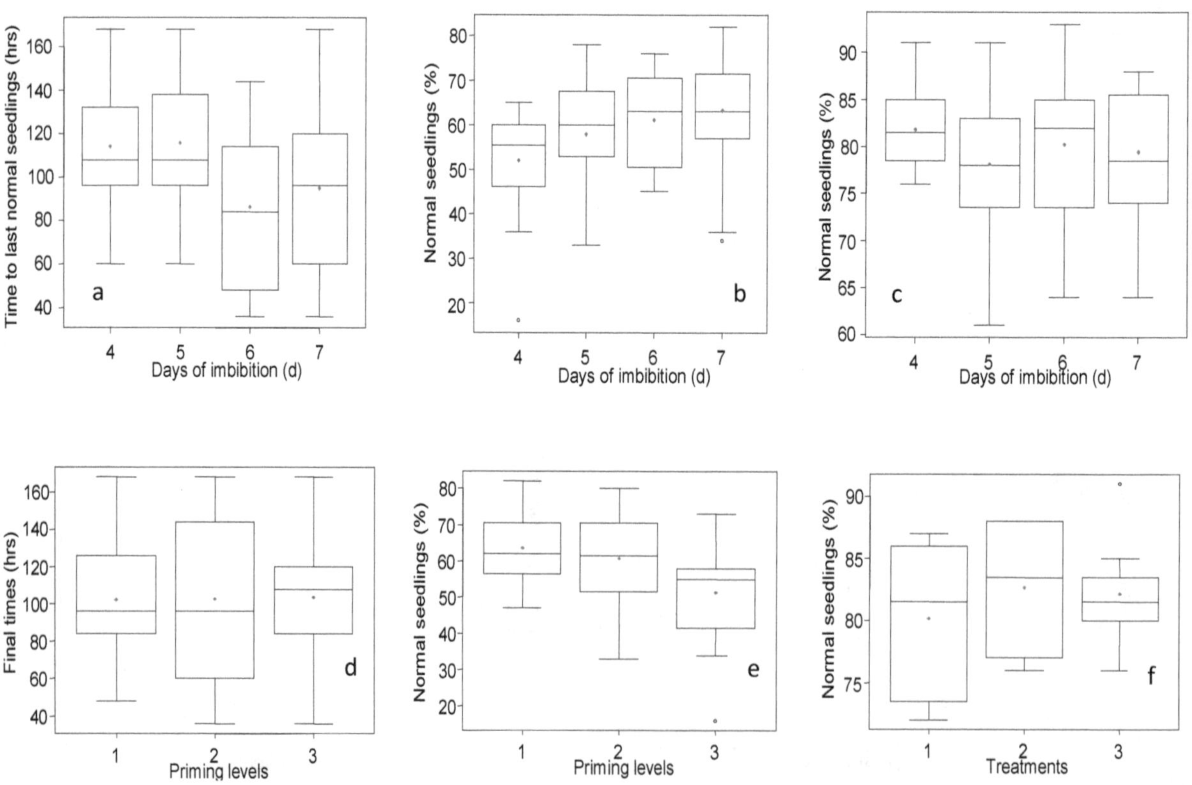

Figure 3. Performance of *Stevia rebaudiana* (Bert) Bertoni seeds germinated after treatments at 25°C for four, five, six, and seven days (a, b, and c) in the control (1), -0.8 (2), and -1.0 MPa (3) (d, e, and f) under laboratory conditions.

Conclusion

Osmotic priming increased the final percentage of normal *Stevia rebaudiana* seedlings and reduced the time to the first and last germination events.

References

Banzatto, D. A., & Kronka, S. N. (2008). *Experimentação Agrícola* (4a. ed.). Jaboticabal, SP: Funep.

Bewley, J. D., & Black, M. (1985) *Seeds*: physiology of development and germination. New York, US: Plenum Press.

Brandle, J. E., Starratt, A. N., & Gijzen, M. (1998). *Stevia rebaudiana*: Its agricultural, biological, and chemical properties. *Canadian Journal of Plant Science*, *78*(4), 527-536.

Brasil. (2009). Ministério da Agricultura, Pecuária e Abastecimento. *Regras para análise de sementes*. Brasília, DF: Mapa/ACS.

Carneiro, J. W. P. (1990). *Stevia rebaudiana* (Bert) *Bertoni: produção de sementes*. Maringá, PR: Imprensa Universitária.

Carneiro, J. W. P. (1996). Determinação do número de sementes para avaliar o desempenho germinativo de Stevia (*Stevia rebaudiana* (Bert) Bertoni). *Revista Brasileira de Sementes*, *18*(1), 1-5.

Carneiro, J. W. P. (2007). *Stevia rebaudiana* (Bert) Bertoni: Stages of plant development. *Canadian Journal of Plant Science*, *87*(4), 861-865.

Carneiro, J. W. P., & Guedes, T. A. (1992a). Influencia do contato das sementes de Stevia (*Stevia rebaudiana* (Bert) Bertoni) no substrato, avaliada pela função de distribuição de Weibull. *Revista Brasileira de Sementes*, *14*(1), 65-68.

Carneiro, J. W. P., & Guedes, T. A. (1992b). Influência da temperatura no desempenho germinativo de sementes de cenoura (*Daucus carota* L.) avaliada pela função de distribuição de Weibull. *Revista Brasileira de Sementes*, *14*(2), 207-213.

Carneiro, J. W. P., Muniz, A. S., & Guedes, T. A. (1997). Greenhouse bedding plant production of *Stevia rebaudiana* (Bert) Bertoni. *Canadian Journal of Plant Science*, 77(3), 473-474.

Geuns, J. M. C. (2003). Molecules of interest: Stevioside. *Phytochemistry*, *64*(5), 913-921.

Heydecker, W., Higgins, J., & Turner, Y. J. (1975). Invigoration of seeds. *Seed Science and Technology*, *3*(3/4), 881-888.

International Seed Testing Association [ISTA]. (2014). *International rules for seed testing*. Bassersdorf, SW: ISTA.

Krzywinski, M., & Altman, N. (2014). Visualizing samples with box-plots. *Nature Methods*, *11*(2), 119-120.

Lemus-Mondaca, R., Vega-Galvez, A., Zura-Bravo, L., & Ah-Hen, K. (2012). *Stevia rebaudiana* Bertoni, source of a high potency natural sweetener: A comprehensive review on the biochemical, nutritional and functional aspects. *Food Chemistry* *132*(3), 1121-1132.

Liu, Z., Koh, G. Y., Zhang, F., Jeansonne, D., Stout, R., Dong, L., & Enright, F. (2011). Medicinal plants and cancer: solubility-enhanced reformulation of paclitaxel, an old drug. *Louisiana Agriculture, Winter*, *54*(1), 6-7.

Michel, B. E., & Kaufmann, M. R. (1973). The osmotic potential of polyethylene glycol 6000. *Plant Physiology*, *51*(6), 914-916.

Moraes, R. M., Donega, M. A., Cantrell, C. L., Mello, S. C., & McChesney, J. D. (2013). Effect of harvest timing on leaf production and yield of diterpene glycosides in *Stevia rebaudiana* Bert: A specialty perennial crop for Mississippi. *Industrial Crops and Products*, *51*(11), 385-389.

Oliveira, A. J. B., Gonçalves, R. A. C., Chierrito, T. P. C., Santos, M. M., Souza, L. M., Gorin, P. A. J., Sassaki, G. L., & Iacomini, M. (2011). Structure and degree of polymerisation of fructo-oligosaccharides present in roots and leaves of *Stevia rebaudiana* (Bert) Bertoni. *Food Chemistry*, *129*(2), 305-311.

Ramesh, K., Singh, V., & Megeji, N. W. (2006). Cultivation of Stevia (*Stevia rebaudiana* (Bert) Bertoni): A comprehensive review. *Advances in Agronomy*, *89*(3), 138-179.

Streit, M., & Gehlenborg, N. (2014). Bar charts and box plots. *Nature Methods*, *11*(2), 117.

Takahashi, L., Melges, E., & Carneiro, J. W. P. (1996). Desempenho germinativo de sementes de *Stevia rebaudiana* (Bert) Bertoni sob diferentes temperaturas. *Revista Brasileira de Sementes*, *18*(1), 1-5.

Villela, F. A., Doni Filho, L., & Siqueira, E. L. (1991). Tabela de potencial osmótico em função da concentração de polietileno glicol 6000 e da temperatura. *Pesquisa Agropecuária Brasileira*, *26*(11/12), 1957-1968.

Yadav, A. K., Singh, S., Dhyani, D., & Ahuja, P. S. (2011). A review on the improvement of Stevia [*Stevia rebaudiana* (Bertoni)]. *Canadian Journal of Plant Science*, *91*(1), 1-27.

3

Pochonia chlamydosporia promotes the growth of tomato and lettuce plants

Rosangela Dallemole-Giaretta[1], Leandro Grassi de Freitas[2], Everaldo Antônio Lopes[3*], Marliane de Cássia Soares da Silva[4], Maria Catarina Megumi Kasuya[4] and Silamar Ferraz[2]

*[1]Departamento de Ciências Agrárias, Universidade Tecnológica Federal do Paraná, Pato Branco, Paraná, Brazil. [2]Departamento de Fitopatologia, Universidade Federal de Viçosa, Viçosa, Minas Gerais, Brazil. [3]Instituto de Ciências Agrárias, Universidade Federal de Viçosa, Campus de Rio Paranaíba, Rod. MG-230, Km 7, 38810-000, Rio Paranaíba, Minas Gerais, Brazil. [4]Departamento de Microbiologia, Universidade Federal de Viçosa, Viçosa, Minas Gerais, Brazil.*Author for correspondence. E-mail: everaldolopes@ufv.br*

ABSTRACT. The fungus *Pochonia chlamydosporia* is one of the most studied biological agents used to control plant-parasitic nematodes. This study found that the isolates Pc-3, Pc-10 and Pc-19 of this fungus promote the growth of tomato and lettuce seedlings. The isolate Pc-19 colonized the rhizoplane of tomato seedlings in only 15 days and produced a large quantity of chlamydospores. This isolate was able to use cellulose as a carbon source, in addition to glucose and sucrose. Scanning electron microscopy (SEM) revealed that hyphae of the *P. chlamydosporia* isolate Pc-10 penetrated the epidermal cells of the tomato roots. These three *P. chlamydosporia* isolates promote the growth of tomato and lettuce.

Keywords: *Lactuca sativa*, nematophagous fungus, rhizosphere colonization, *Solanum lycopersicum*, *Verticillium chlamydosporium*.

Introduction

Pochonia chlamydosporia Zare and Gams is one of the most studied biological agents used to control plant-parasitic nematodes. It can produce chlamydospores, which enable it to survive in the soil without additional energy sources (MANZANILLA-LÓPEZ et al., 2013). However, the ability of *P. chlamydosporia* to control endoparasitic nematodes, such as *Meloidogyne* spp. Goeldi, depends on rhizosphere colonization (MANZANILLA-LÓPEZ et al., 2013).

The *P. chlamydosporia* fungus can colonize the rhizoplane and the internal tissues of the roots of some plants, in the manner of an endophyte (BORDALLO et al., 2002; ESCUDERO; LOPEZ-LLORCA, 2012; MACIÁ-VICENTE et al., 2009b; MONFORT et al., 2005). Once inside the root tissues, this antagonistic fungus shares the same niche as root-knot nematodes and is less subject to competition from soil microorganisms (ESCUDERO; LOPEZ-LLORCA, 2012). From the perspective of the biocontrol of nematodes, the endophytic colonization of roots by the fungus is a competitive advantage. The majority of the putatively highly expressed *P. chlamydosporia* genes are related to its endophytic behaviour, including the production of hydrolytic enzymes, transporters, proteases, chitinases and a large number of secondary metabolites (LARRIBA et al., 2014). As a result of this interaction with its host, the fungus can enhance the tolerance of the crop to biotic and abiotic stresses and promote plant growth (ESCUDERO; LOPEZ-LLORCA, 2012; MACIÁ-VICENTE et al., 2009b; MANZANILLA-LÓPEZ et al., 2013; MONFORT et al., 2005).

Pochonia chlamydosporia acts as a growth-promoting agent of monocot and dicot crops, such as barley (MACIÁ-VICENTE et al., 2009b), wheat

(MONFORT et al., 2005), lettuce (DIAS-ARIEIRA et al., 2011), pistachio (EBADI et al., 2009) and tomato (ESCUDERO; LOPEZ-LLORCA, 2012). However, the extent to which plant growth is promoted depends on the combination of the crop and the particular *P. chlamydosporia* isolate (MANZANILLA-LÓPEZ et al., 2011).

In a previous study, we screened different isolates of *P. chlamydosporia* for the ability to manage the root-knot nematode *Meloidogyne javanica* (DALLEMOLE-GIARETTA et al., 2012). However, little is known about the effects of these isolates on plant growth. Thus, the aims of this study were to evaluate the colonization of tomato and lettuce roots by three isolates of *P. chlamydosporia* var. *chlamydosporia* and to characterize the effects of these organisms on plant development.

Material and methods

Fungal isolates

Pochonia chlamydosporia var. *chlamydosporia* Pc-3, Pc-10 and Pc-19 were isolated from soil naturally infested with *Meloidogyne* spp. and cultivated with vegetables in Viçosa, Minas Gerais State, Brazil (DALLEMOLE-GIARETTA et al., 2012). The fungal isolates were recovered by incubating pieces of colonized filter paper (SMITH; ONIONS, 1994) on corn meal agar (CMA, Difco, Detroit, MI, USA) at 25°C in darkness for 21 day.

Promotion of the growth of tomato seedlings

Two assays were performed to assess the potential of Pc-3, Pc-10 and Pc-19 to promote the growth of tomato cv. Santa Clara under laboratory conditions. Glass tubes (23 mm in diameter x 150 mm in length) were filled with 1.5 g of coconut fiber (Table 1), which was moistened with 7.5 mL (assay 1) or 6.5 mL (assay 2) of distilled water. After 12h, the tubes were capped and autoclaved at 121°C for 30 min. This process of sterilization was performed twice with an interval of 24h.

Table 1. Chemical characterization of the coconut fiber used as a substrate for the growth of tomato and lettuce plants.

Macronutrients (g kg^{-1})						
Nitrogen	Phosphorus	Potassium	Calcium	Magnesium	Sulphur	Sodium
7.3	3.3	11.7	11.6	4.8	5.1	0.4
Micronutrients (mg kg^{-1})						
Zinc	Iron	Manganese		Copper		Boron
502	3668	250		228		126
C/N				pH		
26.49				5.01		

Fungal inoculum was applied to the substrate either as mycelial discs or as a conidia suspension. In assay 1, four discs (5 mm in diameter) were cut from the edge of a 21-day-old culture of *P. chlamydosporia* on CMA, and each disc was placed at a depth of 2 cm in a tube containing the substrate. In assay 2, *P. chlamydosporia* conidia were collected and suspended in sterile water, and the suspension was adjusted to provide 1.52 x 10^6, 1.83 x 10^5 and 1.92 x 10^6 conidia per tube of the isolates Pc-3, Pc-10 and Pc-19, respectively. Sterile PDA discs and distilled water were used as controls. One pre-germinated tomato seed was added in each tube at a depth of 0.5 cm. The tomato seeds were previously disinfected with 70% alcohol (v v^{-1}) for 1 min., then immersed for 10 min. in sodium hypochlorite diluted to 1% active chlorine (v v^{-1}) and finally rinsed with sterile water. The seeds were then pre-germinated for 2 d in the culture medium described by Bordallo et al. (2002). For controls, tomato seeds were placed in the substrate without fungus. The tubes were transferred into a growth chamber at 25°C with a 16-h photoperiod.

The height and the fresh mass of the aboveground parts of the seedlings were recorded after 15 day. The experiment was carried out in a completely randomized design with 10 (assay 1) or eight (assay 2) replicates.

Root colonization and the use of carbon sources by *P. chlamydosporia* isolate Pc-19

Ten root segments, 0.5 cm in length, subjected to colonization by Pc-19 in assay 1 were incubated in Petri dishes on semi-selective medium (GASPARD et al., 1990) at 25°C, and the number of colonized segments was recorded after three days. The remaining root segments were stored in a solution containing formaldehyde, lactic acid and alcohol (FAA 50; 5:5:90, v v^{-1} v^{-1}) to study rhizospheric colonization. The root segments were prepared according to a protocol used for the observation of mycorrhizal associations (BRUNDRETT et al., 1996).

To investigate which carbon sources can be used by the isolate Pc-19, the culture medium was prepared using glucose (control), sucrose, cellulose or pectin and contained 8 g L^{-1} of carbon, 2 g L^{-1} of $NaNO_3$, 1 g L^{-1} of K_2HPO_4, 0.5 g L^{-1} of $MgSO_4$, 0.5 g L^{-1} of KCl, 0.01 g L^{-1} of $FeSO_4$ and 17 g L^{-1} of agar in distilled water (SUN; LIU, 2006). A mycelial disc 5 mm in diameter was cut from the edge of a 15-day-old culture of the fungus and was placed at the centre of each 9-cm-diameter Petri dish. The fungal biomass was recorded after 15 d of culturing at 25°C in the dark. To measure the fungal biomass, the culture medium was dissolved in 200 mL of boiling water, and the mycelium was transferred to a crucible and held at 105°C until a constant mass was reached. The assay was arranged in a randomised design with seven replicates per treatment.

Root colonization of tomato plants by the Pc-10 isolate as determined by scanning electron microscopy (SEM)

Root segments from assay 1 and colonized by Pc-10 were fixed in phosphate buffer solution (0.05 mol. L^{-1}, pH 7.1) with 2.5% of glutaraldehyde for one hour at room temperature. They were washed six times in phosphate buffer solution (0.05 mol. L^{-1}, pH 7.1) for ten min. each time, dehydrated in an ethanol series of 30 and 50% for 10 min. at each concentration and placed in 70% ethanol in the refrigerator until the following day. Dehydration was then continued with an additional ethanol series of 80 and 95% for 10 min. each, followed by 100% ethanol for 15 min., three times. The samples were dried in a critical point dryer (Bal-Tec, model CPD 030, Germany) and mounted on aluminium stubs, sputter-coated in gold and viewed under a scanning electron microscope (Zeiss, model LEO 1430 VP, England).

Promotion of the growth of lettuce seedlings

The potential of the isolate Pc-10 to promote the growth of lettuce seedlings was assessed using the cultivars Verônica, Americana, Regina and Manteiga (Isla Sementes, Porto Alegre, Rio Grande do Sul State, Brazil). Each cultivar was evaluated separately, using the methodology described for assay 1. Root segments subjected to colonization by Pc-10 were incubated in Petri dishes on semi-selective medium (GASPARD et al., 1990) at 25°C, and the number of colonized segments was recorded after three days.

Statistical analysis

The data for each variable were tested using the Kolmogorov-Smirnov test to determine whether they were normally distributed. The Bartlett test was used to determine whether variances were homogeneous. The data were subjected to a one-way analysis of variance (ANOVA, F-test, $p < 0.05$). Treatment means were compared using tukey's HSD test ($p < 0.05$). Statistical analyses were performed using R software, version 2.12.2 (R DEVELOPMENT CORE TEAM, 2011).

Results and discussion

Promotion of tomato seedling growth

All isolates of *P. chlamydosporia* increased the height (from 1,179 to 1,404%) and the aboveground mass (from 1,350 to 1,650%) of tomato seedlings in assay 1, when mycelial discs were used as the source of inoculum, and from 72 to 103, and 100 to 148%, respectively, in assay 2, when the fungus was added to the substrate via conidia suspension (Table 2). No significant difference among the isolates was observed in either assay. Despite the differences between the assays, we concluded that the growth of tomato seedlings was promoted by the *P. chlamydosporia* var. *chlamydosporia* isolates, excluding the sole effect of the nutrients from the culture medium (CMA).

Microorganisms can stimulate plant growth both directly (i.e., via the production of plant hormones, biological nitrogen fixation, phosphorus solubilization, acceleration of the process of mineralization and siderophore synthesis) and indirectly (i.e., via the induction of systemic resistance, the production of antibiotics and antagonism in relation to pathogens) (HAYAT et al., 2010). Pochonia chlamydosporia can promote the growth of wheat seedlings via direct mechanisms, such as the production of growth regulators associated with the activity of peroxidase (MONFORT et al., 2005). In this study, it is likely that the increase in plant growth was related to an increase in the area of absorption of the roots, allowing them to overcome possible water stresses in the tubes.

Table 2. Height and mass of the aboveground portion of tomato seedlings cv. Santa Clara grown on a coconut fiber substrate infested using mycelial discs (assay 1) or a conidia suspension of *Pochonia chlamydosporia* var. *chlamydosporia* isolates Pc-3, Pc-10 and Pc-19.

Treatments	Plant height (cm)	Increase (%)	Aboveground mass (g)	Increase (%)
Assay 1				
Control	0.69 (± 1.21) b	-	0.004 (± 0.006) b	-
Isolate Pc-3	9.45 (± 3.15) a	1,269	0.063 (± 0.028) a	1,475
Isolate Pc-10	10.38 (± 1.36) a	1,404	0.070 (± 0.021) a	1,650
Isolate Pc-19	8.83 (± 2.68) a	1,179	0.058 (± 0.022) a	1,350
Assay 2				
Control	2.87 (± 1.69) b	-	0.023 (± 0.022) b	-
Isolate Pc-3	5.83 (± 1.07) a	103	0.057 (± 0.023) a	148
Isolate Pc-10	5.00 (± 1.52) a	74	0.046 (± 0.020) a	100
Isolate Pc-19	4.95 (± 1.81) a	72	0.048 (± 0.018) a	109

Means (± standard deviations) within a column in each assay followed by the same letter do not differ by tukey's HSD test ($p < 0.05$).

Root colonization and the use of carbon sources by *P. chlamydosporia* isolate Pc-19

The fungus colonized all root segments of the tomato plants and also the coconut fiber used as a substrate for the plants (Figure 1A). A large quantity of chlamydospores was produced in the rhizoplane (Figure 1B), and the hyphae of the fungus penetrated into cells of the root cortex (Figure 1C). The fungus was able to use cellulose as a carbon source, as much as glucose and sucrose, with an average mycelial production (dry mass) of 0.135, 0.111 and 0.138 g using each of these carbon sources, respectively. However, poor mycelial growth was recorded when the fungus was grown in culture medium amended with pectin as the carbon source (0.009 g of dry mass).

This study provides evidence that Pc-19 can colonize tomato roots and use cellulose as a carbon source. Different species of plants differ in their abilities to allow the growth of *P. chlamydosporia* in the rhizosphere. Tomato, cabbage, crotalaria, beans, corn, kale, potato and pea are plants that favour the development of the fungus, whereas soybean, wheat, sorghum, aubergine and okra are plants that do not permit satisfactory development in the rhizosphere (MANZANILLA-LÓPEZ et al., 2013). In addition to corroborating earlier findings of the ability of tomato to support *P. chlamydosporia* colonization (MANZANILLA-LOPEZ et al., 2013), we found that the fungus can form a tangle of hyphae similar to those formed by arbuscular mycorrhizae. It has been suggested that this nematode antagonist may act as an endophyte (MACIÁ-VICENTE et al., 2009a) and may increase the ability of the tomato to germinate, absorb water from the substrate and resist stress due to the deleterious effects of salts present in the coconut fiber substrate (DOMEÑO et al., 2009; HERNÁNDEZ-APAOLAZA et al., 2005). In addition, enhanced development of the plants may be associated with the chemical degradation of cellulose from the coconut fiber by enzymes produced by the fungus (ESTEVES et al., 2009; LARRIBA et al., 2014) and the subsequent translocation of carbohydrates, as well as nutrients like nitrogen and phosphorus, to the roots of the tomato plants, as has been observed with arbuscular mycorrhizal fungi (SCHREINER, 2007).

Root colonization of tomato plants by Pc-10 as determined by scanning electron microscopy (SEM)

After 15 days, *P. chlamydosporia* var. *chlamydosporia* isolate Pc-10 colonized the roots of tomato seedlings, forming a mantle of hyphae and producing conidia in some parts of the rhizosphere (Figure 1D and E). Thus, like Pc-19, this isolate was able to colonize tomato roots; this information can be used to guide the choice of crop rotation system to support good rhizosphere development. Greater development and reproduction of the fungus in the rhizosphere can improve the control of plant-parasitic nematodes (DALLEMOLE-GIARETTA et al., 2011; VIGGIANO et al., 2014).

Although the extent to which isolate Pc-3 colonizes rhizosphere has not been evaluated, based on the results of the growth promotion assays it is possible to speculate that, like Pc-10 and Pc-19, this isolate may efficiently colonize the rhizosphere of tomato plants (Table 2).

Promotion of the growth of lettuce seedlings

Pc-10 increased the aboveground mass of all the lettuce cultivars from 100 to 330% and the height of the cultivar Regina by 80% (Table 3). Colonization of the roots by the isolate was observed in 100% of the root system segments of the lettuce seedlings, for all the cultivars tested, when plated on semi-selective medium.

In addition to the growth promotion of the plants, the ability of *P. chlamydosporia* isolates to colonize tomato and lettuce roots may protect the hosts against nematodes and other soil pathogens. MONFORT et al. (2005) reported that the *P. chlamydosporia* isolate 4624 improved barley growth and reduced root colonization by the phytopathogenic fungus *G. graminis* var. *tritici*. The protection of the root system by *P. chlamydosporia* against fungal pathogens probably occurs via supplementation with nutrients and competition for infection sites, as has been observed with arbuscular mycorrhizal fungi (BORGES et al., 2007). Moreover, *P. chlamydosporia* induces a defense reaction in the plant, with the formation of papillae in the cell wall of the root system (BORDALLO et al., 2002). Further studies are needed to understand the mechanisms of action of the *P. chlamydosporia* isolates Pc-3, Pc-10 and Pc-19.

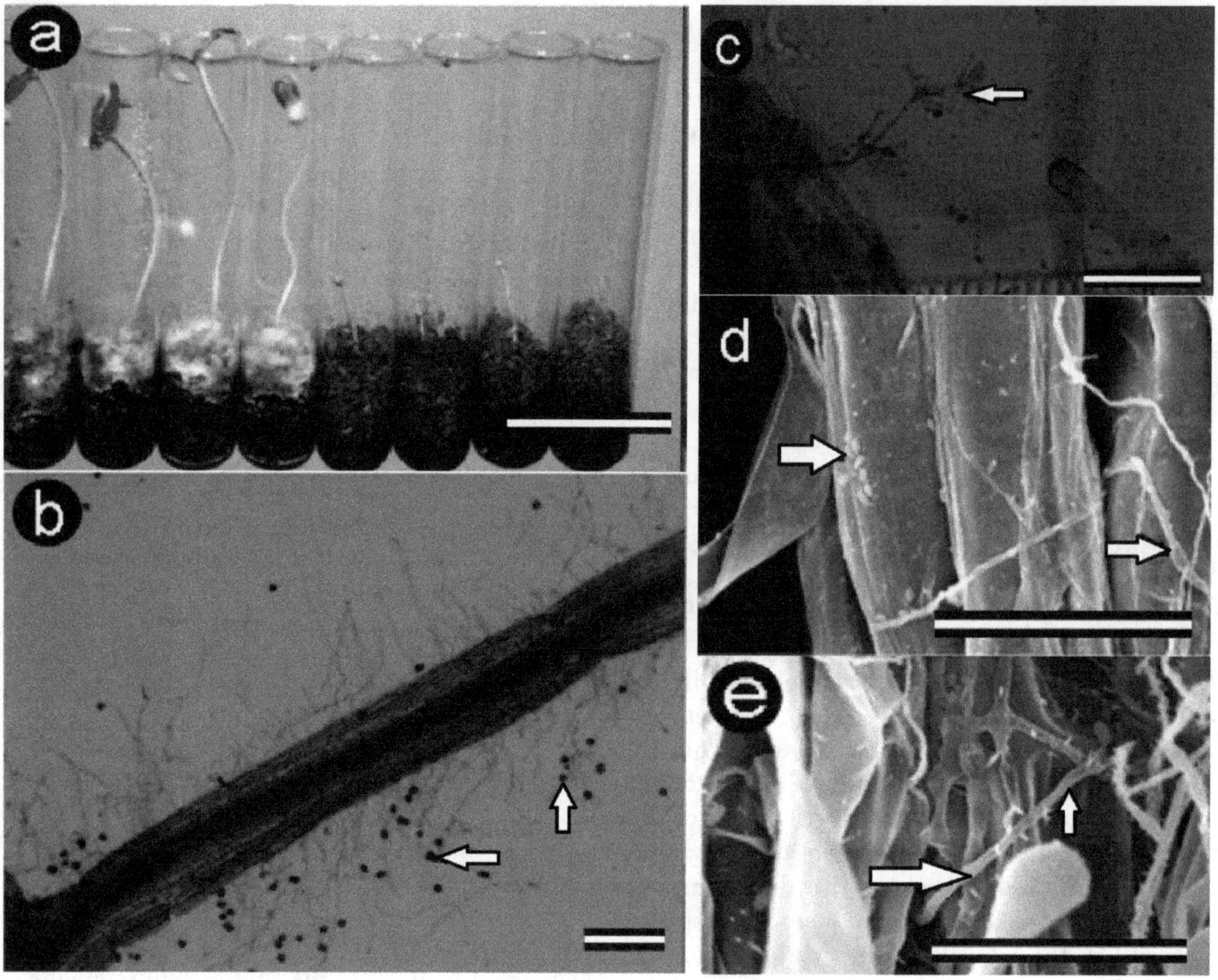

Figure 1. Tomato seedlings grown on a coconut fiber substrate infested with *P. chlamydosporia* var. *chlamydosporia* isolate Pc-19. (A) The four culture tubes on the left contain the isolate Pc-19, while the four on the right do not contain the antagonist (scale bars represent 1 cm). (B) Rhizoplane colonization of tomato seedlings by Pc-19, with the formation of chlamydospores (scale bar represents 250 μm). (C) Phialides of Pc-19 formed on the surface of the tomato root (scale bar represents 100 μm). (D) and (E) Scanning electron microscopy images of tomato roots colonized by *P. chlamydosporia* var. *chlamydosporia* isolate Pc-10. Hyphae and conidia of the fungus in the rhizoplane of the roots (D) and hyphae penetrating into cells of the root cortex of tomatoes (E; scale bars represent 50 μm).

Table 3. Height and fresh mass of the aboveground portion of the lettuce cultivars* Verônica, Americana, Regina and Manteiga grown on a coconut fiber substrate infested with mycelial discs of *Pochonia chlamydosporia* var. *chlamydosporia* isolate Pc-10 (Pc).

Treatments	Plant height (cm)	Increase (%)	Aboveground mass (g)	Increase (%)
Verônica + Pc	1.23 (± 0.678) a	-	0.02 (± 0.013) a	100
Verônica - Pc	0.83 (± 0.453) a	-	0.01 (± 0.007) b	-
Americana + Pc	1.45 (± 0.626) a	-	0.03 (± 0.017) a	200
Americana - Pc	1.34 (± 0.637) a	-	0.01 (± 0.005) b	-
Regina + Pc	1.78 (± 0.646) a	80	0.03 (± 0.016) a	200
Regina - Pc	0.99 (± 0.321) b	-	0.01 (± 0.006) b	-
Manteiga + Pc	1.08 (± 0.436) a	-	0.043 (± 0.016) a	330
Manteiga - Pc	1.21 (± 1.115) a			-

*Each cultivar was evaluated in a separate assay. Means (± standard deviation) within a column in each cultivar followed by the same letter do not differ by F-test ($p < 0.05$).

Conclusion

The isolates Pc-3, Pc-10 and Pc-19 of *P. chlamydosporia* promote the growth of tomato and lettuce seedlings. This fungus colonizes the roots of both plant species and produces hyphae and chlamydospores in the rhizoplane of tomato plants.

Acknowledgements

The authors would like to thank the National Council for Scientific and Technological Development (CNPq), the Foundation for Research Support of the State of Minas Gerais (Fapemig) and the Coordination for the Improvement of Higher

Education Personnel (Capes) for their financial support.

References

BORDALLO, J. J.; LOPEZ-LLORCA, L. V.; JANSSON, H. B.; SALINAS, J.; PERSMARK, L.; ASENSIO, L. Colonization of plant roots by egg-parasitic and nematode-trapping fungi. **New Phytologist**, v. 154, n. 2, p. 491-499, 2002.

BORGES, A. J. S.; TRINDADE, A. V.; MATOS, A. P.; PEIXOTO, M. F. S. Redução do Mal-do-Panamá em bananeira-maçã por inoculação de fungo micorrízico arbuscular. **Pesquisa Agropecuária Brasileira**, v. 42, n. 1, p. 35-41, 2007.

BRUNDRETT, M.; BOUGHER, N.; DELL, B.; GROVET, T.; MALAJCZUK, N. **Working with mycorrhizas in forestry and agriculture**. Canberra: Australian Centre for International Agricultural Research, 1996.

DALLEMOLE-GIARETTA, R.; FREITAS, L. G.; LOPES, E. A.; FERRAZ, S.; PODESTÁ, G. S.; AGNES, E. L. Cover crops and *Pochonia chlamydosporia* for the control of *Meloidogyne javanica*. **Nematology**, v. 13, n. 8, p. 919-926, 2011.

DALLEMOLE-GIARETTA, R.; FREITAS, L. G.; LOPES, E. A.; PEREIRA, O. L; ZOOCA, R. J. F.; FERRAZ, S. Screening of *Pochonia chlamydosporia* Brazilian isolates as biocontrol agents of *Meloidogyne javanica*. **Crop Protection**, v. 42, n. 1, p. 102-107, 2012.

DIAS-ARIEIRA, C. R.; SANTANA, S. M.; FREITAS, L. G.; CUNHA, T. P. L.; BIELA, F.; PUERARI, H. H.; CHIAMOLERA, F. M. Efficiency of *Pochonia chlamydosporia* in *Meloidogyne incognita* control in lettuce crop (*Lactuca sativa* L.). **Journal of Food, Agriculture and Environment**, v. 9, n. 3-4, p. 561-563, 2011.

DOMEÑO, I.; IRIGOYEN, N.; MURO, J. Evolution of organic matter and drainages in wood fibre and coconut fibre substrates. **Scientia Horticulturae**, v. 122, n. 2, p. 269-274, 2009.

EBADI, M.; FATEMY, S.; RIAHI, H. Evaluation of *Pochonia chlamydosporia* var. *chlamydosporia* as a control agent of *Meloidogyne javanica* on pistachio. **Biocontrol Science and Technology**, v. 19, n. 7, p. 689-700, 2009.

ESCUDERO, N.; LOPEZ-LLORCA, L. V. Effects on plant growth and root-knot nematode infection of an endophytic GFP transformant of the nematophagous fungus *Pochonia chlamydosporia*. **Symbiosis**, v. 57, n. 1, p. 33-42, 2012.

ESTEVES, I.; PETEIRA, B.; ATKINS, S. D.; MAGAN, N.; KERRY, B. Production of extracelular enzymes by different isolates of *Pochonia chlamydosporia*. **Mycological Research**, v. 113, n. 8, p. 867-876, 2009.

GASPARD, J. T.; JAFFEE, B. A.; FERRIS, H. Association of *Verticillium chlamydosporium* and *Paecilomyces lilacinus* with root-knot nematode infested soil. **Journal of Nematology**, v. 22, n. 2, p. 207-213, 1990.

HAYAT, R.; ALI, S.; AMARA, U.; KHALID, R.; AHMED, I. Soil beneficial bacteria and their role in plant growth promotion: a review. **Annals of Microbiology**, v. 60, n. 4, p. 579-598, 2010.

HERNÁNDEZ-APAOLAZA, L.; GASCÓ, A. M.; GASCÓ, J. M.; GUERRERO, F. Reuse of waste materials as growing media for ornamental plants. **Bioresource Technology**, v. 96, n. 1, p. 125-131, 2005.

LARRIBA, E.; JAIME, M. D. L. A.; CARBONELL-CABALLERO, J.; CONESA, A.; DOPAZO, J.; NISLOW, C.; MARTÍN-NIETO, J.; LOPEZ-LLORCA, L. V. Sequencing and functional analysis of the genome of a nematode egg-parasitic fungus, *Pochonia chlamydosporia*. **Fungal Genetics and Biology**, v. 65, n. 1, p. 69-80, 2014.

MANZANILLA-LÓPEZ, R. H.; ESTEVES, I.; POWERS, S. J.; KERRY, B. R. Effects of crop plants on abundance of *Pochonia chlamydosporia* and other fungal parasites of root-knot and potato cyst nematodes. **Annals of Applied Biology**, v. 159, n. 1, p. 118-129, 2011.

MANZANILLA-LÓPEZ, R. H.; ESTEVES, I.; FINETTI-SIALER, M. M.; HIRSCH, P. R.; WARD, E.; DEVONSHIRE, J.; HIDALGO-DÍAZ, L. *Pochonia chlamydosporia*: Advances and challenges to improve its performance as a biological control agent of sedentary endo-parasitic nematodes. **Journal of Nematology**, v. 45, n. 1, p. 1-7, 2013.

MACIÁ-VICENTE, J. G.; JANSSON, H. B.; LOPEZ-LLORCA, L. V. Assessing fungal root colonization for plant improvement. **Plant Signaling and Behaviour**, v. 4, n. 5, p. 445-447, 2009a.

MACIÁ-VICENTE, J. G.; ROSSO, L. C.; CIANCIO, A.; JANSSON, H. B.; LOPEZ-LLORCA, L. V. Colonisation of barley roots by endophytic *Fusarium equiseti* and *Pochonia chlamydosporia*: effects on plant growth and disease. **Annals of Applied Biology**, v. 155, n. 3, p. 391-401, 2009b.

MONFORT, E.; LOPEZ-LLORCA, L. V.; JANSSON, H. B.; SALINAS, J.; PARK, J. O; SIVASITHAMPARAM, K. Colonization of seminal roots of wheat and barley by egg-parasitic nematophagous fungi and their effects on *Gaeumannomyces graminis* var. *tritici* and development of root-rot. **Soil Biology and Biochemistry**, v. 37, n. 7, p. 1229-1235, 2005.

R DEVELOPMENT CORE TEAM. **R: A language and environment for statistical computing**. Vienna: R Foundation for Statistical Computing, 2011. Available from: <http://www.R-project.org>. Access on: Apr. 10, 2012.

SCHREINER, R. P. Effects of native and nonnative arbuscular mycorrhizal fungi on growth and nutrient uptake of 'Pinot noir' (*Vitis vinifera* L.) in two soils with contrasting levels of phosphorus. **Applied Soil Ecology**, v. 36, n. 2-3, p. 205-215, 2007.

SMITH, D.; ONIONS, A. H. S. **The preservation and maintenance of living fungi**. Wallingford: CAB International, 1994.

SUN, M. H.; LIU, X. Z. Carbon requirements of some nematophagous, entomopathogenic and mycoparasitic hyphomycetes as fungal biocontrol agents. **Mycopathologia**, v. 161, n. 5, p. 295-305, 2006.

VIGGIANO, J. R.; FREITAS, L. G.; LOPES, E. A. Use of *Pochonia chlamydosporia* to control *Meloidogyne javanica* in cucumber. **Biological Control**, v. 69, n. 1, p. 72-77, 2014.

4

Harvest time and agronomic performance of maize hybrids with contrasting growth cycles

Fernando Panison*, Luís Sangoi, Daniel Fernando Kolling, Cileide Maria Medeiros Coelho and Murilo Miguel Durli

*Centro de Ciências Agroveterinárias, Universidade do Estado de Santa Catarina, Avenida Luiz de Camões, 2090, 88520-000, Lages, Santa Catarina, Brazil. *Author for correspondence. E-mail: fernandopanison@hotmail.com*

ABSTRACT. The delay of the maize harvest may cause qualitative and quantitative grain losses. This work evaluated the effects of the harvest time on the agronomic performance of maize hybrids with contrasting growth cycles. The experiment was set in Lages, Santa Catarina State, South Brazil, during the 2012/2013 and 2013/2014 growing seasons. A randomized block design, arranged in split plots was used. Five single-cross hybrids were tested in the main plots: P1630H and P32R22H (hyper-early), P2530 (super early) and P30F53YH and P30R50YH (early). Five harvest times were assessed in the split plots: 0 (grain physiological maturity), 10, 20, 30 and 40 days after physiological maturity. The sowing was performed on 12/5/2012 and 10/5/2013. The percentage of lodged and broken stems, grain mass, potential and real productivity were determined. The delay of the maize harvest increased the percentage of lodged and broken stems. The harvest time did not affect the weight of 1,000 grains. The harvests performed after physiological maturity decreased the real grain productivity, especially for the hyper-early hybrids P1630H and P32R22H. In the regions with high altitude and latitude, the productivity losses derived from late harvests are mainly due to the increase in the amount of broken and lodged stems.

Keywords: *Zea mays*, maturation, grain productivity, lodging.

Introduction

During the last 20 years, Brazilian soybean breeding programs have emphasized the development of early indeterminate cultivars that do not interrupt their vegetative growth after flowering. This allows farmers to sow them earlier than the genotypes that present a determinate growth habit (Stülp, Braccini, Albrecht, Ávila, & Scapim, 2009).

The anticipation of the sowing date and the earlier maturation of the cultivars have several advantages for soybean production, such as the reduction of stink bug attacks, the lower incidence of Asian rust, the decreased risk of low temperature and water restrictions during grain filling and a better sowing period for maize after the soybean harvest in the states that use this cropping system, such as Mato Grosso and Paraná States (Thomas & Costa, 2010).

However, in the south of Brazil, where soybean and maize are sown during the spring in a crop rotation system, the early ripening of the soybean cultivars, associated with sowing at the end of September/beginning of October, has accelerated the soybean harvest. This operation is currently concentrated in February, coinciding with the harvest of mature maize. Because soybeans are more profitable and more sensitive to harvest delay, growers prefer to harvest the legume first, leaving maize in the field for more than 30 days after its physiological maturity.

The delay of the maize harvest may promote quantitative and qualitative losses to the grain yield, caused by physiological and morphological factors. The grain yield and seed quality are potentially higher when the harvest is carried out right after physiological maturity (Araujo, Araujo, Sofiatti, & Silva, 2006; Marques, Dalpasquale, Vidigal Filho, Scapim, & Reche, 2011; Galvão, Conceição, Araújo, Karstein, & Finger, 2014). However, when maize grains are stored in the field and harvested with less than 20% moisture content, they may lose weight due to respiration (Kaaya, Warren, Kyamanywa, & Kyamuhangire, 2005; Lauren, Smith, & Di Menna, 2007). In addition, the removal of grain with low moisture from the field promotes mechanical damage at harvest. Such injuries may favour the occurrence of insects and diseases during grain storage (Santin, Reis, & Matsumura, 2004; Marqui, Menten, Moraes, & Cícero, 2006).

Morphologically, late harvest increases the susceptibility of maize stems to lodging and breakage. The intensity of the lodging and breakage depends on the hybrid's agronomic traits, the management practices used on the crop (fertilization, plant density and row spacing) and the meteorological conditions at the end of grain filling, especially the wind speed and rain fall distribution (Gomes, Brandão, Brito, Moraes, & Lopes, 2010).

Maize harvest delay may also compromise grain quality and productivity because it favours the incidence of fungi that cause ear and grain rot, such as *Stenocarpella maydis, S. macrospora, Fusarium verticillioides* e *F. graminearum* (Marques, Vidigal Filho, Dalpasquale, Scapim, Princinoto, & Machisnski Junior, 2009). These pathogens use starch from the endosperm as an energy source for their growth and decrease the quality of the grain, its nutritional value and its visual appearance (Pinto, Vargas, & Preis, 2007).

The problems caused by maize harvest delay can be mitigated by the choice of cultivar. Traits such as the hybrid's growth cycle, ear husk coverage, ear decumbence after grain physiological maturity, plant height and ear insertion height may affect the magnitude of morphological and physiological damage caused by late harvests (Gomes et al., 2010).

This work was carried out to evaluate the effects of harvest time on the agronomic performance of maize hybrids with contrasting growth cycles.

Material and methods

This study was conducted in the city of Lages, Santa Catarina State, in the highlands of southern Brazil, during the growing seasons of 2012/2013 and 2013/2014. The experimental site is located at 27° 52' latitude south, 50° 18' longitude west and 900 m above sea level. The climate of the region is classified by Köppen-Geiger, mentioned by Kottek (2006), as Cfb, presenting mild summers, cold winters and adequate rainfall during the whole year. The temperature and precipitation information were collected from a meteorological experimental station located 20 km from the experimental site.

The soil at the study site was an Oxisol (Hapludox), according to Embrapa (Empresa Brasileira de Pesquisa Agropecuária, 2006), having the following chemical characteristics: clay content - 560 g kg^{-1}; organic matter content - 60.0 g kg^{-1}; water pH - 5.2; SMP pH - 5.7; phosphorus – 4.4 mg dm^{-3}; potassium - 186 mg dm^{-3}; calcium – 5.79 cmolc dm^{-3}; magnesium – 2.47 cmolc dm^{-3}; aluminium – 0.2 cmolc dm^{-3}; CTC – 8.94 cmolc dm^{-3}.

A randomized block design arranged in split plots was used, with four replicates per treatment. Five single-cross hybrids with contrasting growth cycles were assessed in the main plots: two hyper-early hybrids (P32R22H and P1630H) that require 1282 and 1220 heat units (HU) to reach physiological maturity; one super-early hybrid (P2530) that requires 1390 HU to attain physiological maturity; and two early hybrids (P30R50YH and P30F53YH) that achieve physiological maturity with 1493 and 1556 HU. Five harvest times were evaluated in the split plots: 0 (grain physiological maturity), 10, 20, 30 and 40 days after physiological maturity. The first harvest time for each hybrid was carried out when there was a visible black layer in the grain insertion point on the cob (R6 stage of the growth scale proposed by Ritchie, Hanway, and Benson (1993). Each split plot comprised four rows, 0.7 apart and 7 m long. All measurements were taken from the two central rows, leaving borders of 0.5 m at the end of each row.

The experiment was set using a no-tillage system under a dead coverage of black oat (*Avena strigosa*).

The experimental area was planted with maize for three consecutive years before installing the trial. The soil fertilization was determined based on the goal of a grain yield of 18,000 kg ha^{-1}. Fertilization was performed at the sowing date by applying 30 kg ha^{-1} of N, 295 kg ha^{-1} of P_2O_5 and 170 kg ha^{-1} of K_2O. The fertilizers were superficially placed close to the sowing rows. Nitrogen was also side-dressed, applying 250 kg of N divided into three equal parts when the plants were at the V4, V8 and V12 growth stages, according to Ritchie et al. (1993). Urea was used as the N source.

The experiment was hand planted on 12/5/2012 and 10/5/2013. The plots were over-sown, dropping three seeds per hill and thinned to the desired density (80,000 pl ha^{-1}) when the plants had three expanded leaves.

The weeds were controlled with two herbicide applications. The first was conducted immediately after sowing and prior to plant emergence with a combination of atrasine (1,400 g a.i. per hectare) and metolachlor (2,100 g a.i. per hectare). The second application was performed after the emergence of the maize, when the plants were at V4, using tembotriona (100 g ha^{-1} de i.a.). Army worm (*Spodoptera frugiperda*) was controlled by spraying the insecticides lufenuron + lambdacyhalothrin (15 + 7.5 g de i.a. ha^{-1}) when the crop reached the V6 and V12 growth stages, according to Ritchie et al. (1993).

For each treatment, the percentage of lodged and broken stems was determined on the harvest day. In 2012/2013, harvest time 0 (grain physiological maturity) was carried out on 5/01/2013 and 5/10/2013 for the hyper-early and the other hybrids, respectively. In 2013/2014, harvest time 0 was performed on 3/06/2014, 3/16/2014 and 3/26/2014 for the hyper, super and early hybrids, respectively. The other harvests for each hybrid were performed at constant intervals of 10 days from harvest time 0.

At all harvest times, the ears were collected manually. Two harvest operations were carried out. In the first, the ears from the lodged and broken stems were harvested. Subsequently, the ears from erect plants were collected. The ears from both harvests were shelled and weighed separately. The kernel moisture was determined, and the grain yield per hectare was calculated and expressed at the standard moisture of 13%.

The grain yield was estimated in two ways. The first yield determination considered all harvested plants from each split plot, including those that were lodged and broken, simulating what would happen if the crop were manually harvested (potential yield). The second yield estimate subtracted from the potential yield a value equivalent to 50% of the weight of the grains harvested in the lodged and broken stems of each split plot, assuming that such kernels would not be removed from the field if the harvest were mechanically performed (real yield).

The data were statistically evaluated by a variance analysis, using the F test, at the significance level of 5%. The data from the lodged and broken stems were transformed before carrying out the variance analysis using the expression $(x + 1)^{1/2}$. When the F values were significant, the values were compared using Tukey's test. The effect of delaying harvest time was also assessed by polynomial regression analysis, testing the linear and quadratic models. Both evaluations were conducted at the significance level of 5%.

Results and discussion

The percentage of lodged and broken stems ranged from 0 to 33.4% in 2012/2013 and from 0 to 95.4% in 2013/2014 (Table 1). This variable was affected by the harvest time. In both growing seasons, the percentage of the lodged and broken plants increased in proportion to the delay in harvest time. The numeric values of the lodged and broken plants were greater in 2013/2014 than in 2012/2013. According to the regression analysis adjusted for the average values of the five hybrids, the percentage of lodged and broken stems presented a linear increase of 4.3 and 10.8% for each 10 days of delay in harvest time for the first and second growing seasons, respectively (Figure 1a).

Table 1. Percentage of lodged and broken stems of maize hybrids with contrasting growth cycles as affected by harvest time. Lages, South Brazil.

Days after physiological maturity	Hybrids					Average	CV (%)
	P1630H	P32R22H	P2530	P30F53YH	P30R50YH		
	Lodged and broken stems						
	Growing season 2012/2013						
0	0.0	0.0	0.0	2.1	3.6	1.1 b*	
10	4.5	5.7	0.4	2.5	8.2	4.3 b	
20	4.8	4.0	4.8	6.5	5.9	5.2 b	79.1
30	17.9	25.1	10.4	10.2	10.6	14.9 a	
40	17.3	33.4	14.5	10.4	12.4	17.6 a	
Averages	8.9NS	13.6	6.0	6.3	8.1		
CV (%)			96.2				
	Growing season 2013/2014						
0	0.0	3.1	10.8	0.5	0.0	2.9 c*	
10	2.4	10.7	13.1	0.5	0.0	5.3 c	
20	4.5	19.9	19.8	6.7	5.3	11.2 bc	80.7
30	8.3	33.7	61.0	3.9	18.6	25.1 b	
40	58.4	95.4	50.4	15.8	17.0	47.4 a	
Averages	14.7 ab*	32.6 a	31 a	5,5 b	8,2 b		
CV (%)			81.7				

[NS]Differences among averages are not significant in the row ($p < 0.05$). *Averages followed by the same lower case letter in the row or in the column do not differ significantly by Tukey's test ($p < 0.05$).

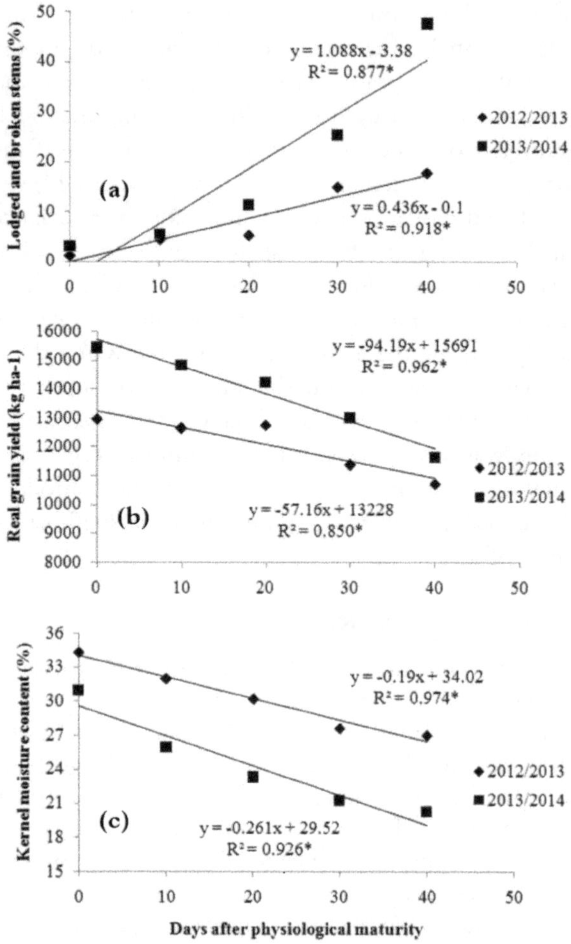

Figure 1. Percentage of lodged and broken stems (a), real grain yield (b) and kernel moisture content (c) of maize as affected by harvest time, in the average of five hybrids. Lages, South Brazil. *Determination coefficient significant at the 5% error probability level ($p < 0.05$).

In 2012/2013, there was no difference in the percentage of the lodged and broken stems among the hybrids. Conversely, in 2013/2014 the hyper- and super-early hybrids P32R22H, P1630H and P2530 had a higher percentage of lodged and broken stems than the early hybrids P30F53YH and P30R50YH. Even when the data transformation was performed prior to the variance analysis, the variation coefficients for this variable were high, ranging from 79.1 to 80.7 % (Table 1).

The delay of the maize harvest time is a risky management strategy because it favours stem lodging due to wind and rain (Gomes et al., 2010). This tendency was confirmed in the present work for both growing seasons. Maize allocates more than 50% of the plant biomass to the grains at physiological maturity (Sangoi, Silva, Argenta, & Rambo, 2010). Therefore, when the harvest is postponed, tissue senescence at the stem base and the constant presence of rainfall and temperatures below 17°C that occur during April, May and June in the highlands of South Brazil (Table 2) increase the ear weight, favouring stem lodging and breaking. This behaviour is accentuated by the crop's early ripening because maize hybrids with shorter growth cycles remobilize greater amounts of the stem-stored carbohydrates to the kernels during grain filling (Blum, Sangoi, Amarante, Arioli, & Guimaraes, 2003). Other factors that may have favoured the increase in the lodged and broken stems when the harvest was delayed were the cropping system with maize for three consecutive years during the spring/summer season and the presence of black oat preceding maize during the Fall/Winter season. This combination of grass species stimulates the presence of fungi, such as *Fusarium graminearum* and *Stenocarpella macrospora*, which promote maize stem rot (Casa, Moreira, Bogo, & Sangoi, 2007; Casa, Reis, Kuhnen Junior, & Bolzan, 2009). Additionally, the tendency during the last 20 years to increase plant density increases maize susceptibility to lodging and breaking when the harvest is delayed (Sangoi et al., 2010).

Table 2. Monthly average temperatures and pluvial precipitation during maize growing seasons in Lages, South Brazil.

Growing season 2012/2013[1]								
Month	Dec.	Jan.	Feb.	Mar.	Apr.	Mai.	Jun.	Average
Average temperature (°C)	21.4	18.9	19.5	17.4	15.3	12.9	12.0	16.7
Precipitation (mm)	203.5	195.4	214.7	179.2	62.3	86.3	199.3	162.9
Growing season 2013/2014								
Month	Oct.	Nov.	Dec.	Jan.	Feb.	Mar.	Apr.	
Average temperature (°C)	15.1	17.4	20.2	21.4	21.0	18.5	16.3	18.5
Precipitation (mm)	136.9	146.1	107.7	182.9	210.9	121.3	94.2	142.9

[1]Temperature and precipitation data were collected from a meteorological experimental station located 20 km from the experimental site.

The potential maize grain yield ranged from 11,320 to 13,686 kg ha^{-1} in 2012/2013 and from 13,165 to 17,047 kg ha^{-1} in 2013/2014 (Table 3). In the first growing season, there was a significant effect of the harvest time on the potential yield. There was a 9.8% (1,285 kg ha^{-1}) decrease on the average potential yield of the five hybrids when the harvest was delayed 40 days in relation to the physiological maturity. The harvest time did not significantly affect the potential yield in the second growing season. There were no differences among the hybrids regarding the potential yield in 2012/2013. On the other hand, the potential yield was lower for hybrid P32R22H than for hybrids P2530 and P30F53YH in 2013/2014, for the average of the five harvest times.

Table 3. Potential grain yield of maize hybrids with contrasting growth cycles as affected by harvest time. Lages, South Brazil.

Days after physiological maturity	Hybrids					Averages	CV (%)
	P1630H	P32R22H	P2530	P30F53YH	P30R50YH		
Potential grain yield (kg ha^{-1})[1]							
Growing season 2012/2013							
0	12,903	12,665	13,541	12,391	13,686	13,037 a*	
10	12,875	12,267	13,233	12,515	13,575	12,893 a	
20	12,377	11,818	13,777	13,451	13,369	12,958 a	7.3
30	11,875	11,629	11,964	12,684	13,176	12,266 ab	
40	11,577	11,320	11,746	12,104	12,012	11,752 b	
Averages	12,321 NS	11,940	12,852	12,629	13,164		
CV (%)			12.1				
Growing season 2013/2014							
0	16,845	13,687	16,675	15,755	15,196	15,631 NS	
10	16,041	13,483	16,108	15,661	14,815	15,221	
20	16,534	14,359	15,050	14,346	14,886	15,035	6.7
30	14,918	13,598	15,642	15,432	14,354	14,788	
40	14,771	13,165	17,047	15,228	15,162	15,074	
Averages	15,821 ab	13,658 b	16,104 a	15,284 a	14,883 ab*		

NS Differences among averages are not significant in the row or in the column ($p < 0.05$)[1]. Calculated values considering manual harvest of all ears in the two central rows of each split-plot. *Averages followed by the same lower case letter in the row or in the column do not differ significantly by the Tukey's test ($p < 0.05$).

The maintenance of the maize kernels in the field for long periods after physiological maturity may increase their respiratory activity, favouring grain weight loss and decreasing crop productivity when the harvest is delayed (Pinto et al., 2007). The magnitude of the losses derived from kernel respiration depends on several factors, such as air temperature and relative humidity, the maize cultivar, and the size and shape of the kernel (Guiscem, Bicudo, Nakagawa, Zanotto, Sansigolo, & Zucarelli, 2002; Cella, Silva, & Azevedo, 2014).

The losses in the potential yield detected in the present work were lower than those reported by Araujo, Araujo, Sofiatti, and Silva (2006) and Marques et al. (2009). There was no significant reduction in the mass of 1,000 grains when the harvest was delayed up to 40 days after physiological maturity (Table 4). Such behaviour is an indication that the kernel respiratory activity from physiological maturity to harvest was not high enough to degrade the endosperm storage compounds and to reduce kernel mass. An important factor that probably contributed to mitigate potential yield losses with harvest delay is the thermal regime that characterizes the region where the experimental site is located. The grain physiological maturity and harvest occurred from March to June (Table 2). The average air temperature at this time of the year in the highlands of South Brazil was low, ranging from 12 to 18°C. The kernel respiratory activity is directly proportional to the temperature (Galvão et al., 2014). Therefore, the low temperatures recorded at the end of the maize-growing season probably reduce plant respiration, avoiding a significant decrease of kernel mass and mitigating losses in potential yield.

Table 4. Mass of 1,000 grains of maize hybrids with contrasting growth cycles as affected by harvest time. Lages, South Brazil.

Days after physiological maturity	Hybrids					Average	CV (%)
	P1630H	P32R22H	P2530	P30F53YH	P30R50YH		
Mass of 1,000 grains (g)							
Growing season 2012/2013							
0	304	340	382	315	338	331 NS	
10	309	336	374	324	337	331	
20	301	337	358	313	342	327	4.5
30	321	345	369	316	335	331	
40	311	324	360	294	344	323	
Average	309 d*	336 bc	369 a	312 cd	339 b		
CV (%)			6.4				
Growing season 2013/2014							
0	289	318	329	331	357	325 NS	
10	287	313	342	313	348	321	
20	295	313	329	321	334	318	4.3
30	301	315	331	325	329	320	
40	303	314	356	329	355	331	
Average	295 d*	315 c	337 ab	324 bc	345 a		
CV (%)			3.8				

NS Differences among averages are not significant in the column ($p < 0.05$). *Averages followed by the same lower case letter in the row do not differ significantly by Tukey's test ($p < 0.05$).

The real productivity, considering a mechanical harvest and assuming that 50% of the grains produced by the lodged and broken plants would not be removed from the field, was affected by the harvest time in 2012/2013 and by the interaction between the harvest time and the hybrid type in 2013/2014 (Table 5). Considering the average of the five hybrids, the grain yield was 2,230 kg ha^{-1} lower when the crop was harvested 40 days after physiological maturity, representing a 17.2% yield loss, in comparison to the productivity registered for the first harvest time. In the second year, hybrids' grain yields reacted differently to the delay in harvest time. The hyper-early hybrids P1630H and P32R22H were more sensitive to late harvests than the early hybrids P30F53YH and P30R50YH. The average real productivity of the two hyper-early hybrids at the last harvest time was only 56% of the value registered when they were harvested at physiological maturity. However, at the last harvest date, the early hybrids showed a real yield equivalent to 90% of the productivity recorded at physiological maturity. The regression analysis adjusted to the average values of the real yield of the five hybrids showed a linear decrease of 57 and 94 kg ha^{-1} for each 10 days of delay in the harvest time after physiological maturity, during the first and second growing season, respectively (Figure 1b).

Kernel moisture content at harvest ranged from 23.5 to 35% in 2012/2013 and from 18 to 34% in 2013/2014 (Table 6). During both growing seasons, this variable was affected by the harvest time and hybrid cycle. The delay of the harvest reduced the kernel moisture content. Considering the average values of the five hybrids, the rates of decrease in the kernel moisture content for each 10 days of harvest delay were 1.9 and 2.6% in 2012/2013 and 2013/2014, respectively (Figure 1c). The lower rate of decrease in the kernel moisture content registered in the first growing season was due to the later sowing date of the trial (12/5/2012). Because maize was sowed in the beginning of December, the period between physiological maturity and harvest was concentrated in May and June, which coincided with the presence of low air temperatures, which slowed the drying of the kernel (Daltro, Albuquerque, França Neto, Guimarães, & Gazziero, 2010).

Table 5. Real grain yield of maize hybrids with contrasting growth cycles as affected by harvest time. Lages, South Brazil.

Days after physiological maturity	Hybrids					Average	CV (%)
	P1630H	P32R22H	P2530	P30F53YH	P30R50YH		
	Real grain yield (kg ha^{-1})[1]						
	Growing season 2012/2013						
0	12903	12665	13541	12260	13432	12960 a*	
10	12578	11948	13205	12355	13026	12622 a	
20	12709	11595	13438	13008	12974	12745 a	8.1
30	10816	10150	11376	12028	12459	11366 b	
40	10560	9464	10894	11465	11265	10730 b	
Average	11913NS	11164	12491	12223	12631		
CV (%)			13.5				
	Growing season 2013/2014						
0	A16845 a*	A13484 a	A15818 a	A15715 a	A15196 a	15412	
10	A15856 a	A12796 a	A14988 a	A15623 a	A14815 a	14816	
20	A16139 a	A13007 a	A13523 a	A13872 a	A14487 a	14206	11.3
30	A14305 ab	B11419 ab	AB12807ab	A15136 a	A13054 a	12985	
40	B10505 b	B 6887 b	AB11013 b	A14019 a	A13872 a	11618	
Average	14730	11519	13630	14873	14285		
CV (%)			16.8				

[1]Calculated values considering that 50% of grains produced by lodged and broken plants would not be removed from the field if harvest was performed mechanically. [NS]Differences among averages are not significant in the row ($p < 0.05$). *Averages preceded by the same upper case letter in the row and followed by the same lower case letter in the column do not differ significantly by Tukey's test ($p < 0,05$).

Table 6. Kernel moisture content of maize hybrids with contrasting growth cycles as affected by harvest time. Lages, South Brazil.

Days after physiological maturity	Hybrids					Average	CV (%)
	P1630H	P32R22H	P2530	P30F53YH	P30R50YH		
	Kernel moisture content (%)						
	Growing season of 2012/2013						
0	34.0	33.0	34.6	35.0	34.3	34.3 a*	
10	27.8	25.5	32.8	35.3	34.0	32.0 b	
20	30.1	28.9	29.2	30.2	30.8	30.2 c	3.7
30	25.8	26.0	26.2	29.1	28.1	27.6 d	
40	24.6	23.5	26.4	29.1	28.4	27.0 d	
Average	28.5 cd*	27.4 d	28.8 bc	31.7 a	31.1 ab		
CV (%)			4.3				
	Growing season 2013/2014						
0	34.0	30.5	30.9	34.0	29.7	30.9 a*	
10	27.5	24.8	25.6	27.5	26.3	25.9 b	
20	23.4	21.6	21.9	23.4	24.9	23.3 c	3.2
30	19.6	18.9	21.2	19.6	23.2	21.2 d	
40	18.7	18.0	19.8	18.7	22.4	20.2 e	
Average	24.6 ab*	22.7 c	24.0 bc	24.6 ab	25.3 a		
CV (%)			4.1				

[NS]Differences among averages are not significant in the column ($p < 0.05$). *Averages followed by the same lower case letter in the row or in the column are not significantly different by Tukey's test ($p < 0,05$).

Collectively, the results of this work show that when the harvest is carried out after physiological maturity, maize grain yield might be reduced. The decrease in crop productivity is caused by the direct effects derived from kernel respiration or by the indirect factors related to the increase in stem lodging and breaking. In regions with high altitude and latitude, such as the highlands of South Brazil, the direct effects of harvest delay on the kernel mass were not significant due to the low air temperatures recorded from March to June, a period that included the end of grain filling, physiological maturity and the harvest. Conversely, the indirect effects of postponing harvest were relevant, mainly for the hyper-early hybrids. These cultivars usually produce a smaller leaf area and a greater carbohydrate mobilization from the stem to the kernels during grain filling (Blum et al., 2003; Brito, Silveira, Brandão, Gomes, & Lopes, 2011). Such traits enhanced their susceptibility to stem lodging and breaking when the harvest was delayed, especially because maize was grown continuously for several years in the same area without rotation with other crops and was preceded by a plant species of the same family (black oat).

Because there is an increasing tendency of growers to use hyper- and very-early maize hybrids in Brazil, which are usually sown at high plant densities, special care must be taken to avoid a delay in harvest time to prevent stem lodging and yield losses.

Conclusion

The delay in harvest time after physiological maturity increases the percentage of lodged and broken plants, reducing real grain yield, especially for hyper- and super-early hybrids.

The delay in harvest time does not affect grain mass in regions with high altitude and latitude.

Acknowledgements

The authors thank the financial aid of Fapesc (Fundação de Amparo à Pesquisa e Inovação do Estado de Santa Catarina).

References

Araujo, E. F., Araujo, R. F., Sofiatti, V., & Silva, R. F. (2006). Physiological quality of maize seeds harvested in different times. *Bragantia, 65*(4), 687-692.

Blum, L. E. B., Sangoi, L., Amarante, C. V. T., Arioli, C. J., & Guimaraes, L. S. (2003). Defoliation, plant populations and hybrid earliness affect the incidence and severity of maize stem rots. *Ciência Rural, 33*(5), 805-811.

Brito, C. S., Silveira, D. L., Brandão, A. M., Gomes, L. S., & Lopes, M. T. G. (2011). Maize leaf area reduction in Brazilian tropical regions and its effect on agronomic traits. *Interciência, 36*(4), 291-295.

Casa, R. T., Moreira, E. N., Bogo, A., & Sangoi, L. (2007). Incidence of stem rot, rotten grains and grain yield of maize hybrids submitted to the increase in plant density. *Summa Phytopathologica, 33*(4), 353-357.

Casa, R. T., Reis, E. M., Kuhnen Junior, P. R., & Bolzan, J. M. (2009). Maize disease control in no tillage system. *Plantio Direto, 112*(1), 15-21.

Cella, V., Silva, J. F., & Azevedo, P. H. (2014). Dissecation effect in anticipated growth stages of soybean. *Bioscience Journal, 30*(3), 1364-1370.

Daltro, E. M. F., Albuquerque, M. C. F., França Neto, J. B., Guimarães, S. C., & Gazziero, D. L. P. (2010). Dissecation application before harvesting: effects on soybean seed quality. *Revista Brasileira de Sementes, 32*(1), 111-122.

Empresa Brasileira de Pesquisa Agropecuária. (2006). *Brazilian system of soil classification* (2a ed.). Rio de Janeiro, RJ: Embrapa CNPS.

Galvão, J. C. C., Conceição, P. M., Araújo, E. F., Karstein, J., & Finger, F. L. (2014). Physiological and enzymatic alterations in maize seeds submitted to different harvest times and shelling methods. *Revista Brasileira de Milho e Sorgo, 13*(1), 14-23.

Gomes, L. S., Brandão, A. M., Brito, C. H., Moraes, D. F., & Lopes, M. T. G. (2010). Resistance to plant lodging and stem breaking in tropical maize. *Pesquisa Agropecuária Brasileira, 45*(2), 140-145.

Guiscem, J. M., Bicudo, S. J., Nakagawa, J., Zanotto, M. D., Sansigolo, C., & Zucarelli, C. (2002). Morphological and physiological traits of maize that affect its grain water loss. *Revista Brasileira de Milho e Sorgo, 1*(2), 28-37.

Kaaya, A. N., Warren, H. L., Kyamanywa, S., & Kyamuhangire, W. (2005). The effect of delayed harvest on moisture content, insect damage, moulds and aflatoxin contamination of maize in Mayuge district of Uganda. *Journal of the Science of Food and Agriculture, 85*(1), 2595-2599.

Kottek, M. (2006). World map of the Köppen-Geiger climate classification updated. *Meteorologische Zeitschrift, 15*(3), 259-263.

Lauren, D. R., Smith, W. A., & Di Menna, M. (2007). Influence of harvest date and hybrid on the mycotoxin content of maize (*Zea mays*) grain grown in New Zealand. *New Zeland Journal of Crop and Horticultural Science, 35*(2), 331-340.

Marques, O. J., Dalpasquale, V. A., Vidigal Filho, P. S., Scapim, C. A., & Reche, D. L. (2011). Grain mechanical damage of maize commercial hybrids as affected by moisture content at harvest. *Semin: Ciências Agrárias, 32*(2), 565-576.

Marques, O. J., Vidigal Filho, P. S., Dalpasquale, V. A., Scapim, C. A., Princinoto, L. F., & Machisnski Junior, M. (2009). Fungi incidence and contamination by grain toxins in maize commercial hybrids as affected by moisture content at harvest. *Acta Scientiarum. Agronomy, 31*, 667-675.

Marqui, J. L., Menten, J. O. M., Moraes, M. H. D., & Cícero, S. M. (2006). Relation among mechanical damage, fungicide treatment and pathogens incidence in maize seeds. *Revista Brasileira de Milho e Sorgo, 5* (3), 351-358.

Pinto, N. F. J. A., Vargas, E. A., & Preis, R. A. (2007). Sanitary quality and B1 fumosinane production in maize kernels before harvest. *Summa Phytopathologica, 33*(3), 304-306.

Ritchie, J., Hanway, J. J., & Benson, G. O. (1993). *How a corn plant develops*. Ames, IA: Iowa State University of Science and Technology.

Sangoi, L., Silva, P. R. F., Argenta, G., & Rambo, L. (2010). *Physiology of maize for high yields*. Lages, SC: Graphel.

Santin, J. A., Reis, E. M., & Matsumura, A. T. S. (2004). Effect of delaying maize harvest on the incidence of rotten grains and pathogenic fungi. *Revista Brasileira de Milho e Sorgo, 3*(3), 182-192.

Stülp, M., Braccini, A. L., Albrecht, L. P., Ávila, M. R., & Scapim, C. A. (2009). Agronomic performance of three soybean cultivars at different sowing dates and two growing seasons. *Ciência e Agrotecnologia, 33*(5), 1240-1248.

Thomas, A. L., & Costa, J. A. (2010). *Soja: manejo para alta produtividade de grãos*. Porto Alegre, RS: Evangraf.

5

Vegetative growth of Conilon coffee plants under two water conditions in the Atlantic region of Bahia State, Brazil

Crescimento vegetativo de plantas de café Conilon sob dois regimes hídricos, na região Atlântica da Bahia, Brasil

André Monzoli Covre[1*], Fábio Luiz Partelli[1], Robson Bonomo[1], Heder Braun[2] and Cláudio Pagotto Ronchi[3]

*[1]Universidade Federal do Espírito Santo, Rodovia BR-101, Km 60, 29932-540, São Mateus, Espírito Santo, Brazil. [2]Universidade Estadual do Maranhão, São Luiz, Maranhão, Brazil. [3]Universidade Federal de Viçosa, Viçosa, Minas Gerais, Brazil. *Author for correspondence. E-mail: andre-covre@hotmail.com*

ABSTRACT. Extreme temperatures and persistent water stress stand out among the main factors that restrict the vegetative growth and productivity of *Coffea canephora*. The objective of this study was to evaluate the vegetative growth of orthotropic and plagiotropic branches of *C. canephora* under non-irrigated and irrigated conditions, and their correlation with climatic factors in the Atlantic region of Bahia State, Brazil. The experiment was established with two treatments (non-irrigated and irrigated) in a completely random design with 14 replicates. One orthotropic and four plagiotropic branches were labelled on each plant. During the two-year experimental period, the growth of these branches was evaluated at 14-day intervals. Two harvests were performed to obtain productivity data. In summary, it was confirmed that irrigation resulted in an increased productivity of Conilon coffee in the Atlantic region of Bahia, Brazil. The growth rate of the orthotropic and plagiotropic branches was higher in irrigated plants. The growth rate of the plagiotropic branches was limited by the fruit load capacity. The growth rate of *C. canephora* branches was not limited by the minimum average air temperature in the Atlantic region of Bahia, Brazil.

Keywords: *Coffea canephora*, water deficit, temperature, plagiotropic branches.

Introduction

The genus *Coffea* comprises at least 124 species, of which *Coffea arabica* and *C. canephora* are economically important (Davis, Tosh, Ruch, & Fay, 2011). Despite *C. arabica* being the most widely farmed species in the world, the cultivation of *C. canephora* (Conilon coffee) has significantly contributed to coffee production worldwide. In 2015, world coffee production exceeded 144.7 million bags, of which 41.5% were Conilon coffee produced in countries considered as emerging, with Brazil being the second largest producer of this species worldwide (International Coffee Organization [ICO], 2016).

The species *C. canephora* is cultivated in the Atlantic region of Brazil, where it represents an important source of income in many counties. With the availability and implementation of efficient technologies for coffee production, this region has shown a considerable increase in production and productivity in the last few years. However, despite technological advances and apparently favourable climatic conditions for coffee farming within this region, coffee growth, development and productivity have been jeopardized by water deficits and extreme or inadequate temperatures that differ from those considered suitable for its cultivation (DaMatta & Ramalho, 2006).

Temperatures below 13°C and accentuated water deficits affect diverse components of the photosynthetic process because they reduce stomatal conductance, net photosynthesis, the photochemical efficiency of photosystem II, the efficiency of the electron transport chain in the thylakoid membrane, enzymatic activity and carbon metabolism, in addition to modifying the composition and structure of the photosynthetic pigment complexes, lipids and fatty acids, with different intensities recorded for different genotypes and species (Batista-Santos et al., 2011; Ferreira, Partelli, Didonet, Marra, & Braun, 2013; Partelli, Marré, Falqueto, Vieira, & Cavatti, 2013; Scotti-Campos, Pais, Partelli, Batista-Santos, & Ramalho, 2014). On the other hand, high air temperatures may cause the denaturation and aggregation of proteins as well as an increase in the production of reactive oxygen species (DaMatta & Ramalho, 2006), ethylene synthesis and evaporative demand, which may result in stomatal closing and a reduction in the supply of CO_2, which consequently reduces net photosynthesis (Vara Prasad, Allen Júnior, & Boote, 2005) and the production of coffee beans (Craparo, Asten, Laderach, Jassogne, & Grab, 2015; Ovalle-Rivera, Laderach, Bunn, Obersteiner, & Schroth, 2015). When cultivated at temperatures below 17°C or above 31°C, Conilon coffee shows an expressive reduction in the growth rate, which negatively impacts production (Partelli et al., 2013).

A water deficit may be considered a limiting factor for coffee plant growth because the majority of cultivated areas are located in regions with water restrictions (Araújo, Reis, Moraes, Garcia, & Nazário, 2011). Consequently, the farming of Conilon coffee has been conducted predominantly under irrigation because this condition enhances the production of flower buds (Carvalho, Colombo, Scalco, & Morais, 2006), increases the number of plagiotropic branches per plant (Nazareno et al., 2003), enhances the number of flowers per plant (Massarirambi, Chingwara, & Shongwe, 2009) and results in better development and grain formation (Pezzopane, Castro, Pezzopane, Bonomo, & Saraiva, 2010). Therefore, irrigation guarantees high productivity (Bonomo et al., 2013; Sakai et al., 2015) and a final product with a better beverage quality (Fernandes, Partelli, Bonomo, & Golynski et al., 2012).

Climate change represents another limiting factor for coffee production and is responsible for the loss of areas suitable for the cultivation of *C. arabica* and *C. canephora* worldwide (Bunn, Laderach, Rivera, & Kirschke, 2015). The loss of suitable areas for Conilon is significant in Brazil and Vietnam. According to Bunn et al. (2015), Conilon coffee may substitute the farming areas of *C. arabica* due to its tolerance to high temperatures. Nevertheless, this scenario may only be feasible in some regions where Conilon coffee develops better (regions with low intra-seasonal variability). The Furthermore, the sensitivity of Conilon coffee to climate changes is prominent at lower latitudes and altitudes (Bunn et al., 2015).

By studying the physiological and biochemical responses of photosynthesis to elevated atmospheric CO_2 concentration and/or temperature in several *C. arabica* and *C. canephora* genotypes, Rodrigues et al. (2016) demonstrated that predictions concerning the impacts of climate change and global warming should consider the role of CO_2 as a key player in coffee heat tolerance. They also demonstrated a relevant heat resilience of coffee species and showed that an elevated CO_2 concentration remarkably mitigated the impact of heat on coffee physiology. In this regard, and fortunately, future perspectives on the sustainability of the coffee crop that are based on increasing temperature scenarios should not be as catastrophic as previously predicted (Rodrigues et al., 2016).

An understanding of the seasonal fluctuations of *C. canephora* vegetative growth under non-irrigated and irrigated conditions, associated with the climatic conditions within particular growth areas, is an important tool for plant evaluation because it has direct implications for the irrigation management and planning of nutritional programmes of coffee plantations. Unfortunately, references that are currently available on this subject are scarce.

Consequently, the objective of this study was to evaluate the vegetative growth of orthotropic and plagiotropic branches of *C. canephora* under irrigated and non-irrigated conditions, and associate this with the climatic factors of the Atlantic region of Bahia State, Brazil.

Material and methods

The experiment was performed in a commercial plantation in Itabela county, Southern Bahia State, Brazil, at an altitude of 108 m, a latitude of 16°42'13" South and a longitude of 39°25'28" West. The climate was classified as Aw, i.e., tropical with a dry winter season winter and a rainy summer season (Alvares et al., 2013). Three-year-old *Coffea canephora* plants from the clonal variety Emcapa 8111, 'genotype 02' (Bragança, Carvalho, Fonseca, & Ferrão, 2001), which were cultivated in full sunlight, with a 3.5 x 1.0-m spacing and supplied with fertigation since the seedling transplant, were used in the experiment. The plantation was established with four productive stems per plant under a cyclic pruning programme. However, the plantation was evaluated before the pruning stage was reached. Additional cultural treatments were performed according to the technical recommendations for coffee plantations.

The plants were grown in soil classified as a Yellow Latosol (Oxisoil), originally dystrophic with a sandy loam texture (Empresa Brasileira de Pesquisa Agropecuária [Embrapa], 2013) and the following chemical and physical characteristics at a depth of 0-20 cm: pH = 6.25; available P, K, S, B, Cu, Fe, Mn, and Zn contents of 28.5, 105, 15.5, 1.49, 1.9, 450, 20, and 4.2 mg dm^{-3}, respectively; Ca and Mg contents of 4.15 and 1.55 $cmol_c$ dm^{-3}, respectively; potential acidity (H+Al) of 2.85 $cmol_c$ dm^{-3}; organic matter content of 4.55 dag kg^{-1}; sand, silt and clay contents of 730, 110, and 160 g kg^{-1}, respectively; and a volumetric content at field capacity and a permanent wilting point of 0.19 and 0.13 cm^3 cm^{-3}, respectively.

The experimental design was completely random and was based on a split-plot design in time, with 14 replicates. The treatments consisted of irrigated and non-irrigated coffee plants within the main plot and branches measured at different periods within the split-plots. The non-irrigated treatment was established by withholding irrigation, four months before the evaluations were started, to allow the plants to acclimate to the water deficit. In the irrigated treatment, a surface dripping system was used, with one line of emitters per plant row, with a 0.5-m spacing and a flow rate of 2.0 L h^{-1}. Both treatments received dosages of 500, 100, and 400 kg ha^{-1} $year^{-1}$ of N, P and K, respectively, according to the needs of each phenological stage of the coffee plant. In the irrigated treatment, fertigation was performed weekly, while in the non-irrigated treatment, the fertilizers were applied on the soil surface under the coffee plant canopy and the total amount was partitioned over five application times within one year (September, November, January, March and June).

The maximum, average and minimum temperature, global solar radiation, rainfall and relative humidity of the air were collected at an automatic weather station located at a distance of 800 m from the experimental area (Figure 1). The meteorological data were used to determine the reference evapotranspiration (ETo) according to the Penman-Monteith model (Allen, Pereira, Raes, & Smuth, 1998). To manage irrigation, the daily water balance was used based on the crop evapotranspiration (ETc = ETo $_x$ Kc), the measured local rainfall and the soil water storage characteristics. A daily soil water balance (for both irrigated and non-irrigated conditions) was calculated to identify water deficit periods (Figure 1).

On July 28th 2011, one orthotropic (its terminal portion) and two plagiotropic branches per plant were randomly labelled: an old plagiotropic-bearing branch with 12 productive nodes and two completely expanded leaves per node (PlagBB) and the last-released plagiotropic branch within the orthotropic branch (PlagN1). The orthotropic branch (Orto) was marked from its insertion in the branch PlagBB to the apex. In a similar way, in the next year (2012), the two newest plagiotropic branches from the upper orthotropic branch were also labelled: one on January 28th (PlagN2) and the other on April 29th (PlagN3). These two additional plagiotropic branches were, measured due to the loss of strength of the previously–labelled branches (Partelli, Vieira, Silva, & Ramalho, 2010). The length of these branches was recorded at intervals of approximately 14 days during a two-year period (until July 30th 2013). The total growth of each branch was calculated as the difference between the last and the first measurements. In addition, the daily absolute growth ratio was calculated using a relationship based on the difference between the final and initial branch measurements divided by the number of days between the measurements.

To assess the coffee production per plant, two manual harvests were performed in April 2012 and 2013, using the 14 labelled plants within each treatment. The average production of the coffee grains was quantified in litres (L) of fruit *in natura* per plant. After the first harvest, the PlagBB branch was eliminated by pruning and therefore once it lost its bearing capacity for the next harvest period. In a similar manner, the PlagN1 branch was eliminated after the second harvest.

The vegetative growth and production data were subjected to variance analysis ($p < 0.05$) and the mean values were compared using a t test ($p < 0.05$). The statistical analysis was performed using the ASSISTAT 7.7 Beta software (Silva, 2015). The mean value and standard error were calculated for each variable to be included in the graphics.

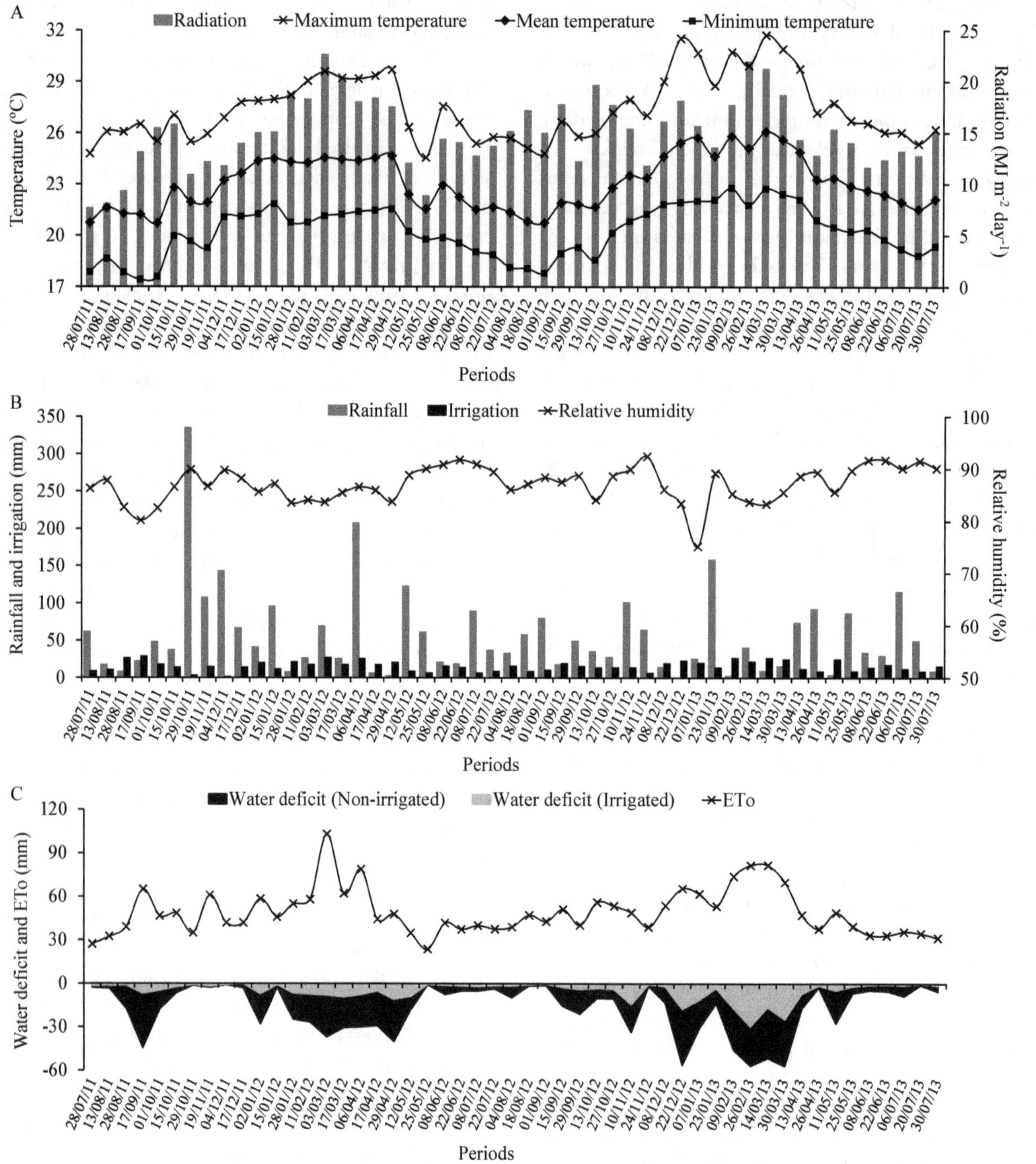

Figure 1. Global solar radiation, maximum, mean and minimum air temperatures (A); rainfall, irrigation and relative humidity (B); water deficit (in non-irrigated and irrigated treatments) and total reference evapotranspiration (ETo) (C), determined during the experimental period between July 2011 and July 2013.

Results and discussion

Under the irrigated treatment, the coffee plants showed greater total vegetative growth in all branch categories, with the exception of PlagN3, which showed a growth pattern similar to the non-irrigated plants (Table 1). The total growth of the orthotropic branches of the irrigated coffee plants was 31.4% higher than that of the non-irrigated coffee plants. Irrigation increased the vegetative growth of PlagBB, PlagN1 and PlagN2 by 68.1, 35.1 and 34.5%, respectively, compared with the non-irrigated coffee plants (Table 1). The lowest total growth was observed for PlagBB, within the non-irrigated treatment (Table 1). The total growth of the plagiotropic branches (PagN1 and PlagN2) was similar to that of the orthotropic branches.

In the Atlantic region of Bahia, Brazil, the growth pattern of the orthotropic and plagiotropic branches of *C. canephora* was not determined by the minimum air temperature (Figures 2, 3 and 4). This is in contrast to the States of Espírito Santo (Partelli et al., 2013) and Rio de Janeiro (Partelli et al., 2010) because the average minimum air temperature, even during the winter period, always remained above 17°C in the present study (Figure 1A). It has been demonstrated that *C. canephora* plants growing at latitudes higher than 15°S show higher vegetative growth rates during periods with longer, hotter days and higher rainfall, while low vegetative growth rates are observed during periods with colder and shorter days (Nazareno et al., 2003; Partelli et al., 2010). *Coffea canephora* plants can endure higher temperatures than *C. arabica* (Ramalho et al., 2014); however, they are less adapted to low temperatures (DaMatta & Ramalho, 2006; Batista-Santos et al., 2011; Ramalho et al., 2014; Scotti-Campos et al., 2014).

The irrigated plants showed a higher growth rate with respect to the orthotropic branch compared with the non-irrigated plants, and the growth pattern of this branch type differed slightly between the evaluation years (Figure 2). The whole area was originally irrigated and the application of mineral nutrients was achieved via fertigation. Consequently, the plants needed to adjust to the water deficit condition resulting from the management modifications imposed to evaluate the non-irrigated condition (Figure 1C). This probably jeopardized the growth rate, even under temperatures that were adequate for growth.

Different coffee genotypes have different adaptation mechanisms for water stress, for example, an increase in stomatal control and the efficiency of water extraction from the soil (DaMatta, & Ramalho, 2006), a deepening of the root system (Pinheiro et al., 2005), and a reduction in leaf area (Cai et al., 2007) and vegetative growth (DaMatta, & Ramalho, 2006), which consequently results in a reduction in photoassimilate production (Sakai, Barbosa, Silveira, & Pires, 2015).

Table 1. Total length of the orthotropic branch (Orto), plagiotropic bearing branch (PlagBB) and new plagiotropic branches (PlagN1, PlagN2 and PlagN3) between July 2011 and July 2013.

Treatment	Total growth (cm)				
	Orto	PlagBB	PlagN1	PlagN2	PlagN3
Non-irrigated	46.1 b*	11.6 b	41.0 b	38.6 b	35.1 a
Irrigated	60.6 a	19.5 a	55.4 a	51.9 a	41.4 a
CV (%)	18.3	43.3	27.6	18.7	27.7

*Mean values followed by the same letter within a column are not significantly different according to the *t* test at a 5% probability level.

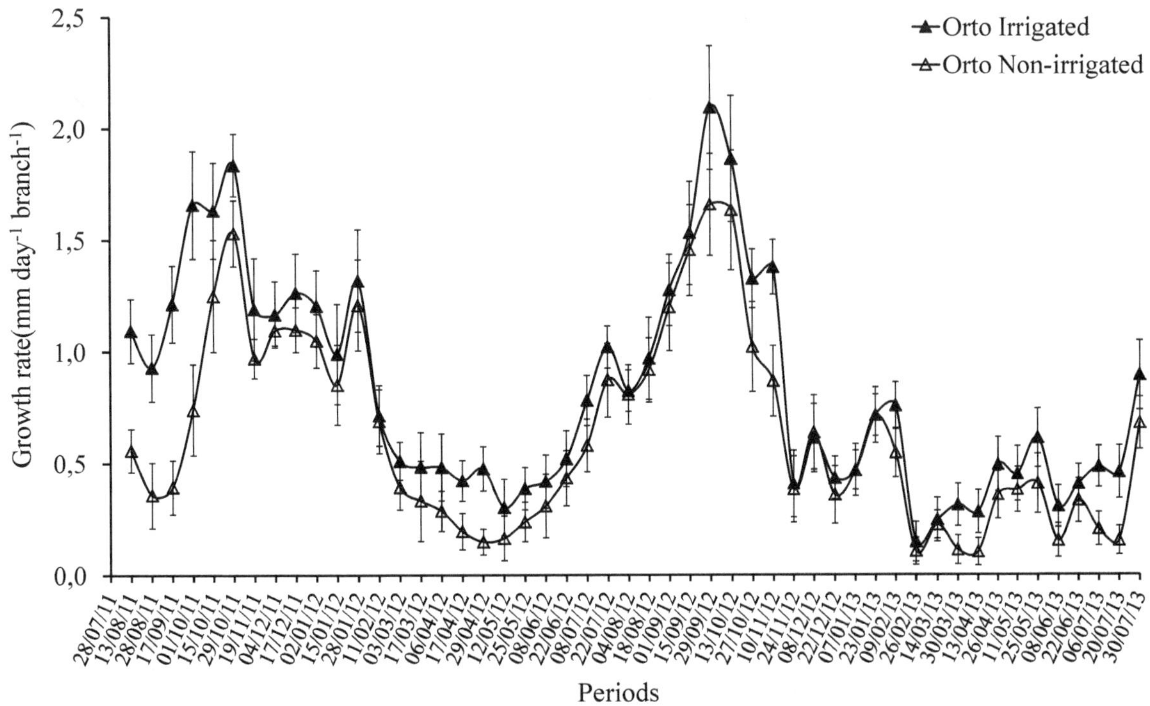

Figure 2. Absolute growth rate of orthotropic branches (Orto) of *Coffea canephora* as a function of the water regime (non-irrigated and irrigated) between July 2011 and July 2013. The graph was plotted using the mean ± mean standard error.

During the first year, higher growth rates were observed during the period from October 2011 to January 2012, while during the second year, higher growth rates were observed during the period from July/August to November 2012 (Figure 2). Within these periods, it is possible to observe a temperature increase in the region, i.e., 3.0 to 5.0°C in the average maximum and minimum temperatures, respectively (Figure 1A), compared with the winter temperatures. According to Partelli et al. (2010), the vegetative growth rate of *C. canephora* cultivated in the North Fluminense region of the Rio de Janeiro State indicates a linear increment as a function of the temperature increase (end of winter and during spring), a condition that is also observed in different genotypes in the Northern region of the Espírito Santo State (Partelli et al., 2013).

In the Atlantic region of the Bahia State, an average maximum temperature above 29.0°C and a water deficit of -30.0 mm (Figure 1) coincided with periods of low vegetative growth of the coffee plants, from February until April 2012, and during December 2012 and April 2013, but without stopping branch growth. During the second harvest, daily maximum temperatures between 34 and 36°C were observed (Figure 1A), mainly on March 11, 13 and 28th, 2013, and April 6th, 2013, corresponding to a period of low vegetative growth (Figure 2). Maximum air temperatures above 31°C result in reduced vegetative growth of *C. canephora* (Partelli et al., 2013).

After the coffee fruits were harvested, which was completed at the end of April 2012 and 2013, the growth rate of the branches remained low (Figures 2, 3, and 4). This may be associated with the stress induced by fruit harvesting, due to the pruning of the non-bearing plagiotropic branches and to sprout thinning procedures.

Higher vegetative growth of the plagiotropic branches occurred between spring and summer (Figure 3 and 4), a period characterized by high rainfall indexes, a gradual increase in temperature within non-limiting levels and a higher availability of light (Figure 1). From the end of October 2011, the PlagBB branch showed a gradual reduction in the net growth rate, until it was approximately equal in both treatments, in the period preceding the harvest (Figure 3). This result was certainly associated with the grain formation stage of the coffee plants, when the requirement for nutrients is high. As fruits constitute preferential sinks for nutrients and photoassimilates (Laviola et al., 2008; Partelli, Espindola, Marré, & Vieira, 2014), vegetative growth is severely jeopardized during this stage, due to an accentuated imbalance in the source:sink ratio of these productive branches, despite these branches (PlagBB) possessing limited autonomy in the production and supply of carbohydrates, as verified for *C. arabica* (Chaves, Martins, Batista, Celin, & DaMatta, 2012). Additionally, the advent of new branches (budding) may reduce the growth of older branches (Ferreira et al., 2013) due to the changes in the source:sink relationship among different branches.

Consistent with the results and discussion above, it was verified that a gradual reduction in the net growth rate from January to March could be observed in the orthotropic branches (Figure 2) and in PlagN1 (Figure 3). However, these branches maintained a considerably high net growth rate, which was not observed in PlagBB (Figure 3). Therefore, the fruiting branches did not lose their growth potential, reinforcing the fact that reproductive growth limits vegetative growth. In fact, Amaral, Rena and Amaral (2006) demonstrated that the growth rate of plagiotropic branches of *C. Arabica* decreased during a period from the end of summer to the end of autumn, for both bearing and non-bearing branches. Nevertheless, these authors also demonstrated that the net growth rates were always higher for the defruited branches.

The PlagN1 branches showed higher net growth rates during the first evaluated period (July 28th, 2011 to July 7th, 2012) compared with the second period (July 22th, 2012 to July 30th, 2013) (Figure 3). This is because these branches did not produce fruit during the first year, in contrast to the second year of evaluation. This behaviour demonstrates that the phenological cycle of coffee has a succession of vegetative and reproductive stages, which occurs within a period of approximately two years (Pezzopane, Pedro Júnior, Camargo, & Fazuoli, 2008).

The growth curves of the PlagN2 and PlagN3 branches indicated similar behaviour during the whole experimental period, with the highest growth values occurring during the spring, between early September and the end of October 2012 (Figure 4). The higher growth values observed from the beginning of November until the end of December 2012, and from the end of January until early April 2013 may be associated with the low amount of rainfall and hence the water deficit, increased temperatures and solar radiation, and the decrease in relative humidity, which was the main driving factor (Figure 1). As the response pattern and the magnitude of the values were similar among the treatments, it may be suggested that the water deficit played less of a role in controlling the growth rates. Nevertheless, with the increase in rainfall amounts recorded in January 2013, the growth rates of the orthotropic and plagiotropic branches showed a small increase (Figures 2, 3 and 4). This means that *C. canephora* responded to the rainfall, a similar behaviour to that observed by Ferreira et al. (2013) for *C. arabica* in the Cerrado Goiano region.

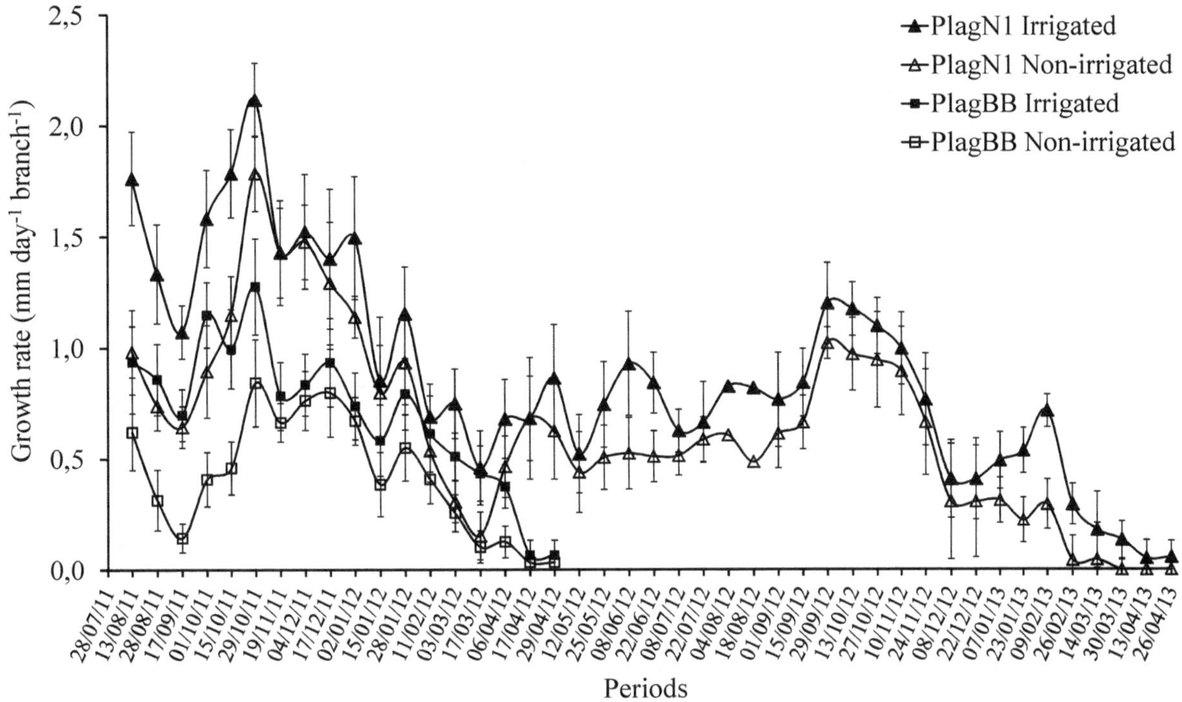

Figure 3. Absolute growth rate of plagiotropic bearing branches (PlagBB) and new plagiotropic branches (PlagN1) of *Coffea canephora*, as a function of the water regime (non-irrigated and irrigated) between July 2011 and July 2013. The graph was plotted using the mean ± mean standard error.

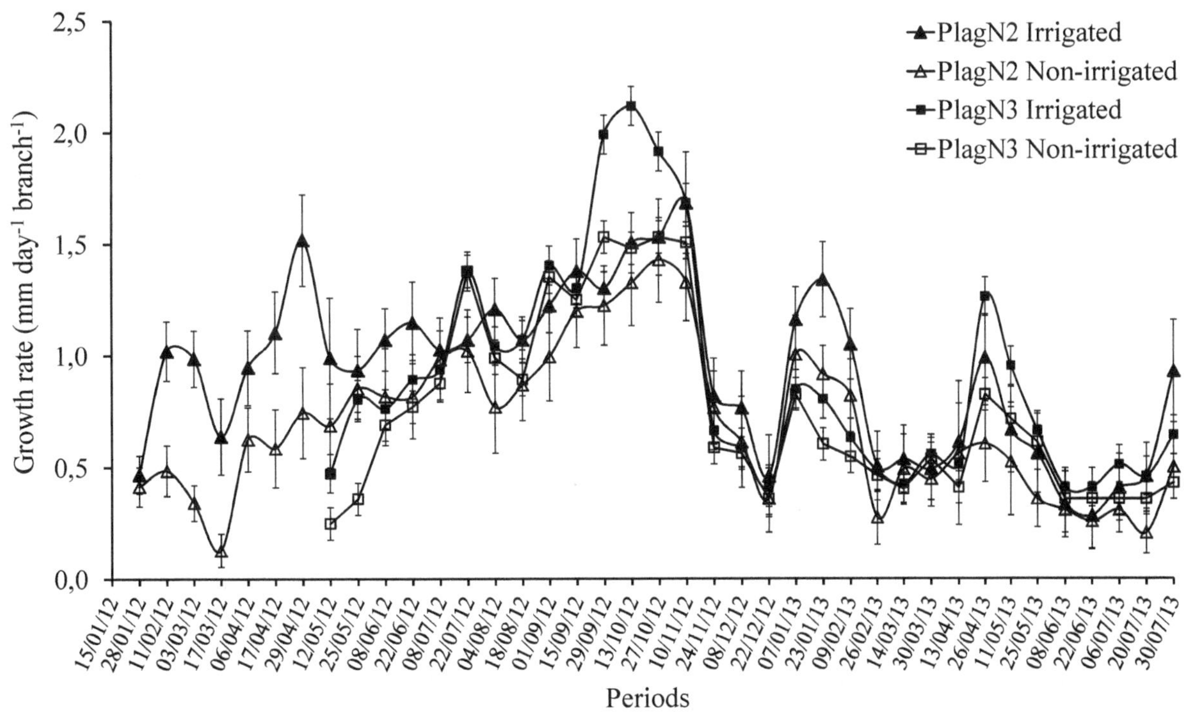

Figure 4. Absolute growth rate of new plagiotropic branches (PlagN2 and PlagN3) of *Coffea canephora*, as a function of the water regime (non-irrigated and irrigated) between July 2011 and July 2013. The graph was plotted using the mean ± mean standard error.

The effect of the water deficit could also be observed on the irrigated plants; however, this effect was lower compared with the non-irrigated plants (Figure 1C). This suggests that the irrigation system used may not have met the water demand of this coffee crop and would need to be configured according to the climatic conditions of the region and the physical characteristics of the soil. According to Soares, Mantovani, Rena, and Soares (2005), a reduction in coffee branch growth results in a lower production of available nodes for bud formation, which consequently reduces fruit production. According to DaMatta (2004), a water deficit affects the development of coffee plant shoots and reduces the leaf area. Further, a reduction in stomatal opening ultimately results in lower CO_2 absorption and a reduction in the photosynthetic rate. A water deficit is the main environmental factor affecting agricultural production in Brazil (DaMatta, Ronchi, Maestri, & Barros, 2010). According to Silva, Cavatte, Morais, Medina, and DaMatta (2013), short water deficit periods may substantially reduce coffee production.

There was no significant difference in the coffee production per plant between the water treatments for the 2012 harvest. Average yields of 22.7 and 22.0 L plant^{-1} were observed in the irrigated and non-irrigated treatments, respectively. However, in 2013, the irrigated plants produced 86.3% more coffee than the non-irrigated plants, resulting in values of 16.1 and 8.6 L plant^{-1}, respectively. It must be emphasized that the branches of non-irrigated plants that produced coffee in the 2012 harvest period were formed in the previous year (this occurs naturally in coffee plants – DaMatta, Ronchi, Maestri, & Barros, 2007), i.e., prior to the suppression of irrigation. As these branches were already completely developed and considering that during the critical period of grain formation, water availability was apparently adequate to guarantee fruit formation (due to adequate rainfall – Figure 1B), the suppression of irrigation did not affect the coffee yield potential. The higher yield of the irrigated plants observed in the second harvest period could be attributed mainly to the high total vegetative growth of PlagN1 (Table 1) because these branches would only produce coffee grains in the second harvest period.

The biennial coffee production effect could be observed in 2013. A biennial cycle is a characteristic of coffee plants that occurs predominantly, and it is associated with excessive fruit yield in the previous year, as evidenced in the 2012 harvest period. The excessive production results in a depletion of plant reserves, affecting growth in the subsequent year (Jaramillo-Botero, Santos, Martinez, Cecon, & Fardin, 2010) and producing alternate cycles of high and low production in coffee plants. Biennial cycling is a naturally occurring process in coffee plants, and not even procedures such as irrigation are able to modify this behaviour (DaMatta et al., 2007), as verified in the present study.

During the period of low production, coffee plants recover their productive capability because the quantity of energy consumed in the process of fructification is small, and the energy supply is focused on vegetative growth and bud differentiation (Sakai, Barbosa, Silveira, & Pires, 2015). In the second harvest period (2013), plants without irrigation showed a coffee reduction of 7.5 L plant^{-1} compared with those that were irrigated. This reinforces the importance of using irrigation with fertilization in *C. canephora* farming in the Atlantic region of the Bahia State.

Irrigation has a positive effect on coffee growth (Nazareno et al., 2003; Ferreira et al., 2013) with respect to the size and quality of the grains (Sakai, Barbosa, Silveira, & Pires, 2013), as well as the yield (Bonomo, Bonomo, Partelli, Souza, & Magiero, 2013; Sakai et al., 2015). Increased coffee plant yield resulting from higher water availability in the soil provided by irrigation is also associated with greater development of the vegetative parts of the plant canopy (Alves, Faria, Guimarães, Muniz, & Silva, 2000), especially the yield components of coffee plants, such as the number of branches per plant, number of nodes per branch, number of fruits per node and dry mass per fruit (DaMatta et al., 2007).

The treatment supplied with irrigation throughout the whole experimental period also showed a reduction in the growth of the orthotropic and plagiotropic branches during periods with high temperature and evapotranspiration, as well as under water deficit conditions that resulted from low rainfall and insufficient irrigation to satisfy the evaporative demand. However, the irrigated plants exhibited the highest rates of vegetative growth compared with the non-irrigated plants during the various evaluations periods. Similar results were reported by Nazareno et al. (2003), Carvalho et al. (2006), Ferreira et al. (2013) and Sakai et al. (2015) for *C. arabica*.

The mean interval period of 14 days between branch growth evaluations was not adequate to conclude which climatic factor is correlated with or explains the behaviour of the vegetative growth of branches of non-irrigated and irrigated Conilon coffee plants because pronounced climatic fluctuations occurred in a short time period in the region under study. Consequently, shorter intervals (weekly or daily) are necessary between branch evaluations to better characterize the behaviour of the seasonal vegetative growth of *C. canephora* as a function of the climatic variables in the Atlantic region of the Bahia State.

Conclusion

Irrigation increased Conilon coffee plant yield in the Atlantic region of Bahia State, Brazil, without changing the biennial pattern.

The growth rate of the orthotropic and plagiotropic branches was higher under irrigation compared with the non-irrigated plants.

The growth rate of the plagiotropic branches depended on their load-bearing capacity.

The maximum growth rates occurred during spring and the minimum growth rates occurred during the endosperm formation period and/or during periods of water deficit, high temperature and low relative humidity.

The growth rate of *C. canephora* branches was not limited by the prevailing average minimum temperatures of the Atlantic region in Bahia, Brazil.

Acknowledgements

The authors thank the Universidade Federal do Espírito Santo (UFES), the Coordenação de Aperfeiçoamento de Pessoal de Nível Superior (CAPES), the Conselho Nacional de Desenvolvimento Científico e Tecnológico (CNPq), the producers Ademir Trevizani, Daniel Trevizani and Luiz Antônio Covre for their support, and the Veracel Celulose company.

References

Allen, R. G., Pereira, L. S., Raes, D., & Smuth, M. (1998). *Crop Evapotranspiration: Guidelines for computing crop water requirements* (Irrigation and Drainage Paper 56). Rome: FAO.

Alvares, C. A., Stape, J. L., Sentelhas, P. C., Gonçalves, J. L. M., Sparovek, G. (2013). Koppen's climate classification map for Brazil. *Meteorologische Zeitschrift*, 22(6), 711-728. doi: 10.1127/0941-2948/2013/0507

Alves, M. E. B., Faria, M. A., Guimarães, R. J., Muniz, J. A., & Silva, E. L. (2000). Crescimento do cafeeiro sob diferentes lâminas de irrigação e fertirrigação. *Revista Brasileira de Engenharia Agrícola e Ambiental*, *4*(2), 219-225. doi: 10.1590/S1415-43662000000200015

Amaral, J. A. T., Rena, A. B., & Amaral, J. F. T. (2006). Crescimento vegetativo sazonal do cafeeiro e sua relação com fotoperíodo, frutificação, resistência estomática e fotossíntese. *Pesquisa Agropecuária Brasileira*, *41*(3), 377-384. doi: 10.1590/S0100-204X2006000300002

Araújo, G. L., Reis, E. F., Moraes, W. B., Garcia, G. de O., & Nazário, A. A. (2011). Influência do déficit hídrico no desenvolvimento inicial de duas cultivares de café conilon. *Irriga*, *16*(2), 15-124. doi: 10.15809/irriga.2011 v16n2p115

Batista-Santos, P., Lidon, F. C., Fortunato, A., Leitão, A. E., Lopes, E., Partelli, F., ... Ramalho, J. C. (2011). The impact of cold on photosynthesis in genotypes of *Coffea* spp. photosystem sensitivity, photoprotective mechanisms and gene expression. *Journal of Plant Physiology*, *168*(8), 792-806. doi: 10.1016/j.jplph. 2010.11.013

Bonomo, D, Z., Bonomo, R., Partelli, F. L., Souza, J. M., & Magiero, M. (2013). Desenvolvimento vegetativo do cafeeiro Conilon submetido a diferentes lâminas de irrigação. *Revista Brasileira de Agricultura Irrigada*, 7(2), 157-169. doi: 10.7127/RBAI.V7N200008

Bragança, S. M., Carvalho, C. H. S., Fonseca, A. F. A., & Ferrão, R. G. (2001). Variedades clonais de café conilon para o estado do Espírito Santo. *Pesquisa Agropecuária Brasileira*, *36*(5), 765-770. doi: 10.1590/S0100-204X200100 0500006

Bunn, C., Laderach, P., Rivera, O. O., & Kirschke, D. (2015). A bitter cup: climate change profile of global production of Arabica and Robusta coffee. *Climatic Change*, *129*(1), 89-101. doi: 10.1007/s10584-014-1306-x

Cai, C. T., Cai, Z. Q., Yao, T. Q., & Qi, X. (2007). Vegetative growth and photosynthesis in coffee plants under different watering and fertilization managements in Yunnan, China. *Photosynthetica, 45*(3), 455-461. doi: 10.1007/s11099-007-0075-4

Carvalho, C. H. M., Colombo, A., Scalco, M. S., & Morais, A. R. (2006). Evolução do crescimento do cafeeiro (*Coffea arabica* L.) irrigado e não irrigado em duas densidades de plantio. *Ciência e Agrotecnologia*, *30*(2), 243-250. doi: 10.1590/S1413-7054200600020 0008

Chaves, A. R. M., Martins, S. C. V., Batista, K. D., Celin, E. F., & DaMatta, F. M. (2012). Varying leaf-to-fruit ratios affect branch growth and dieback, with little to no effect on photosynthesis, carbohydrate or mineral pools, in different canopy positions of field-grown coffee trees. *Environmental and Experimental Botany,* 77(1), 207-218. doi: 10.1016/j.envexpbot.2011.11.011

Craparo, A. C. W., Asten, P. J. A., Laderach, P., Jassogne, L. T. P., & Grab, S. W. (2015). *Coffea arabica* yields decline in Tanzania due to climate change: Global implications. *Agricultural and Forest Meteorology*, *207*(1), 1-10. doi: 10.1016/j.agrformet.2015.03.005

DaMatta, F. M. (2004). Exploring drought tolerance in coffee: a physiological approach with some insights for plant breeding. *Brazilian Journal of Plant Physiology*, *16*(1), 1-6. doi: 10.1590/S1677-04202004000100001

DaMatta, F. M., & Ramalho, J. D. C. (2006). Impacts of drought and temperature stress on coffee physiology and production: a review. *Brazilian Journal of Plant Physiology*, *18*(1), 55-81. doi: 10.1590/S1677-04202006000100006

DaMatta, F. M., Ronchi, C. P., Maestri, M., & Barros, R. S. (2007). Ecophysiology of coffee growth and production. *Brazilian Journal of Plant Physiology*, *19*(4), 485-510. doi: 10.1590/S1677-04202007000400014

DaMatta, F. M., Ronchi, C. P., Maestri, M., & Barros, R. S. (2010). Coffee: environment and crop physiology. In DaMatta, F. M. (Ed.), *Ecophysiology of Tropical Tree Crops* (p. 181-216). New York, USA: Nova Science Publishers.

Davis, A. P., Tosh, J., Ruch, N., & Fay, M. F. (2011). Growing coffee: *Psilanthus* (Rubiaceae) subsumed on the basis of molecular and morphological data implications for the size, morphology, distribution and evolutionary history of *Coffea*. *Botanical Journal of the Linnean Society*, *167*(4), 357-377. doi: 10.1111/j.1095-8339.2011.01177.x

Empresa Brasileira de Pesquisa Agropecuária [Embrapa]. (2013). *Sistema brasileiro de classificação de solos*. Rio de Janeiro, RJ: Centro Nacional de Pesquisa de Solos.

Fernandes, A. L. T., Partelli, F. L., Bonomo, R., & Golynski, A. (2012). A moderna cafeicultura dos cerrados brasileiros. *Pesquisa Agropecuária Tropical*, *42*(2), 231-240. doi: 10.1590/S1983-40632012000200015

Ferreira, E. P. de B., Partelli, F. L., Didonet, A. D., Marra, G. E. R., & Braun, H. (2013). Crescimento vegetativo de *Coffea arabica* L. influenciado por irrigação e fatores climáticos no Cerrado Goiano. *Semina: Ciências Agrárias*, *34*(6), 3235-3244. doi: 10.5433/1679-0359.2013v34n6Supl1p3235

International Coffee Organization [ICO]. (2016). *Total production by all exporting countries*. Retrieved on June 28, 2016 from www.ico.org/prices/po-production.pdf.

Jaramillo-Botero, C., Santos, R. H. S., Martinez, H. E. P., Cecon, P. R., & Fardin, M. P. (2010). Production and vegetative growth of coffee trees under fertilization and shade levels. *Scientia Agricola*, *67*(6), 639-645. doi: 10.1590/S0103-90162010000600004

Laviola, B. G., Martinez, H. E. P., Salomão, L. C. C., Cruz, C. D., Mendonça, S. M., & Rosado, L. (2008). Acúmulo em frutos e variação na concentração foliar de NPK em cafeeiro cultivado em quatro altitudes. *Bioscience Journal*, *24*(1), 19-31.

Massarirambi, M. T., Chingwara, V., & Shongwe, V. D. (2009). The effect of irrigation on synchronization of coffee (*Coffea arabica* L.) flowering and berry ripening at Chipinge, Zimbabwe. *Physical Chemistry Earth*, *34*(13-16), 786-789. doi: 10.1016/j.pce.2009.06.013

Nazareno, R. B., Oliveira, C. A. S., Sanzonowicz, C., Sampaio, J. B. R., Silva, J. C. P., & Guerra, A. F. (2003). Crescimento inicial do cafeeiro Rubi em respostas a doses de nitrogênio fósforo e potássio e a regime hídrico. *Pesquisa Agropecuária Brasileira*, *38*(8), 903-910. doi: 10.1590/S0100-204X2003000800002

Ovalle-Rivera, O., Laderach, P., Bunn, C., Obersteiner, M., & Schroth, G. (2015). Projected shifts in *Coffea arabica* suitability among major global producing regions due to climate change. *Plos One*, *10*(4), 1-13. doi: 10.1371/ journal.pone.0124155

Partelli, F. L., Espindola, M. C., Marré, W. B., & Vieira, D. V. (2014). Dry matter and macronutrient accumulation in fruits of Conilon coffee with different ripening cycles. *Revista Brasileira de Ciência do Solo*, *38*(1), 214-222. doi: 10.1590/S0100-06832014000100021

Partelli, F. L., Marré, W. B., Falqueto, A. R., Vieira, H. D., & Cavatti, P. C. (2013). Seasonal vegetative growth in genotypes of *Coffea canephora*, as related to climatic factors. *Journal of Agricultural Science*, *5*(8), 108-116. doi: 10.5539/jas.v5n8p108

Partelli, F. L., Vieira, H. D., Silva, M. G., & Ramalho, J. C. (2010). Seasonal vegetative growth of different age branches of Conilon coffee tree. *Semina: Ciências Agrárias*, *31*(3), 619-626. doi: 10.5433/1679-0359.2010v31n3p619

Pezzopane, J. R. M., Castro, F. da S., Pezzopane, J. E. M., Bonomo, R., & Saraiva, G. S. (2010). Zoneamento de risco climático para a cultura do café Conilon no Estado do Espírito Santo. *Ciência Agronômica*, *41*(3), 341-348. doi: 10.1590/S1806-66902010000300004

Pezzopane, J. R. M., Pedro Júnior, M. J., Camargo, M. B. P. de, & Fazuoli, L. C. (2008). Exigência térmica do café arábica cv. Mundo Novo no subperíodo florescimento-colheita. *Ciência e Agrotecnologia*, *32*(6), 1781-1786. doi: 10.1590/S1413-70542008000600016

Pinheiro, H. A., DaMatta, F. M., Chaves, A. R. M., Loureiro, M. E., Ducatti, C. (2005). Drought tolerance is associated with rooting depth and stomatal control of water use in clones of *Coffea canephora*. *Annals of Botany*, *96*(1), 101-108. doi: 10.1093/aob/mci154

Ramalho, J. C., DaMatta, F. M., Rodrigues, A. P., Scotti-Campos, P., Pais, I., Batista-Santos, P., ... Leitão A. E. (2014). Cold impact and acclimation response of *Coffea* spp. plants. *Theoretical and Experimental Plant Physiology*, *26*(1), 5-18. doi: 10.1007/s40626-014-0001-7

Rodrigues, W. P., Martins, M. Q., Fortunato, A. S., Rodrigues, A. P., Semedo, J. N., Simões-Costa, M. C., ... Ramalho, J. C. (2016). Long-term elevated air [CO_2] strengthens photosynthetic functioning and mitigates the impact of supra-optimal temperatures in tropical *Coffea arabica* and *C. canephora* species. *Global Change Biology*, *22*(1), 415-431. doi: 10.1111/gcb.13088

Sakai, E., Barbosa, E. A. A., Silveira, J. M. C., & Pires, R. C. de M. (2015). Coffee productivity and root systems in cultivation schemes with different population arrangements and with and without drip irrigation. *Agricultural Water Management*, *148*, 16-23. doi: 10.1016/j.agwat.2014.08.020

Sakai, E., Barbosa, E. A. A., Silveira, J. M. C., & Pires, R. C. de M. (2013). *Coffea arabica* (cv. Catuaí) production and bean size under different population arrangements and soil water availability. *Engenharia Agrícola*, *33*(1), 145-156. doi: 10.1590/S0100-69162013000100015

Scotti-Campos, P., Pais, I. P., Partelli, F. L., Batista-Santos, P., & Ramalho, J. C. (2014). Phospholipids profile in chloroplasts of *Coffea* spp. genotypes differing in cold acclimation ability. *Journal of Plant Physiology*, *171*(3-4), 243-249. doi: 10.1016/j.jplph. 2013.07.007

Silva, F. A. S. (2015). *ASSISTAT - Statistical Assistance versão 7.7 beta* [Software]. Campina Grande:

Universidade Federal de Campina Grande. Retrieved on Dez 13 2015 from www.assistat.com /indexi.html.

Silva, P. E. M., Cavatte, P. C., Morais, L. E., Medina, E. F., & DaMatta, F. M. (2013). The functional divergence of biomass partitioning, carbon gain and water use in *Coffea canephora* in response to the water supply: implications for breeding aimed at improving drought tolerance. *Environmental and Experimental Botany*, *87*(1), 49-57. doi: 10.1016/j.envexpbot.2012. 09.005

Soares, A. R., Mantovani, E. C., Rena, A. B., & Soares, A. A. (2005). Irrigação e fisiologia da floração em cafeeiros adultos na região da zona da mata de Minas Gerais. *Acta Scientiarum. Agronomy*, *27*(1), 117-125. doi: 10.4025/actasci agron.v27i1.2128

Vara Prasad, P. V., Allen Júnior, L. H., & Boote, K. J. (2005). Crop responses to elevated carbon dioxide and interaction with temperature. *Journal of Crop Improvement*, *13*(1), 113-155. doi: 10.1300/J411v13n01_07

6

Growing Princesa apples under semiarid conditions in northeastern Brazil

Paulo Roberto Coelho Lopes[1], Inez Vilar de Morais Oliveira[1], Raissa Rachel Salustriano da Silva[2] and Ítalo Herbert Lucena Cavalcante[3*]

*[1]Empresa Brasileira de Pesquisa Agropecuária, Embrapa Semiárido, Petrolina, Pernambuco, Brazil. [2]Programa de Pós-graduação em Solos e Nutrição de Plantas, Universidade Federal do Piauí, Bom Jesus, Piauí, Brazil. [3]Departmento de Engenharias, Campus "Profa. Cinobelina Elvas", Universidade Federal do Piauí, Bom Jesus, Piauí, Brazil. *Author for Correspondence. E-mail: italohlc@ufpi.edu.br*

ABSTRACT. Phenological observations are some of the most sensitive data for the identification of plant species growth in regional climate conditions. Thus, an experiment was conducted from October 2008 to December 2009 to characterize the phenological stages, fruit set and fruit yield of Princesa apples grown under a semiarid climate in northeastern Brazil. Phenological data (stages) were determined in the orchard by daily observations from bud breaking until fruit ripening. The following variables were evaluated: i) fruit set, which is the relation between the number of flowers and the number of fruits (%); ii) the number of fruit per plant; iii) the fruit production per plant (kg); and iv) the fruit yield (t ha^{-1}). Under semiarid conditions, the phenological cycle of Princesa apples is completed in 123 days and that it is possible to obtain apple yield under semiarid conditions with the pollinator cultivar. Furthermore, additional studies and more evaluation years are necessary to generate an apple production system that is reliable under semiarid conditions.

Keywords: *Malus domestica*, climatic conditions, phenology, cold hours requirement.

Introduction

Apples are a fruit of worldwide horticultural importance, and Brazil is the eleventh top apple producer in the world (FAO, 2010).

Apple fruit trees grow well in temperate climate zones where most commercial varieties satisfy their required chilling temperature, which is often expressed as hours at less than 7°C (TROMP, 2005). In Brazil, the main apple varieties cultivated are Gala and Fuji, and the favorable climatic conditions for cultivation of these varieties are observed only in southern Brazil. More than 90% of the Brazilian apple fruit production comes from these two varieties planted only in the colder regions with altitudes above 900 m in southern Brazil (NACHTIGALL, 2004).

The main obstacle in the production of temperate fruit crops in tropical areas is the lack of effective accumulated chilling because warm winters result in prolonged dormancy leading to poor blooming, strong apical dominance, unsynchronized growth patterns and, consequently, low yields (COOK; JACOBS, 2000). One of the possible alternatives to avoid such problems is using low chilling requirement varieties, such as Anna (NJUGUNA et al., 2004) and Princesa apples, which have a low requirement for chilling from 350 to 450 chill units for satisfactory budding and blooming (NACHTIGALL, 2004). Another alternative is bringing the trees into an artificial dormancy by stopping the irrigation (JONES, 1987) and defoliating them by hand followed or not followed by chemical treatment to break dormancy (using oils or other chemicals).

Due to the worldwide economic importance of apples, some studies have been developed aiming to grow apples under different conditions from those required by this crop. Ashebir et al. (2010) studied apple growth under tropical mountain climate conditions in northern Ethiopia, and they concluded that it is possible to develop new apple production in the northern region of Tigray in Ethiopia where there is an average of 148 hours at less than 7°C per year during the dormant season. Moreover, the beginning of the apple tree blossom stage depends on annual deviations in air temperature, i.e., years with temperatures above normal in late winter and early spring are clearly related to negative anomalies in the date of the beginning of the blossom stage as observed by Chmielewski et al. (2004).

This study may lead to new solutions for crop diversification in irrigated areas of northeastern Brazil to meet fruit producer demands and to ensure the sustainability of fruit production. In addition, the strategy used in this study will enable the market to offer various products in different seasons.

The study of the phenological behavior of crops as a part of a well-characterized environment is important both to obtain satisfactory production results and to determine the most suitable agronomic techniques (VALENTINI et al., 2001), specifically for apples. Moreover, the study of phonological behavior is a useful system to study blooming in woody angiosperms (FOSTER et al., 2003).

The present work aimed at characterizing the phenological stages, fruit set and yield of Princesa apples grown under semiarid climate conditions in northeastern Brazil.

Material and methods

Plant material and growth conditions

Princesa apple (*Malus domestica*) trees propagated by grafting (with 'M9 filter and Marubakaido rootstock) and transplanted in 2007 were used in this study.

The study was conducted from October 2008 to December 2009 on the experimental orchard located in Bebedouro Experimental Station, which belongs to the Brazilian Agricultural Research Corporation (Embrapa Tropical SemiArid) in Petrolina (9°09' S and 40°22' W; at an altitude of 365.5 m above sea level), Pernambuco State, Brazil.

The climate of this region is classified as Bswh (Köeppen), which corresponds to a semiarid region with an average annual temperature of 26°C and average minimum and maximum temperatures of 21.2 and 32.7°C, respectively. The average annual rainfall is 481.7 mm, and the majority of the rainfall is observed during the months of February to April. The dry season occurs from June to November, and the average relative humidity is 67%. During the execution of the experiment, the climatic data were collected by a meteorological station installed inside the experimental station.

Apple trees were defoliated with 1.0% copper sulfate. After defoliation (August 5, 2008 and July 11, 2009), the apple trees were treated with 0.08% hydrogen cyanamide and 3.0% mineral oil (Assist®) following the recommendation of Petri and Palladini (1999). A completely randomized block design was used with treatments constituted by years (2008 and 2009), and the evaluations were subdivided into parcels of time. Four branches from 25 trees were selected at random and examined in 2008 (from October 5 to December 29) and 2009 (from July 11 to November 25) for daily monitoring.

The orchard conduction system was set up with a central leader, and it was composed of Eva, Condessa, Daiane, Gala and Princesa apple cultivars. The Princesa apple cultivar was the pollinator cultivar. In the orchard, one Princesa apple tree was planted for every five trees of the varieties cited. The spacing between the rows was 4.0 m, and the spacing between the trees was 1.3 m. The trees were drip-irrigated each day with ten self-regulated emitters per tree with a flow of 2 L hour^{-1} based on daily evapotranspiration registers recorded by the local Bebedouro Meteorological Station and corrected according to the apple culture coefficient (Kc).

The soil from the study area was classified as a Red-Yellow Oxisol. The fertilization of the orchard was performed according to the recommendations of Nachtigall et al. (2004).

Variables recorded

Phenological data (stages) were determined in the orchard by daily observations from bud break (stage B) (Figure 1) until fruit ripening following

Fleckinger stages (GAUTIER, 1988). The duration of each phenological stage was recorded in number of days.

The following fruit variables were also recorded: i) fruit set, which is the relation between the number of flowers and number of fruits (%); ii) number of fruits per plant; iii) fruit production per plant, which was measured using a Filizola® CF15 brand precision scale (0.5 g precision) and expressed in kilograms (kg); and iv) yield, which was obtained by multiplying the fruit production per plant by the total number of plants in one hectare.

Experimental design and statistical analysis

Statistical analyses included descriptive and correlation analysis between the duration of each plant stage (in days) and each climatic variable in 2008 and 2009, and a t-test was used to compare years. All the calculations were performed using Assistat software.

Results and discussion

The phenological stages of the Princesa apple tree from bud break (stage B) until fruit ripening (GAUTIER, 1988) are shown in Figure 1.

The apple tree vegetative stage lasted from June to July (D_2 stage) with a difference of five days between 2008 and 2009 (Table 1).

In the 2008 and 2009 cycles, the F_2 phase reached 47 and 38 days, respectively, which was longer than that reported by Valentini et al. (2001), who characterized the phenological behavior of 15 different apple varieties and recorded an average duration of 35 days between the dormant bud and full bloom phases. Additionally, Valentini et al. (2001) reported ranges from 30 (Bouras cultivar) to 41 days (Rave V.G. variety) showing that some apple cultivars are compatible to the Princesa apple cultivar grown under a semiarid climate.

Table 1. Days needed to change the phenophases of Princesa apple at 2008 and 2009 cycles in Petrolina, Pernambuco State, Brazil.

Phenophases	Days		Bud —— % ——	
	2008	2009	2008	2009
A	1	1	100.0	100.0
B	15	8	76.1	60.9
C	29	23	3.5	18.3
C_3	30	24	2.1	12.2
D	33	27	3.5	7.8
D_2	34	29	4.2	8.7
E	37	31	2.8	15.6
E_2	40	33	4.2	13.0
F	41	36	4.9	15.6
F_2	47	38	4.2	13.0
G	48	39	5.6	18.3
H	49	41	24.6	24.4
I	51	44	7.7	23.5
J	54	46	4.9	33.0
L	146	126	2.33	10.2

A: dormant buds; B: beginning of swelling; C: green tips; C_3: half-inch green; D: half-inch green leafless; D_2: half-inch green leaves, E: green button, E_2: pink button, F: flowering start; F_2: full bloom; G: end of flowering; H: petal fall, I: fruit set; J: green fruit and L: fruit ripe.

The duration of each phenological stage differed between years studied (Table 1) showing that all stages in 2008 were longer than in 2009. The beginning of flowering (F stage; first flower) occurred five days earlier in 2009 than in 2008 (Table 1). The shorter phenological cycle from 2008 to 2009 (20 days in total) was caused by climatic factors, such as lower average temperature, lower global radiation, higher levels of relative humidity and higher levels of rainfall (Figure 2).

Figure 1. Phenological stages of Princesa apple tree, Petrolina, Pernambuco State, Brazil, 2009. A: dormant buds; B: beginning of swelling; C: green tips; C_3: half-inch green; D: half-inch green leafless; D_2: half-inch green leaves, E: green button, E_2: pink button, F: flowering start; F_2: full bloom; G: end of flowering; H: petal fall, I: fruit set; J: green fruit and L: fruit ripe.

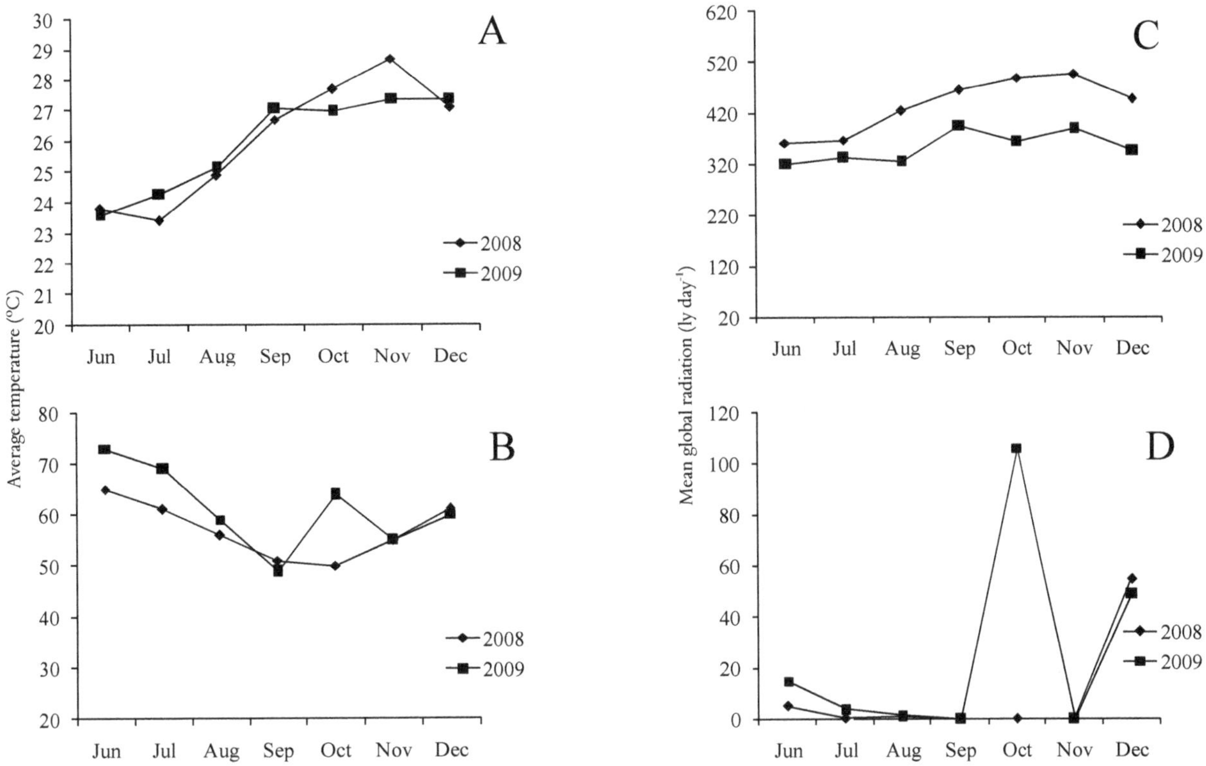

Figure 2. Average temperature (A), mean global radiation (B), relative humidity (C) and rainfall (D), from June to December 2008 and 2009 in Petrolina, Pernambuco State, Brazil.

Independently of the year (Tables 2 and 3), the apple phenophases were directly and positively ($p < 0.01$) correlated with air temperature with a range between 0.52 (2009) and 0.90 (2008). These results demonstrated that higher air temperatures generated longer phenological stages. Apple trees are commonly cultivated around the world in temperate zones where the relationship between air temperature and phenological stage duration is inverse. In a field study on apples in Germany, Chmielewski et al. (2004) observed that all phenophases are well-correlated to the average air temperature, thus indicating that higher temperatures after the winter accelerate developmental processes and ultimately lead to an advanced timing of spring events.

The results of the present study agreed with the findings of Sparks et al. (2000) and Chmielewski and Rötzer (2001), who explained that there are positive correlations between phenological phases and temperature, thus confirming the relationship between changes in air temperature and changes in the timing of the phenological stages.

With regard to the difference of apple tree phenological behavior in temperate and semiarid climates, it is important to remember that apple is a temperate crop and that it grows well in temperate climate zones where most commercial varieties satisfy their required chilling temperature, which is often expressed as the number of hours at less than 7°C (TROMP, 2005).

Chilling units accumulated during the cold season enable the plant to overcome dormancy (LEGAVE et al., 2008). Moreover, Iuchi (2006) explain that plant growth, phenology and fruit production are influenced by climatic conditions, such as air temperature, which may explain the disagreement of results from temperate and semiarid climates. According to Chmielewski et al. (2004), plant development and phenology strongly depend on air temperature after the release of dormancy. In the present study, it was not possible to accumulate cold hours, which highlighted the importance of irrigation efficiency. Chmielewski et al. (2004) explained that if temperature is increased, the biochemical reactions are accelerated up to a threshold where enzyme systems are destroyed and cells die. In the present work, this acceleration did not happen because the air temperatures during the time studied did not reach the threshold value for the Princesa apple cultivar. Furthermore, phenology is a sensitive biosphere indicator of climate change (WALTHER et al., 2002).

In addition to air temperature, relative humidity and rainfall were also positively correlated with the duration of apple phenophases in 2008 (Table 2), with an extremely high correlation coefficient (r = 0.86) for rainfall. According to this result, the rainfall in October of 2008 had more influence on the apple phenophases than relative humidity.

Table 2. Simple correlation coefficients (*r*) between climatic factors and phenophases of Princesa apples in 2008. Petrolina, Pernambuco State, Brazil.

	Phenophases[1]	Air Temperature[2]	RH[3]	Radiation[4]	Rainfall[5]
Phenophases[1]	-	0.5997*	0.5289*	-0.3868ns	0.8596**
Temperature[2]	-	-	0.4083ns	-0.3697ns	0.2424ns
UH[3]	-	-	-	-0.1134ns	0.3016ns
Radiation[4]	-	-	-	-	-0.3530ns
Rainfall[5]	-	-	-	-	-

1: Duration of phenological stages; 2: Average air temperature; 3: Relative Humidity; 4: Average Global Radiation; ★★: significant at 1% probability; ★: significant at 5% probability; ns: not significant.

Table 3. Simple correlation coefficients (*r*) between climatic factors and phenophases of Princesa apples in 2009. Petrolina, Pernambuco State, Brazil.

	Phenophases[1]	Air Temperature[2]	RH[3]	Radiation[4]	Rainfall[5]
Phenophases[1]	-	0.5188*	-0.2465ns	0.3254ns	0.4130ns
Temperature[2]	-	-	-0.5501*	0.2635ns	0.1778ns
UH[3]	-	-	-	- 0.4543ns	-0.0191ns
Radiation[4]	-	-	-	-	0.0353ns
Rainfall[5]	-	-	-	-	-

1: Duration of phenological phases; 2: Average air temperature; 3: Relative Humidity; 4: Average Global Radiation; ★: significant at 5% probability; ns: not significant.

Table 4 shows that all evaluated fruit characteristics were drastically and statistically higher in 2009 than in 2008, which may have been caused by the climatic conditions during this period. The fruit set average of Princesa apples in 2009 was higher than in 2008. This average (Table 4) was considered high because apples have a high rate of abscission; thus, they retain only 4-10% of their potential fruits according to Iuchi (2006).

The number of fruits per plant increased by 64.96% from 2008 to 2009 (Table 4). Independently of the year studied, all plants produced a lower number of fruits as compared to the number of fruits reported by others. Biffi and Rafaeli Neto (2008) studied Fuji apples in Brazil and reported a range of 394.43 - 293.09 for the number of fruits, and Fallahi (2007) studied Ryan Spur Delicious apples in the USA and reported a range of 624-630 for the number of fruits. In addition to the climate differences, this inferiority may be explained by the age of the plants in the present study because they were young and not at their full capacity for fruit production.

Fruit production per plant presented the same tendency of other fruit variables, i.e., results from 2009 were extremely higher than those from 2008 (Table 4). Fruit production per plant of the present study was lower than the results of Biffi and Rafaeli Neto (2008) in Brazil and Liu et al. (2008) in China. However, the fruit production results of the present study were higher than the 8.57 kg per plant reported by Santos et al. (2007) for the Clone 217 genotype in Portugal.

Yield is one of the most important characteristics of apple crops. From 2008 to 2009, the yield was increased by 9.16 t ha^{-1} (Table 4). The average values shown in Table 4 were lower than those presented by Gul (2005) in Turkey (24.25 t ha^{-1}) and by Nava and Dechen (2009) for Fuji apples in Brazil (16.1 t ha^{-1}). Moreover, the average yield values of the present study were compatible to the values reported by Naor et al. (2008) for 15-year-old Smoothie (a Golden delicious strain) apples grown in Israel (13.1 t ha^{-1}) and superior to the range reported by Di Vaio et al. (2009) for Annurca Rossa del Sud apples in Italy (7.4 - 25.1 t ha^{-1}).

Importantly, apple trees take four to five years to produce their first fruit (FERREE; WARRINGTON, 2003). In the present study, the trees were transplanted in 2007, so the fruit production was anticipated in three or four years. Moreover, it is common that the first harvest presents a low yield. Therefore, the continuous evaluation of the next harvests has crucial importance to determine and consolidate the viability of apple crops under semiarid conditions in northeastern Brazil because the preliminary results are satisfactory.

Table 4. Fruit set, number of fruits per plant (NFP), fruit production per plant (P/P) and yield of Princesa apples. Petrolina, Pernambuco State, Brazil.

	Fruit Set (%)		NFP		P/P (kg)		Yield (t ha^{-1})	
	2008	2009	2008	2009	2008	2009	2008	2009
Magnitude	2.86	9.00	47.00	139.00	5.12	17.35	10.46	33.37
VC (%)	6.28	20.83	16.17	45.35	14.64	45.92	14.64	45.92
Variance	1.94	4.34	234.92	302.71	2.41	9.24	10.73	34.19
Average	2.33 b	10.20 a	20.48 b	58.45 a	1.54 b	6.62 a	3.57 b	12.73 a
SD (%)	1.08	2.08	9.50	26.50	0.35	3.04	1.28	5.84

VC=variation coefficient; SD=standard deviation.

Accordingly, most apple cultivars require cross-pollination using a compatible cultivar to obtain desirable set fruit, and the Princesa apple cultivar is used for cross-pollination and not for yield. In comparison to apple cultivars traditionally grown for yield in productive regions, Princesa apple results were previously expected to be lower, but according to the present work, this apple cultivar (Princesa) has potential for apple production under semiarid climate conditions. However, further studies are necessary.

Conclusion

Thus, the results of this study indicated that the phenological cycle of Princesa apples is completed in 123 days and that it is possible to obtain apple yield under semiarid conditions with the pollinator cultivar. Furthermore, additional studies and more evaluation years are necessary to generate an apple production system that is reliable under semiarid conditions.

References

ASHEBIR, D.; DECKERS, T.; NYSSEN, J.; BIHON, W.; TSEGAY, A.; TEKIE, H.; POESEN, J.; HAILE, M.; WONDUMAGEGNEHEU, F.; RAES, D.; BEHAILU, M.; DECKER, J. Growing apple (*Malus domestica*) under tropical mountain climate conditions in Northern Ethiopia. **Experimental Agriculture**, v. 46, n. 1, p. 53-65, 2010.

BIFFI, L. J.; RAFAELI NETO, L. S. Comportamento espacial de variáveis agronômicas da maçã 'Fuji' durante dois anos de observações no planalto serrano de Santa Catarina. **Revista Brasileira de Fruticultura**, v. 30, n. 4, p. 975-980, 2008.

CHMIELEWSKI, F.-M.; MÜLLER, A.; BRUNS, E. Climate changes and trends in phenology of fruit trees and field crops in Germany, 1961-2000. **Agricultural and Forest Meteorology**, v. 121, n. 1-2, p. 69-78, 2004.

CHMIELEWSKI, F.-M.; RÖTZER, T. Response of tree phenology to climate change across Europe. **Agricultural and Forest Meteorology**, v. 108, n. 2, p. 101-112, 2001.

COOK, N.; JACOBS, G. Progression of apple (*Malus* × *domestica* Borkh.) bud dormancy in two mild winter climates. **Journal of Horticultural Science and Biotechnology**, v. 75, n. 2, p. 233-236, 2000.

DI VAIO, C.; CIRILLO, C.; BUCCHERI, M.; LIMONGELLI, F. Effect of interstock (M.9 and M.27) on vegetative growth and yield of apple trees (cv "Annurca"). **Scientia Horticulturae**, v. 119, n. 3, p. 270-274, 2009.

FAO-Food and Agriculture Organization of the United Nations. **FAOSTAT**: statistics database. Available from: <http://www.apps.fao.org/>. Access on: Nov. 18, 2010.

FALLAHI, E. Influence of 1-aminoethoxyvinylglycine hydrochloride and α-naphthalene acetic acid on fruit retention, quality, evolved ethylene, and respiration in apples. **International Journal of Plant Production**, v. 1, n. 1, p. 53-61, 2007.

FERREE, D. C.; WARRINGTON, I. J. **Apples**: botany, production, and uses. Wallingford: CABI Publishing, 2003.

FOSTER, T.; JOHNSTON, R.; SELEZNYOVA, A. A morphological and quantitative characterization of early floral development in apple (*Malus domestica* Borkh.). **Annals of Botany**, v. 92, n. 2, p. 199-206, 2003.

GAUTIER, M. **Les productions fruitières**. Paris: Lavoisier, 1988.

GUL, M. Technical efficiency and productivity of apple farming in Antalya province of Turkey. **Pakistan Journal of Biological Sciences**, v. 8, n. 11, p. 1533-1540, 2005.

IUCHI, V. L. Botânica e fisiologia. In: EPAGRI. **A cultura da macieira**. Florianópolis: Epagri, 2006. p. 59-104.

JONES, H. G. Repeat flowering in apple caused by water stress or defoliation. **Trees – Structure and Function**, v. 1, n. 3, p. 135-138, 1987.

LEGAVE, J. M.; FARRERA, I.; ALMERAS, T.; CALLEJA, M. Selecting models of apple flowering time and understanding how global warming has had an impact on this trait. **Journal of Horticultural Science and Biotechnology**, v. 83, n. 1, p. 76-84, 2008.

LIU, C.; HAN, M.; ZHANG, L. The effects of fertilizer application at early summer on growth, yield and quality of Fuji apple in Weibei Highland. **Agricultural Research in the Arid Areas**, v. 26, n. 1, p. 124-137, 2008.

NACHTIGALL, G. R. **Maçã**: produção. Brasília: Embrapa Informação Tecnológica, 2004. (Frutas do Brasil, 37).

NAOR, A.; NASCHITZ, S.; PERES, M.; GAL, Y. Responses of apple fruit size to tree water status and crop load. **Tree Physiology**, v. 28, n. 2, p. 1255-1261, 2008.

NAVA, G.; DECHEN, A. R. Long-term annual fertilization with nitrogen and potassium affect yield and mineral composition of 'Fuji' apple. **Scientia Agricola**, v. 66, n. 3, p. 377-385, 2009.

NJUGUNA, J. K.; LEONARD, S. W.; TEDDY, E. M. Temperate fruits production in the tropics: a review on apples in Kenya. **HortScience**, v. 39, n. 4, p. 841-841, 2004.

PETRI, J. L.; PALLADINI, L. A. Eficiência de diferentes volumes e concentrações de calda para quebra de dormência na macieira cultivar 'Gala'. **Pesquisa Agropecuária Brasileira**, v. 34, n. 8, p. 1491-1495, 1999.

SANTOS, A.; LOPES, A.; SÁ, M.; LOUSADA, J. L. Relações de produtividade, área folhear e alternância na macieira Bravo de Esmolfe. **Revista de Ciências Agrárias**, v. 31, n. 1, p. 132-138, 2007.

SPARKS, T. H.; JEFFREE, E. P.; JEFFREE, C. E. An examination of the relationship between flowering times and temperature at the national scale using long-term phenological records from the UK. **International Journal of Biometeorology**, v. 44, n. 2, p. 82-87, 2000.

TROMP, J. Dormancy. In: TROMP, J.; WEBSTER, A. D.; WERTHEIM, S. J. (Ed.). **Fundamentals of temperate zone tree fruit production**. Leiden: Backhuys Publishers BV, 2005. p. 65-73.

VALENTINI, N.; ME, G.; FERRERO, R.; SPANNA, F. Use of bioclimatic indexes to characterize phenological phases of apple varieties in Northern Italy. **International Journal of Biometeorology**, v. 45, n. 4, p. 191-195, 2001.

WALTHER, G.-R.; POST, E.; CONVEY, P.; MENZEL, A.; PARMESAN, C.; BEEBEE, T. R. C.; FROMENTIN, J.-M.; HOEGH-GULDBERG, O.; BAIRLEIN, F. Ecological responses to recent climate change. **Nature**, v. 416, p. 389-395, 2002.

7

Germination and development of *Amburana cearensis* seedlings as influenced by seed weight, light and temperature

João Paulo Nobre de Almeida[1*], Bruno Franca da Trindade Lessa[2], Charles Lobo Pinheiro[3], Fernanda Melo Gomes[3], Sebastião Medeiros Filho[3] and Cydianne Cavalcante Silva[1]

*[1]Universidade Federal Rural do Semi-Árido, Cx. Postal 137, 59625-900, Mossoró, Rio Grande do Norte, Brazil. [2]Universidade Federal do Vale do São Francisco, Petrolina, Pernambuco, Brazil. [3]Universidade Federal do Ceará, Fortaleza, Ceara, Brazil. *Author for corresponde. E-mail: joaopaulonobre@yahoo.com.br*

ABSTRACT. The objective of this work was to study the seed germination and seedling growth of *A. cearensis* (Allemão) A. C. Sm. influenced by seed weight and under different temperature and light conditions. Seeds were individually weighed and divided into three weight classes (light, medium and heavy), and each was subjected to germination tests under two light conditions (presence or absence) and six temperature regimes (20, 25, 30, 35, 40, and 20-30°C). The experiment was established with a completely randomized design in a split-split plot arrangement with four replicates. We determined the first count, final germination percentage and germination speed index. To assess seedling growth, the length and dry mass of the epicotyl and root + hypocotyl axis were also determined. The data were subjected to Tukey's test ($p \leq 0.05$). The optimal conditions for seed germination occurred at a temperature of 30°C, with light and medium seeds showing greater vigour. Temperatures of 35 and 40°C were detrimental to the growth of the species, whereas the best conditions occurred with light and medium seeds at temperatures of 25 and 30°C in the presence of light.

Keywords: cumaru, abiotic factors, seed vigour.

Introduction

The predatory extraction suffered by some important plant species native to the Caatinga biome is putting them at risk of extinction. The survival of *Amburana cearensis* (Allemão) A. C. Sm., commonly known in Brazil as 'cumaru' or 'imburana-de-cheiro', is threatened, given that the simple removal of its bark (medicinal part) causes great damage to the plant (Canuto, Silveira, Bezerra, Leal, & Viana, 2008). It is necessary to preserve this species as, despite its unquestionable economic and social importance, due to logging and the wide use of ther parts of the plant in folk medicine, its predatory use continues to decimate large areas (Almeida et al., 2010). However, primary information about *A. cearensis*is limited and scattered, particularly regarding its propagation and ecology, which makes studies on the germination and seedling establishment of this species necessary to enable agronomic management including appropriate techniques for its proliferation.

The classification of seeds by weight is a strategy that can be used to standardize seedling emergence in the field and to obtain seedlings with similar sizes and/or greater vigour (Carvalho & Nakagawa, 2012). Seed weight is a key variable in the production process, as it is directly related to seed quality (Silva, Maia, &

Moraes, 2007). This is because heavier seeds were better nourished during their development and are therefore more vigorous due to well-formed embryos and larger amounts of reserves (Carvalho & Nakagawa, 2012).

Several studies have been conducted to assess the influence of seed weight on the germination and physiological quality of many plant species. However, studies are still quite scarce for forest species, with some species such *Hymenaea stigonocarpa*, *Tabebuia heptaphylla*, *A. cearensis* and *Enterolobium contortisiliquum* being citedamong them (Pereira, Giraldelli, Laura, & Souza, 2011; Ribeiro, Costa, Senna, & Caliman, 2012; Almeida, Pinheiro, Lessa, Gomes, & Medeiros Filho, 2014; Lessa, Almeida, Pinheiro, Gomes, & Medeiros Filho, 2015).

The ecological significance of the relationships among seed weight, light requirements and temperature fluctuation seems to be related to the need to avoid germination in deep soil, where it is difficult for small seeds to emerge (Thompson & Grime, 1983). In some species, the light requirement for seed germination is strongly influenced by the temperature range over which seeds can germinate, which is characteristic of each species (Bewley, Bradford, Hilhorst, & Nonogaki, 2013). Temperature directly influences the germination percentage, affecting both the water absorption rate and biochemical reactions (Amato et al., 2007). According to their response to the presence of light, seeds can be classified as positively photoblastic (benefited by light), negatively photoblastic (negatively affected by light) ornon-photoblastic or indifferent (Marcos Filho, 2015).

Some researchers have studied the temperature and type of light suitable for the seed germination of native species such as *Licania rigida* Benth. (Diniz, Moreira, Silva, & Medeiros Filho, 2008) and *Luetzelburgia auriculata* (Nogueira, Silva, Bezerra, & Medeiros Filho, 2012). Nevertheless, there is a lack of information regarding the direct effects of seed reserves and environmental factors on the germination process of forest species.

Thus, this work aimed at studying the germination and development of *A. Cearensis* seedlings from different seed weight classes under different temperature and light conditions.

Material and methods

We randomly collected seeds of *A. cearensis* from the ripe fruit of 20 adult plants at the Experimental Farm Vale do Curu in Pentecoste, Ceará State, 110 km from Fortaleza. The harvested fruit were packed in polyethylene bags and sent to the Seed Analysis Laboratory at the Federal University of Ceará (UFC) in Fortaleza, Ceará State, where the seeds were extracted, manually processed and dried under shade conditions for six days. Thereafter, the seeds were packed in polyethylene bags and stored at ±12°C and 50% relative humidity.

Four hundred seeds were separated from the total seed sample and individually weighed using an analytical balance (0.0001 g) to classify the sample into three distinct classes based on their specific weight (Figure 1), defined as light (< 0.25 g), medium (0.25 - 0.35 g) and heavy (> 0.35 g). Thereafter, the seeds from the submitted sample were individually weighed and classified into their respective classes.

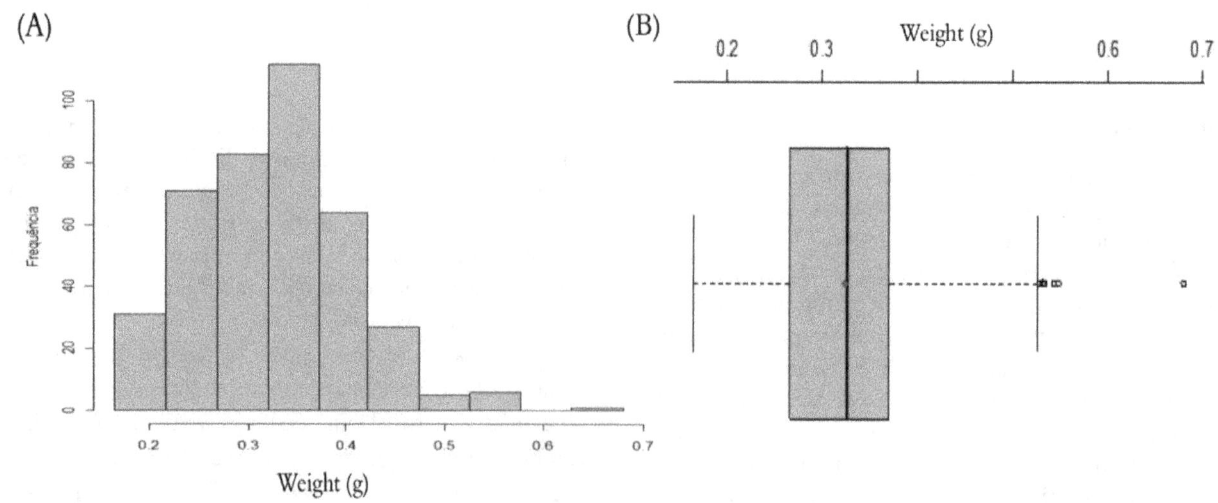

Figure 1. Frequency histogram (A) and box plots (B) for the weight distribution of a sample of 400 *Amburana cearensis* (Allemão) A.C. Sm. seeds.

The water content and weight of a thousand seeds from every class were determined based on the recommendations of the Rules for Seed Analysis (Brasil, 2009). Seed sterilization was performed by immersing all the seeds from every class in 5% sodium hypochlorite for 15 minutes and a 0.4% nystatin® solution to prevent fungal growth.

The effects of temperature, light and weight were studied using a completely randomized experiment witha split-split plot design, with temperature regimes applied to the whole plots (20, 25, 30, 35, 40, or 20-30°C) and light conditions applied to the subplots (presence or absence). The sub-subplots consisted of seed weight classes (light, medium and heavy) with four replicates, and every experimental unit was one paper roll containing 25 seeds.

Seeds from every weight class were placed in rolls of blotting paper moistened with distilled water (2.5 times the paper weight) to germinate. These rolls were packed in transparent plastic bags to reduce water loss in B.O.D. (Biochemical Oxigen Demand) incubators regulated at constant (20, 25, 30, 35, and 40°C) and alternating temperatures (20 - 30°C) with a photoperiod of 12 hours under two light conditions (presence or absence). The treatments under darkness included paper rolls with seeds wrapped in aluminium foil and black bags, and the first count was performed under green safe light.

Germination was assessed on a daily basis, and the substrate was moistened when necessary. The germinated seeds showed radicle protrusion equal to or greater than 2 mm. At the end of the test, we recorded the germination percentage (first and final count at 7 and 21 days, respectively) and the germination speed index (GSI), which was determined based on the daily counts of the number of germinated seedlings up to 21 days after sowing, based on Maguire (1962).

After 21 days, the length of normal seedlings was assessed by the portion above (epicotyl length - EPL) and below (root + hypocotyl axis - RHL) the cotyledon insertion; these results were expressed in cm.seedling^{-1}. After removing the cotyledons, the epicotyl and root + hypocotyl fractions of every experimental unit were dried in a forced air oven (at 80°C for 24 hours) to determine the epicotyl dry mass (EPDM) and root + hypocotyl dry mass (RHDM). Thereafter, the seedlings were weighed using an analytical balance with an accuracy of 0.0001 g; these results were expressed in g seedling^{-1}. Furthermore, the total length and dry biomass of the seedlings were determined by the following calculations: EPL + RHL and EPDM + RHDM, respectively.

These data were evaluated by the Shapiro-Wilk and Kolmogorov tests to assess their normality. Analysis of variance (ANOVA) and Tukey's test (5%) were then conducted for the normally distributed data. The Box-Cox transformation was used for the non-normal data, and the non-parametric Mann-Whitney test (5%) was conducted when the normal distribution assumption was not met. Assistat 7.6 beta software was used to perform the ANOVA and Tukey's test, whereas the normality and Mann-Whitney tests were performed using the Action 2.4 software.

Results and discussion

Seeds from the three weight classes had a water content of approximately 9.4% (Table 1), regardless of their weight, showing that the classes did not influence the hygroscopic equilibrium between the seeds and the environment. A significant increase in the thousand-seed weight due to the increase in weight categories is also observed in Table 1, wherein a marked difference of more than 180 grams between heavy and light seeds stands out. This reveals that *A. cearensis* seeds have great weight variability, which is a critical factor in the perpetuation and survival of this species in highly competitive environments.

Table 1. Water content (WC) and thousand-seed weight (TSW) of Amburana cearensis (Allemão) A.C. Sm. seeds as influenced by weight classes.

Weight Classes	WC	TSW
	---- % ----	--- g ---
Light	9.4	221.4 c
Medium	9.3	285.4 b
Heavy	9.4	403.3 a
CV (%)	-	1.85

Means followed by the same letter do not differ by Tukey's test at the 5% probability level.

There were significant differences within the isolated factors and their interactions for the germination and seedling growth variables (Table 2). The interaction between light and temperature as well as between weight and temperature were highly significant for the germination speed index. For the root + hypocotyl dry mass and the total dry mass, there was only an interaction between temperature and light, whereas for the root + hypocotyl length, there was an interaction between temperature and seed weight. In turn, the mutual influence of the three factors (temperature, light and weight) was critical to the epicotyl length and epicotyl dry mass. The germination percentage, first count and total length data did not meet the normal distribution assumption, even after the data transformation, unlike the length of the root + hypocotyl axis, which met the normality assumption after its transformation.

Table 2. Summary of ANOVA for germination speed index (GSI), epicotyl length (EPL), root + hypocotyl axis (RHL), epicotyl dry mass (EPDM), root + hypocotyl dry mass (RHDM) and total dry biomass (TDB) of *Amburana cearensis* (Allemão) A.C. Sm.

FV	GL	GSI	EPL	RHL	EPDM	RHDM	TDB
		Mean Square					
Temp.	5	16.63**	118.91**	152.74**	$1.17x10^{-3}$**	$1.24x10^{-3}$**	$4.64x10^{-3}$**
error 1	18	0.32	0.70	0.56	0	10^{-5}	$2x10^{-5}$
Light	1	1.15 ns	89.36**	40.78**	$3x10^{-4}$**	$9.1x10^{-4}$**	$2.32x10^{-3}$**
T x L	5	3.62**	6.99**	3.25 ns	$6x10^{-5}$**	$6x10^{-5}$**	$1.7x10^{-4}$**
error 2	18	0.26	0.88	1.72	0	10^{-5}	$2x10^{-5}$
Weight	2	41.33**	18.01**	19.35**	$5x10^{-5}$**	10^{-5} ns	10^{-4}*
T x W	10	2.17**	5.76**	4.21**	$2x10^{-5}$*	10^{-5} ns	$4x10^{-5}$ ns
L x W	2	0.57 ns	0.42 ns	0.23 ns	10^{-5} ns	10^{-5} ns	$3x10^{-5}$ ns
T x L x W	10	0.45 ns	0.26*	0.5 ns	10^{-6}*	10^{-6} ns	10^{-5} ns
error 3	72	0.32	0.88	0.93	10^{-5}	10^{-5}	$3x10^{-5}$
Total	143						
CV 1		19.5	28.57	23.71	22.88	24.14	22.27
CV 2		17.63	32.1	41.56	23.25	23.98	22.46
CV 3		19.45	31.89	30.68	35.01	24.8	28.18

** $p < 0.01$; * $p < 0.05$; ns: not significant, based onthe F-test.

By analysing the germination behaviour of *A. cearensis* seeds, it is possible to see marked differences in the effects of environmental factors and seed reserves onthe first count and final germination percentage. Seed germination was observed both in the presence and absence of light (Figure 2A), which characterizes the seeds as neutrally photoblastic or insensitive to light. These results are similar, in part, to those reported by Lessa et al. (2015), who also found that the germination of *Enterolobium contortisiliquum* seeds occurs regardless of the light conditions.

Means followed by the same letter do not differ by the non-parametric Mann-Whitney test at a 5% probability level.

This insensitivity to light may be due to the quantity of phytochrome present in the seeds in the active form being insufficient to induce the germination process (Bewley et al., 2013). The search for knowledge about the effects of light on the ecological succession of forest species, such as *Luetzelburgia auriculata* (Nogueira et al., 2012), *Clitoria fairchildiana* (Alves et al., 2012), *Guazumaul mifolia, Dimorphand ramolli* sand *Anadenanthera colubrine* (Figliolia, Aguiar, & Silva, 2009), has been frequently reported in the literature.

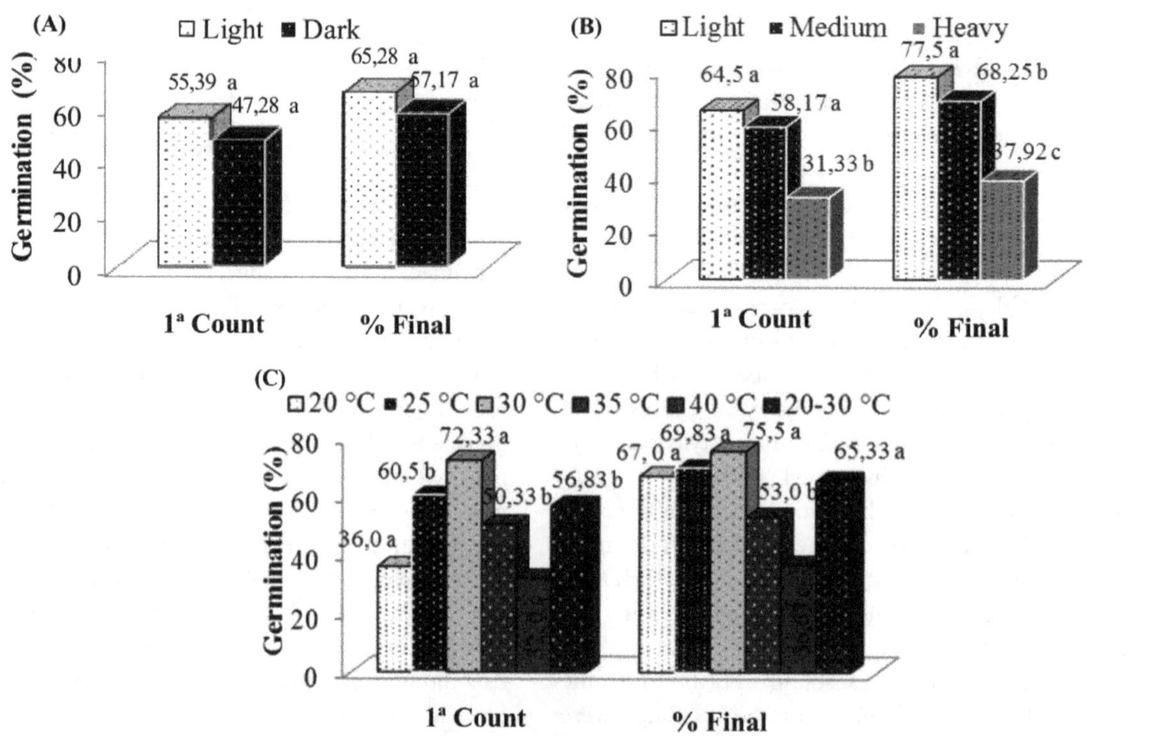

Figure 2. First count and final germination percentage (%) of *Amburana cearensis* (Allemão) A. C. Sm. seeds as influenced by light conditions (A), weight classes (B) and temperature regimes (C).

Regarding the temperature regimes (Figure 2C), evidence of the best temperature for germination of *A. cearensis* seeds was already noted in the first count, for which 30°C was ideal. As for the final germination percentage, the temperature range was increased, with 20, 25, 30 and 20-30°C being considered as appropriate for germination. In both tests, the temperature of 40°C was strongly detrimental. The simple fact that germination occur sunder more than one temperature regime characterizes, for many plant species, adaptation to natural temperature variation in the environment. These results corroborate, in part, those obtained by Guedes et al. (2010), who found 25, 30, and 20-30°C to be some of the most ideal temperatures for the final germination percentage of *A. cearensis* seeds. The temperature range of 20 to 30°C can be considered optimal for the seed germination of forest species (Diniz et al., 2008; Alves et al., 2012; Nogueira et al., 2012).

Significant differences in the first count and the final germination percentage can be further seenby evaluating the influence of seed weight on germination (Figure 2B). Seeds from the light and medium classes showed higher germination performance in the first count in relation to heavy seeds. Nevertheless, this characteristic changed in the final germination percentage, for which anincrease in seed weight adversely affected germination. In this case, light seeds are the best option to increase the germination of *A. cearensis*. These results differ from the reports of Carvalho and Nakagawa (2012), in which heavier seeds were noted to be the most well-nourished, being more vigorous due to the presence of well-formed embryos and larger amounts of reserves.

However, it is important to note that, in some circumstances, heavier seeds might not be the most vigorous. This may be due to factors such as temperature, water and nutrient supply, light intensity and photoperiod length, which may affect seed development and the accumulation of reserves, consequently affecting seed vigour (Carvalho & Nakagawa, 2012). On the other hand, studies demonstrating that light seeds exhibit better germination are scarce.

Taking into account that heavy seeds have a larger amount of reserves, it can be inferred that the various types of substances present in this weight class, as well as their quality, could be somehow detrimental to seed germination. Considering that *A. cearensis* seeds are highly allelopathic due to the coumarin compound, this substance might influence their germination process. Chemicals present in the extract of *A. cearensis* seeds have strong allelopathic potential, which can interfere with seed germination and the development of other plants in the same environment (Silva et al., 2006; Felix, Ono, & Araújo, 2010).

Barbosa (2003) makes reference to the fact that, so far, no inhibition of the germination of *A. cearensis* seeds has been reported and that coumarin might be present in low concentrations in mature seeds of this species, as the inhibition of germination can be related to increasing levels of this substance. Based on these studies, it can be assumed that heavy *A. cearensis* seeds (> 0.35 g) have large amounts of coumarin, which is likely to inhibit their germination, a fact that has not yet been proven in the literature. Hence, future studies should be directed at explaining this gap.

From an ecological point of view, regarding the dispersibility of this species, light seeds tend to travel longer distances than heavy seeds, the latter of which usually remain near the mother plant due to their weight and end up competing for nutrients and space. In terms of survival, it can also be inferred that this species may repel its dispersers by investing in a larger coumarin content in the heavy seeds.

Table 3 shows that the lowest GSI values were obtained at 20 and 40°C. This result was more evident in the absence of light at the temperature of 40°C (1.04), resulting in this being the only treatment for which the absence of light had a significant effect. This shows that under the different temperature regimes, *A. cearensis* seeds are insensitive to light, i.e., are neutrally photoblastic, except at a temperature of 40°C, at which they are preferably positively photoblastic. Certain temperature limits above or below the upper or lower limits of the optimal temperature range tend to decrease the germination speed, thus resulting in a disruption of membranes and leading to a reduction in the total germination (Carvalho & Nakagawa, 2013).

Table 3. Germination speed index (GSI) of *Amburana cearensis* (Allemão) A.C. Sm. seeds subjected to different temperatures in the presence or absence of light.

GSI						
	Temperatures					
Light	20°C	25°C	30°C	35°C	40°C	20-30°C
Presence	2.22 aC	3.00 aB	4.12 aA	3.30 aB	2.74 Abc	2.81 aBC
Absence	2.26 aC	3.41 aB	4.39 aA	3.09 aB	1.04 bD	2.94 aB
lsd for columnn = 0.4552 lsd for lines = 0.6383						

Means followed by the same letter, upper case in the same row and lower case in the same column, do not differ by Tukey's test at a 5% probability level.

The highest germination index occurred at a temperature of 30°C, regardless of the presence of light, differing significantly from the other temperatures (Table 3). These results are similar to those obtained by Diniz et al. (2008), who studied the influence of light and temperature on the germination of *Licania rigida* Benth. seeds and found an interaction between these factors, with the highest germination speed index obtained at a temperature of 30°C in the absence of light.

By analysing the results shown in Table 4, it can be observed that light and medium seeds had higher germination speed indices at all temperatures except for 35°C, at which only light seeds differed from the other weight classes. However, the combination of a temperature of 30°C and light and medium seeds promoted the highest germination speed. Heavy seeds negatively influenced this speed, which can be clearly seen at 40°C. These results differ from the results reported in the literature, which indicate that heavy seeds are usually those with the highest germination speed. Lessa et al. (2015) found that heavy *Enterolobium contortisiliquum* seeds had a higher germination speed index.

Table 4. Germination speed index (GSI) of *Amburana cearensis* (Allemão) A.C. Sm. seeds of different weight classes under different temperature regimes.

	GSI					
Classes	Temperatures					
Light	2.63 aC	3.85 aB	4.99 aA	4.38 aAB	2.22 aC	3.74 Ab
Edium	2.19 abC	3.32 aB	5.27 aA	3.14 bB	2.77 aBC	3.13 Ab
Heavy	1.89 bA	2.45 bA	2.50 bA	2.07 cA	0.67 bB	1.74 bA
lsd for column = 0.6714 lsd for lines = 0.8195						

Means followed by the same letter, upper case in the same row and lower case in the same column, do not differ by Tukey's test at a 5% probability level.

A possible explanation for the negative influence of heavy seeds on the germination speed index is that the ratio between the surface and volume of a seed is reduced with an increase in seed size. This decreases the ability of a seed to obtain enough water to beginthe germination process (Harper & Benton, 1966), thus causing greater delays in germination and seedling establishment. Such findings may provide an appropriate weight class for the proliferation of this species to produce more homogeneous and vigorous seedlings, as light and medium *A. cearensis* seeds are likely to contribute the most to an increase in the propagation of the species.

The following results take into account the growth behaviour and dry biomass of *A. cearensis* seedlings. Temperatures of 40 and 35°C were highly detrimental to the establishment of the species, killing seedlings shortly after germination.

Regarding the total length of *A. Cearensis* seedlings, for the isolated factors, it can be seen from Figure 3 that the best growth occurred when the seeds were germinated in the presence of light (Figure 3A). The weight classes did not influence seedling development (Figure 3B), and the temperatures of 25 and 30°C were considered as ideal for seedling establishment (Figure 3C). With respect to the temperatures, the corresponding expected results were confirmed by Guedes et al. (2010), with 30°C being one of temperatures that promoted greater seedling length in *A. cearensis*. In regard to the fact that there was no influence of seed weight on shoot growth, Lessa et al. (2015) reported similar results in *Enterolobium contortisiliquum* seedlings.

Means followed by the same letter do not differ by the non-parametric Mann-Whitney test at a 5% probability level.

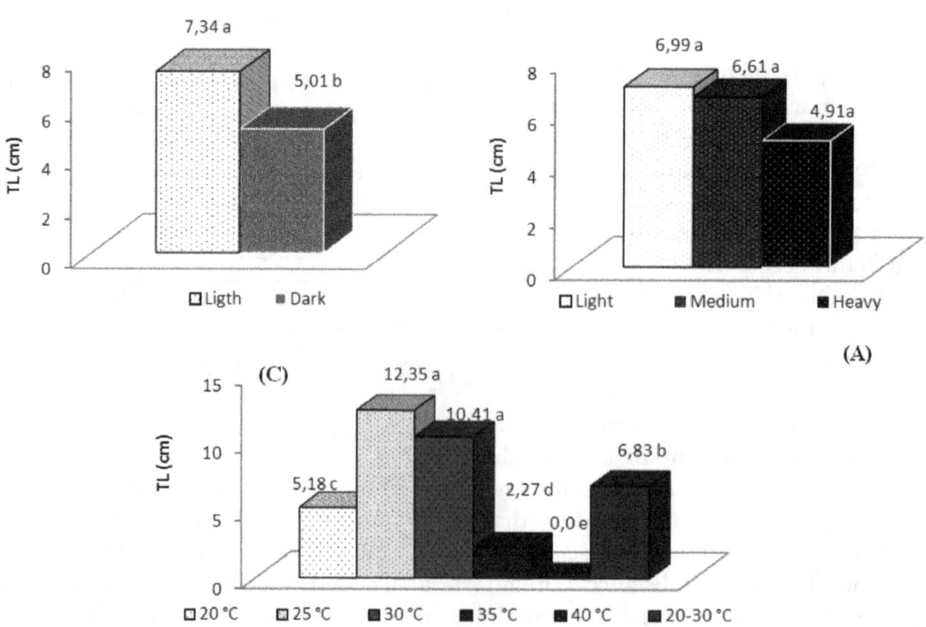

Figure 3. Total length (TL) of *Amburana cearensis* (Allemão) A.C. Sm. seedlings as influenced by light conditions (A), weight classes (B) and temperature regimes (C).

The light requirement of seeds is associated with different ecological groups: pioneer, secondary and climax (Figliolia & Piña-Rodrigues, 1995). *Amburana cearensis* is described in different ways in the literature, being included in more than one ecological group. This has been verified by Barbosa (2003), who describes that *A. cearensis* demonstrates shade tolerance, as it germinates under the forest canopy, and it later shows characteristics of pioneer species during the young plant stage.

Regarding the influence of different temperatures and seed weight on the root + hypocotyl length (Table 5), it can be seen that low (20°C) and high temperatures (35 and 40°C) were not critical in the differentiation among the weight classes. Temperatures of 25, 30, and 20-30°C were essential to the perception of such differences, for which light and medium seeds stood out as the best in terms of the promotion of increased root growth. The optimal temperature for germination and seedling growth is a prime factor that ensures the action of seed reserves in the establishment of the species, with the ideal temperature being responsible for increasing metabolism and the translocation of these reserves to the embryonic axis.

Table 5. Root + hypocotyl length (RHL) of *Amburana cearensis* (Allemão) A. C. Sm. seedlings as influenced by seed weight.

RHL						
	Temperature					
Classes	20°C	25°C	30°C	35°C	40°C	20-30°C
Light	3.70 aC	7.95 aA	5.24 Ab	0.81 aD	0.00 aD	3.66 aC
Mediu	3.87 aC	7.27 aA	5.86 Ab	0.66 aD	0.00 aD	3.28 aC
Heavy	3.24 aB	5.49 bA	2.86 Bb	0.88 aCD	0.13 aD	1.94 bBC
lsd for columns = 1.1606 lsd for lines = 1.314						

Means followed by the same letter, upper case in the same row and lower case in the same column, do not differ by Tukey's test at a 5% probability level.

The process that led to this greater root growth is the result of favourable conditions for germination, which are associated with a complex sequence of biochemical reactions. In these reactions, reserve substances stored in the supporting tissue are broken down, transported and resynthesized in the embryonic axis. The higher the temperature, the faster and more efficient the germination, up to a certain limit (Carvalho & Nakagawa, 2012). Within these limits, the temperature of 25°C enabled the better translocation of reserves, with maximum efficiency, when light and medium seeds were used. A comparison between light and heavy seeds reveals an increase in root growth of more than 44 and 83.22% at temperatures of 25°C and 30°C, respectively.

It can be seen in Table 6 that the factor of light along with temperature showed unique responses in terms ofthe epicotyl length. This growth was certainly evidenced by the perception of light, withits presence being critical for growth. Along with the light factor, temperatures of 25 and 30°C led to increased epicotyl length, with increases of 20.89 and 38.98%, respectively, when compared to that in the absence of light. These results are similar to those reported by Alves et al. (2012), who found that the greatest *Clitoria fairchildiana* length occurred at a temperature of 30°C under a white light regime.

Table 6. Epicotyl length (EPL), epicotyl dry mass (EPDM), root + hypocotyl dry mass (RHDM) and total dry biomass (TDB) of *Amburana cearensis* (Allemão) A.C. Sm. as a function of temperature and light presence or absence.

	Temperature					
Light	20°C	25°C	30°C	35°C	40°C	20-30°C
Presence	2.06 aD	5.96 aA	6.56 aA	3.10 aC	0.16 aE	4.36 aB
Absence	1.08 bC	4.93 bA	4.42 bA	0.00 bC	0.00 aC	2.47 bB
			EPDM (g)			
Presence	0.006 aC	0.017 aAB	0.018 aA	0.007 aC	0.00 aD	0.014 aB
Absence	0.005 bD	0.017 aA	0.014 bB	0.000 bE	0,00 aE	0.008 bC
			RHDM (g)			
Presence	0.013 aC	0.023 aA	0.018 aB	0.007 aD	0.00 aE	0.014 aC
Absence	0.011 bBC	0.015 bA	0.011 bB	0.000 bD	0.00 aD	0.008 bC
			TDB (g)			
Presence	0.019 aC	0.040 aA	0.036 aA	0.014 aC	0.00 aE	0.0284 aB
Absence	0.015 aC	0.033 bA	0.025 bB	0.000 bD	0.00 aD	0.0241 aB

Means followed by the same letter, upper case in the same row and lower case in the same column, do not differ by Tukey's test at a 5% probability level. LSD for EPL, EPDM, RHDM and TDB in columns (0.8083, 0.0018, 0.0020, 0.0036, respectively) and lines (1.0957, 0.0025, 0.0029, 0.0052, respectively).

The temperatures of 30, 25, and 20-30°C (in this order) showed higher mean values of epicotyl dry mass, with the highest results obtained under light conditions, with the exception of 25°C, at which the mean values were not different (Table 6). For the root + hypocotyl dry mass, the greatest increase (53.33%) was observed in the presence of light at a temperature of 25°C.The total dry missal so responded similarly to epicotyl growth, for which the highest results were obtained at temperatures of 25 and 30°C (21.21 and 44%, respectively) when compared to seedlings under light limitation. One of these results is similar to those reported by Pacheco, Matos, Ferreira, and Feliciano (2007), where a constant temperature of 30°C resulted in a greater increase in the hypocotyl dry mass of *Apeibati bourbou* Aubl. seedlings.

The explanation for the shorter seedling length in the dark is the permanence of phytochrome A throughout seedling development, which inhibits etiolation, a phenomenon that only occurs in the presence of phytochrome B (Carvalho & Peres, 2013). Based on this response, it can be deduced that under low light conditions, such as under a very closed canopy, *A. cearensis* suffers strong competition for light, thus requiring a clearing for its growth and survival. Although *A. cearensis* is a species typically adaptable to semi-arid conditions and considering that there was no influence of light on germination, the results are quite enlightening. The presence of light was critical for seedling growth, enabling the normal development of seedling structures, thus indicating that this species might be classified as a pioneer in ecological succession.

Table 7 shows the influence of the factor of seed weight on EPL and EPDM, which showed a significant three-way interaction. Based on the results previously mentioned and described by the statistical breakdown of the temperature × light interaction (Table 6), which revealed the superiority of the combination of 30 or 25°C in the presence of light for these variables, it can be seen from Table 7 that seed weight is a key factor in the development of *A. cearensis* seedlings, affecting even seed photoblastism. The data showed that under the best combination of temperature and light (30 or 25°C in the presence of light) for EPL, seeds classified as light or medium produced more vigorous seedlings when compared to the heavy seeds. For EPDM, this behaviouroccurred again at 30°C, whereas at 25°C, there was no difference among the weight classes. These results are also similar to those reported by Almeida et al. (2014), who observed higher growth of *A. cearensis* seedlings by using light and medium seeds at a temperature of 30°C.

Table 7. Epicotyl length (EPL) and epicotyl dry mass (EPDM) means of *Amburana cearensis* (Allemão) A. C. Sm. seedlings for the three-way interaction of temperature × light × weight.

	EPL (cm)			EPDM (g)		
	Weight Classes			Weight Classes		
Temp. x Light	Light	Medium	Heavy	Light	Medium	Heavy
20°C x Light	2.06 A	2.16 A	1.97 A	0.006 A	0.001 A	0.007 A
20°C x Dark	0.99 A	1.16 A	1.08 A	0.004 A	0.006 A	0.004 A
25°C x Light	7.01 A	5.83 AB	5.03 B	0.018 A	0.016 A	0.016 A
25°C x Dark	5.88 A	4.97 AB	3.93 A	0.019 A	0.017 A	0.015 A
30°C x Light	7.60 A	7.49 A	4.59 B	0.018 AB	0.020 A	0.015 B
30°C x Dark	5.32 A	5.93 A	2.02 B	0.014 AB	0.018 A	0.010 B
35°C x Light	3.94 A	2.90 A	2.88 A	0.008 A	0.006 A	0.006 A
35°C x Dark	0.00 A	0.00 A	0.00 A	0.000 A	0.000 A	0.000 A
40°C x Light	0.00 A	0.00 A	0.49 A	0.000 A	0.000 A	0.000 A
40°C x Dark	0.00 A	0.00 A	0.00 A	0.000 A	0.000 A	0.000 A
20-30°C x Light	5.11 A	4.45 A	3.53 A	0.015 A	0.015 A	0.013 A
20-30°C x Dark	3.35 A	2.52 A	1.53 B	0.010 A	0.009 AB	0.005 B

Uppercase letters in the same rows do not differ by Tukey's test at a 5% probability of error. EPL: LSD = 1.5862. EPDM: LSD = 0.0053.

Overall, light and medium seeds promoted better development of *A. cearensis*, which may indicate that they have higher tolerance to adverse environmental conditions. In terms of the low performance obtained with heavy seeds, these seeds have been influenced since the germination stage, which also causes changes in seedling establishment. A possible explanation for the detrimental effect of heavy seeds on seedling development under optimal temperature and light conditions is most likely associated with higher coumarin levels.

Seed weight also did not influence seedling development under non-optimal temperatures (20, 35, and 40°C), with the exception of alternating temperatures (20 - 30°C) in the absence of light, which also shows the influence of weight on light perception. This reveals once more a superiority of light and medium seeds (Table 7), i.e., the seedlings obtained from heavy seeds under dark conditions do not exhibit good development rates.

Temperature modifies seedling growth according to the amount of reserves in the seeds, indicating changes in the type of phytochrome throughout the development of this species. This shows that environmental factors establish different behaviours in this species depending on the ambient temperature and the type of light. Both extrinsic factors affect responses in the plant according to the amount of reserves in the seeds. These results may provide ecological and agronomic information for a better understanding and the sustainable exploitation of this species, allowing it to be propagated more efficiently for reforestation and preservation purposes.

Conclusion

The germination of *A. cearensis* shows neutral photoblastism, where seeds from light and medium classes have greater vigour, and light seeds have a higher final germination percentage.

The temperature of 30°C is the most suitable for germination, whereas 35 and 40°C are strongly detrimental to seedling growth.

Temperature and seed weight influence the perception of light for seedling development.

Seedlings obtained from light and medium seeds resulted in greater epicotyl growth and dry mass at temperatures of 25 and 30°C in the presence of light.

References

Almeida, J. R. G. S., Guimarães, A. G. G., Siqueira, J. S., Santos, M. R. V., Lima, J. T., Nunes X. P., & Quintans, Júnior, L. J. (2010). *Amburana cearensis* – uma revisão química e farmacológica. *Scientia Plena*, *6*(11), 1-8.

Almeida, J. P. N., Pinheiro, C. L., Lessa, B. F. T. M, Gomes, F. M., & Medeiros Filho, S. (2014). Estresse hídrico e massa de sementes na germinação e crescimento de plântulas de *Amburana cearensis* (Allemão) A.C. Smith. *Revista Ciência Agronômica*, *45*(4), 777-787.

Alves, M. M., Alves, E. U., Bruno, R. L. A., Silva, K. R. G., Moura, S. S. S., Barrozo, L. M., & Araújo, L. R. (2012). Potencial fisiológico de sementes de *Clitoria fairchildiana* R. A. Howard. *Fabaceae* submetidas a diferentes regimes de luz e temperatura. *Ciência Rural*, *42*(12), 2199-2205.

Amato, A. L. P., Maia, F. C., Maia, M. S., Caetano, L. S., Simioni, S. B., Conto, L., & Bonini Filho, R. M. (2007). Estabelecimento de condições de luz e temperatura na germinação de sementes de amendoim forrageiro. *Revista Brasileira de Sementes*, *29*(3), 61-66.

Barbosa, D. C. A. (2003). Estratégias de germinação e crescimento de espécies lenhosas da caatinga com germinação rápida. In I. R. Leal, M. Tabarelli, & J. M. C. Silva (Eds.), *Ecologia e conservação da caatinga* (p. 625-656). Recife, PE: Universidade Federal de Pernambuco.

Bewley, J. D., Bradford, K. J., Hilhorst, H. W. M., & Nonogaki, H. (2013). *Seeds:* Physiology of development, germination and dormancy (3rd ed.). New York, US: Springer.

Brasil. (2009). Ministério da Agricultura, Pecuária e Abastecimento – Secretaria de Defesa Agropecuária. *Regras para análise de sementes*. Brasília, DF: Mapa/ACS.

Canuto, K. M., Silveira, E. R., Bezerra, A. M. E., Leal, L. K. A. M., & Viana, G. S. B. (2008). *Uso de Plantas Jovens de Amburana cearensis A. C. Smith: Alternativa para preservação e exploração econômica da espécie.* Petrolina, PE: Embrapa Semiárido.

Carvalho, N. M., & Nakagawa, J. (2012). *Sementes: ciência, tecnologia e produção* (5a ed.). Jaboticabal, SP: Funep.

Carvalho, R. F., & Peres, L. E. P. (2013). *Fotomorfogênese.* Retrieved from http://www.miniweb.com.br/ciencia/artigos/fotomorfogenese.pdf.

Diniz, F. O., Moreira, F. J. C., Silva, F. D. B., & Medeiros Filho, S. (2008). Influência da luz e temperatura na germinação de sementes de oiticica (*Licania rigida* Benth.). *Revista Ciência Agronômica*, *39*(3), 476-80.

Felix, R. A. Z., Ono, E. O., & Araújo, F. P. (2010). Efeitos alelopáticos de extratos de *Amburana cearensis* na germinação de sementes de rabanete e alface. *Cultivando o Saber*, *3*(1), 18-31.

Figliolia, M. B., Aguiar, I. B., & Silva, A. (2009). Germinação de sementes de três arbóreas brasileiras. *Revista do Instituto Florestal*, *21*(1), 107-115.

Figliolia, M. B., & Piña-Rodrigues, F. C. M. (1995). Considerações práticas sobre o teste de germinação. In *Manual técnico de sementes florestais* (p. 45-59). São Paulo, SP: Instituto florestal. (Série Registros, 14).

Guedes, R. S., Alves, E. U., Gonçalves, E. P., Braga Júnior, J. M., Viana, J. S., & Colares, P. N. Q. (2010). Substratos e temperaturas para testes de germinação e vigor de sementes de *Amburana cearensis* (Allemão) A.C. Smith. *Revista Árvore*, *34*(1), 57-64.

Harper, J. L., & Benton, R. A. (1966). The behaviour of seeds in soil: II. The germination of seeds on the surface of a water supplying substrate. *Journal of Ecology*, *54*(1), 151-166.

Lessa, B. F. T., Almeida, J. P. N., Pinheiro, C. L., Gomes, F. M., & Medeiros Filho, S. (2015). Germinación y crecimiento de plántulas de *Enterolobiumcontortisiliquum* enfunciόndel peso de lasemilla y las condiciones de temperatura y luz. *Agrociencia*, *49*(3), 315-327.

Maguire, J. D. (1962). Speed of germination-aid in selection and evaluation for seedling emergence and vigor. *Crop Science*, *2*(1), 176-177.

Marcos-Filho, J. (2015). *Fisiologia de sementes de plantas cultivadas* (2a ed.). Piracicaba, SP: Fealq.

Nogueira, F. C. B., Silva, J. W. L., Bezerra, A. M. E., & Medeiros Filho, S. (2012). Efeito da temperatura e luz na germinação de sementes de *Luetzelburgia auriculata* (Alemão) Ducke Fabaceae. *Acta Botanica Brasilica*, *26*(4), 772-778.

Pacheco, M. V., Matos, V. P., Ferreira, R. L. C., & Feliciano, A. L. P. (2007). Germinação de sementes de *Apeibati bourbou* Aubl. em função de diferentes substratos e temperaturas. *Scientia Forestalis*, *19*(73), 19-25.

Pereira, S. R., Giraldelli, G. R., Laura, V. A., & Souza, A. L. T. (2011). Tamanho de frutos e de sementes e sua influência na germinação de jatobá-do-cerrado (*Hymenaea stigonocarpa* var. *Stigonocarpa* Mart. ex Hayne, Leguminosae Caesalpinoideae. *Revista Brasileira de Sementes*, *33*(1), 141-148.

Ribeiro, C. A. D., Costa, M. P., Senna, D. S., & Caliman, J. P. (2012). Fatores que afetam a germinação das sementes e a biomassa de plântulas de *Tabebuia heptaphylla*. *Revista Floresta*, *42*(1), 161-168.

Silva, G. M., Maia, M. S., & Moraes, C. O. C. (2007). Influência do peso da semente sobre a germinação e o vigor de cevadilha vacariana (*Bromus auleticus* Trinius). *Revista Brasileira de Agrociência*, *13*(1), 123-126.

Silva, W. A., Nobre, A. P., Leites, A. P., Silva, M. S. C., Lucas, R. C., & Rodrigues, O. G. (2006). Efeito alelopático de extrato aquoso de *Amburana cearensis* A. Smith na germinação e crescimento de sementes de sorgo (*Sorghum bicolor* L.). *Agropecuária Científica no Semi-árido*, *2*(1), 48-54.

Thompson, K., & Grime, J. P. (1983). A comparative study of germination responses to diurnally fluctuating temperatures. *Journal of Applied Ecology*, *20*(1), 141-156.

8

Effects of silicon on the growth and genetic stability of passion fruit

Bárbara Nogueira Souza Costa[1*], Gabrielen de Maria Gomes Dias[1], Irton de Jesus Silva Costa[1], Franscinely Aparecida de Assis[1], Flávia Aparecida da Silveira[1] and Moacir Pasqual[2]

[1]*Programa de Pós-graduação, Departamento de Agricultura, Universidade Federal de Lavras, Av. Doutor Sylvio Menicucci, 1001, Kennedy, 37200-000, Lavras, Minas Gerais, Brazil.* [2]*Departamento de Agricultura, Universidade Federal de Lavras, Lavras, Minas Gerais, Brazil. *Author for correspondence. E-mail: babinogueiraagro@hotmail.com*

ABSTRACT. The objective of this study was to determine the silicon concentration that would provide good growth in passion fruit plants. Passion fruit seeds were sown in polystyrene. After 60 days, when they were approximately 15 cm tall, the plants were transplanted into polyethylene pots containing 1.1 kg Tropstrato® substrate. Treatments consisted of four concentrations (0, 0.28, 0.55, and 0.83 g pot^{-1}) of silicon applied as a silicic acid solution 1%. This solution was applied around the stem of the plants (*drenched*), with the first application being administered 15 days after transplanting. In total, three applications were made at intervals of 15 days. After the last application, the plants were subjected to chemical analysis to determine the silicon concentration and to X-ray microanalysis and flow cytometry. Phytotechnical analyses were performed during the applications. The use of silicon in concentrations of 0.28 and 0.55 g pot^{-1} provides better growth of the passion fruit, and the absorption and deposition of the silicon in the passion fruit leaves are proportional to the availability of this element in the plant. The roots of the passion fruit plant are silicon accumulators, and the DNA stability and amount are preserved in the silicon-treated passion fruit plants.

Keywords: flow cytometry, silicic acid, *Passiflora edulis*.

Introduction

The genus Passiflora is economically important to Brazil, which has 129 known native species from the genus, 83 of which are endemic; these plants can be used as food, medicine and ornaments (Cervi, Azevedo, & Bernacci, 2010).

Brazil is the main producer and consumer of passion fruit, with a planted area in 2011 of 61,631 hectares and an average yield of approximately 923,035 tons (Agrianual, 2014).

Fertilization often increases crop yields due to increased plant vigor (Espindula, Rocha, Souza, Grossi, & Souza, 2010). However, the practice of fertilization should be based on knowledge of the morphological and physiological characteristics of the plant in addition to factors such as the availability of nutrients in the soil and the behavior of these nutrients in the plant (Almeida, Damatto Junior, & Leonel, 2007).

Although silicon (Si) is not considered an essential element for most plants, the benefits of silicate fertilization have been studied and recognized in cultivated species (Epstein, & Bloom, 2006; Ma, & Yamaji, 2008; Richmond, & Sussman, 2003).

The beneficial effect of Si on biomass formation in cultivated plants is associated with alterations in the plant structure, allowing, for example, better solar energy collection and reducing lodging. Although the role of Si in plant metabolism is still unknown (Epstein, & Bloom, 2006), this element is postulated to be solubilized in the plant and to play a role in the synthesis of plant defense molecules (Ma, & Yamaji, 2008; Rodrigues et al., 2004). Thus, silicon can indirectly promote plant growth and production, causing an increase in the chlorophyll content of the leaf tissue, altering the plant architecture, making the plants more upright, preventing excessive auto shading, delaying senescence, increasing the structural tissue rigidity and protecting the plants from abiotic and biotic agents (Epstein, & Bloom, 2006; Ma, & Yamaji, 2008; Marschner, 1995).

Few studies exist on the effects of silicon on plant growth, with the majority of publications dealing with the nutritional aspects and beneficial role of this element in biotic stress resistance, with possible evaluations of final productivity of the crop (Laviola, Martinez, Souza, & Alvarez, 2007; Pozza et al., 2009; Reis, Figueiredo, Guimarães, Botrel, & Rodrigues, 2008). Ma and Yamaji (2008) stated that the beneficial effects of Si on plant growth are commonly observed in plants under stress conditions. From a physiological point of view, for the growth and development of plants, silicon has demonstrated beneficial effects on the growth and development of plants in the increased production of various crops (Gomes, Moraes, & Assis, 2008).

A lack of silicon adversely affects DNA synthesis and chlorophyll in diatoms (Werner, 1977; Raven, 1983). However, no reports exist of excess silicon altering the DNA content of plants. In this sense, the flow cytometry technique has gained particular attention because it allows the relative amount of nuclear DNA of plant cells to be estimated rapidly and with high accuracy (Jin et al., 2008; Bairu, Aremu, & Van staden, 2011; Smulders, & Klerk, 2011).

This work was carried out to evaluate the effect of the addition of silicon on the growth, morphology and genetic stability of passion fruit plants.

Material and methods

Passion fruit (*Passiflora edulis* Sims f. *flavicarpa* Deg.) were sown in polystyrene trays and allowed to grow until they reached a size appropriate for transplantation into pots.

After 60 days, the plants, which were approximately 15 cm in height, were transplanted to polyethylene pots that contained 1.1 kg of Tropstrato® substrate. The plants were randomly arranged on a bench in the greenhouse and irrigated daily to meet their water needs.

Treatments consisted of four concentrations (0, 0.28, 0.55, and 0.83 g pot^{-1}) of silicon in the form of 1% silicic acid solution ($SiO_2.XH_2O$) (Pereira, Moraes, Prado, & Dacosta, 2010). This solution was applied around the stems of the plants (*drenched*), with the first application being administered 15 days after transplanting. All three applications were done at intervals of 15 days with the same concentrations. The control pots received water in the same amount. After the last application, the plants were subjected to chemical analysis to determine the concentration of silicon and to X-ray microanalysis and flow cytometry. The phytotechnical analyses were performed in the course of the applications.

Phytotechnical analyses

All experimental plants were evaluated for the number of leaves, plant height (cm), stem diameter (mm), leaf length (mm), web width (mm), aerial part fresh and dry weight (g) and root fresh and dry weight (g). The dry weight of the plant material was obtained after drying in an oven at 60°C for 72 hours to a constant weight.

Silicon concentration

To determine the silicon concentrations, analyses were carried out in the Fertilizer Laboratory of the Federal University of Uberlândia, Institute of Agricultural Sciences. The leaves, stems and roots of 12 plants were collected. The materials were dried in a forced circulation oven at 60°C for 72 hours to a constant weight and ground separately. For the determination of silicon, we used the methodology proposed by Korndörfer, Pereira, and Nolla (2004).

X-ray microanalysis

This analysis was conducted in the Electron Microscopy Laboratory, Department of Plant Pathology, UFLA. Samples from the middle third of 2 leaves from 3 plants were mounted on "stubs" and maintained in a silica gel desiccator for 72 hours for evaporation of all water. Later, the samples were coated with carbon CED 020 Baltec and analyzed in an LEO-EVO scanning electron microscope, following the protocol of Alves (2004).

Flow cytometry

Flow cytometry was performed at the Plant Tissue Culture Laboratory of the Agriculture Department of UFLA. Approximately 30 mg of leaf samples was collected, along with the same amount of tomato leaf biomass (*Solanum lycopersicon*) (reference standard with a quantity of 1.96 pictograms (pg) of DNA), with the samples being crushed in a petri dish containing 1 mL Marie nuclei extraction buffer (Dolezel, Binarova, & Lucretti, 1989). The extract containing the nuclei was subsequently stained with 25 $\mu L\ mL^{-1}$ of propidium iodide and placed in the equipment; 5,000 nuclei analyses were performed on each sample. Histograms were obtained on a FacsCalibur® cytometer (Becton Dickinson) with the Cell Quest program (Dickinson, 1998). The DNA content (pg) of the plants was obtained with the following equation: DNA content (pg) = the G1 peak position the sample position/G1 peak of the standard x pg. Three repetitions were conducted, and statistical analysis was performed using the WinMDI 2.8 program (Trotter, 2000).

Experimental design and statistical analysis - The design was completely randomized with 4 treatments and 20 repetitions/treatment. All data were submitted to an analysis of variance using the statistical program SISVAR (Ferreira, 2011), followed by data regression or the Scott-Knott test.

Results and discussion

Analysis phytotechnical

No interaction was observed between the treatments and the three application times; therefore, the factors were studied separately. For the regression analysis results, no model fit the curve, and we used the statistical tests to better explain the results.

An increase was observed in the stem diameter with increasing silicon concentrations, with the largest diameter, 2.33 mm, being attained with 0.21 g silicon pot^{-1} and a reduction occurring at higher concentrations (Figure 1a). Similar results were obtained for the plant height, with the maximum height, 9.82 cm, being attained at a concentration of 0.28 g silicon pot^{-1} (Figure 1b).

These results agree with those of Prado and Natale (2005), whose work with calcium silicate and passion fruit also showed that the application of silicon increased the height and stem diameter quadratically.

Similar results were obtained by Prado and Natale (2004), who observed while working with the application of chrome iron slag to passion fruit that a silicon application of rate of 333.4 g kg^{-1} increased the stem diameter and plant height quadratically.

This positive relation between height and stem diameter was also observed by Ferri (1985) and indicates the great morphophysiological importance of vegetative characteristics, as reflected in a practical way in the growth and differentiation of the plant. In addition, several studies have shown that fertilization with silicon can have a positive influence on the plant growth and productivity (Sávio, Silva, Teixeira, & Borém, 2011).

Increasing silicon concentrations resulted in a proportional increase in the fresh and dry leaf biomass and fresh stem biomass (Figure 1c, d and g), reaching a fresh leaf biomass of 12.09 g at levels of 0.66 g silicon pot^{-1}, a dry weight of 2.08 g with 0.57 g pot^{-1} and a fresh stem biomass of 5.29 g at a concentration of 0.60 g silicon pot^{-1}. At higher concentrations, a downward trend in values was recorded for these three variables.

A linear decrease for root fresh and dry biomass was observed with the increasing silicon concentrations (Figure 1e and f). These results agree with those obtained by Ribeiro et al. (2011), who applied calcium silicate to coffee plants and found that the highest calcium silicate dose (6 mg ha^{-1}) provided a reduction in coffee plant root systems without compromising the functionality and development of the plant aerial parts.

Among the benefits related to silicon fertilization are the shoot and root dry matter increase and the importance of silicon for the growth and development of plants (Epstein, 1994). Prado and Natale (2005) observed a quadratic increase in passion fruit plant shoot and roots dry matter in response to increased calcium silicate concentrations.

The same authors, while assessing steel slag in passion fruit plants, observed a quadratic increase in the shoot and root dry matter (Prado & Natale, 2004). Similar data were obtained in the present experiment with respect to the dry and fresh weight of the leaves and stems, thereby showing the positive effect of silicon on plant growth. However, a decrease was observed in the fresh and dry biomass of the root, which may be due to the production of photoassimilates being directed to vegetative production, thus expressing the higher plant growth. Therefore, the production of assimilates for the root was lower, which resulted in lower fresh and dry biomass.

Table 1 presents the significant difference observed for the dry biomass of the stem, and the length and width of the leaf. Concentrations of 0.28, 0.55, and 0.83 g silicon pot^{-1} provided higher dry stem biomass with respect to the control, which reflects the results found for the fresh stem biomass. The silicon accumulates in the stem support and sustaining tissues, substantially strengthening the plant structure (Plucknett, 1971).

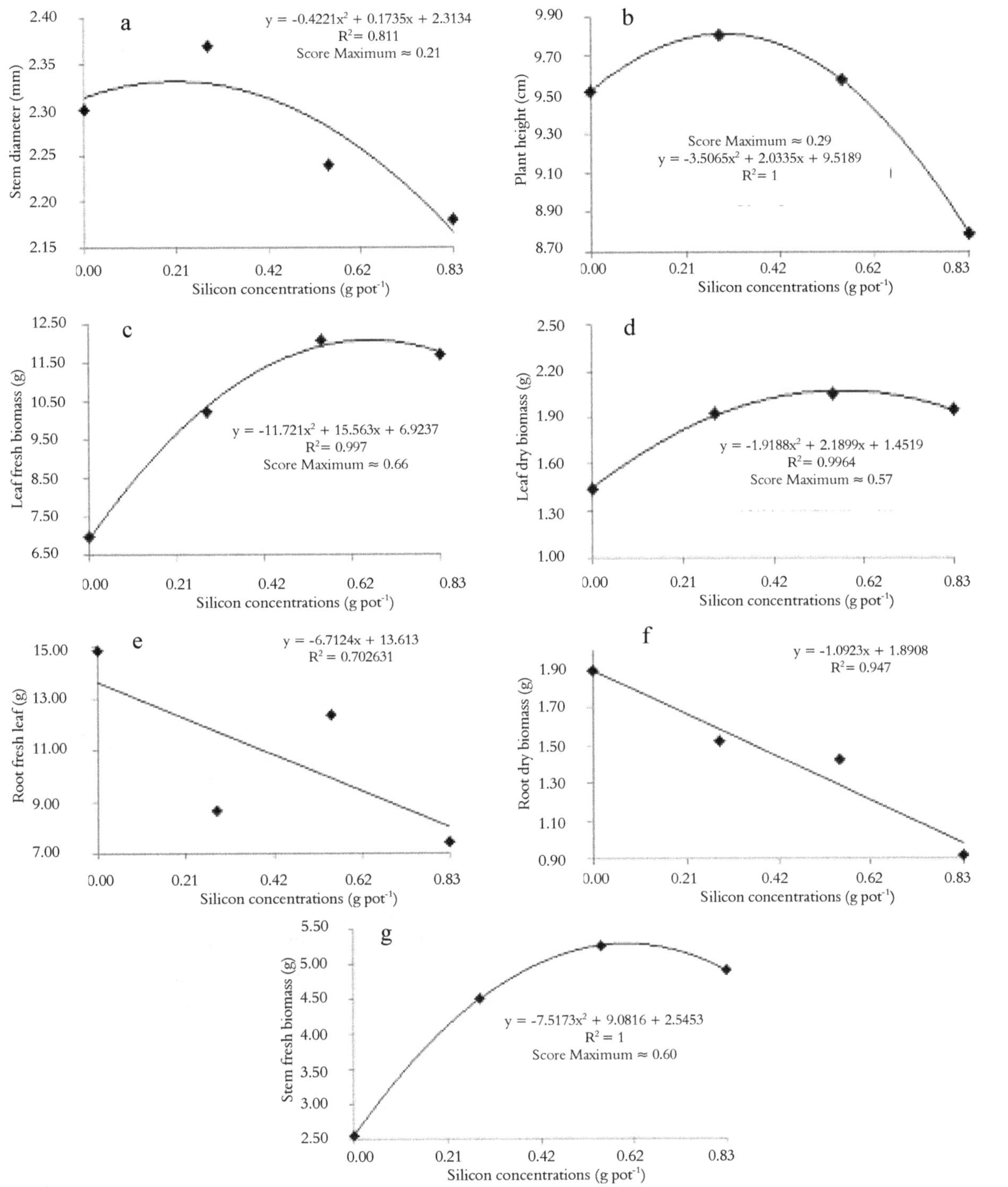

Figure 1. Phytotecnical characteristics of passion fruit (Passiflora edults Sims f. flavicarpa Deg.), submitted to different silicon concentrations.

No difference was observed between the control and the concentration 0.28 g silicon pot^{-1}, which showed greater leaf length and width relative to the other treatments. The leaves are the most important part of the plant with respect to the amount of CO_2 fixed by photosynthesis, a physiological process that results in dry matter accumulation (Ritchie, Hanway, Thompson, & Benson, 1994).

The growth can be measured by the amount of accumulated dry matter, which is directly related to the amount of light absorbed and increases as the leaf area increases (Gomes et al., 2008). However, the amount of dry matter produced depends not only on the amount of radiation received by the plant but also the efficiency in the use of such energy (Shibles, & Weber, 1966). Thus, an increase in the leaf area does not always mean an increase in the amount of dry matter produced by the plants (Gomes et al., 2008). The results obtained in this article show that, despite the control having larger leaf area than the other treatments, the leaf and stem dry weight were higher at concentrations of 0.28, 0.55 and 0.83 g silicon pot^{-1} compared to the control.

Table 1. Dry stem biomass, number, length and width of passion fruit (*Passiflora edulis* Sims f. *flavicarpa* Deg.) leaves submitted to different concentrations of silicon.

Silicon concentrations (g pot^{-1})	Stem dry biomass (g)	Number of leaf	Leaf legth (mm)	Leaf width (mm)
Control	0.81 b	9.97 a	78.83 a	41.92 a
0.28	1.32 a	10.08 a	79.62 a	43.64 a
0.55	1.38 a	9.75 a	69.89 b	38.63 b
0.83	1.12 a	9.87 a	74.48 b	40.33 b
CV	29.38	12.33	18.72	17.07

*Means followed by the same lowercase letters in the same column belong to the same group, by Scott-Knott test ($p \leq 0.05$).

X-ray microanalysis

A linear increase was observed in the percentage of silicon in the passion fruit leaves as the concentrations of the applied silicon increased (Figure 2). Silicon is a mineral element, which is polymerized after being absorbed by the plants and accumulates in the cell wall of the epidermis (Jarvis, 1987).

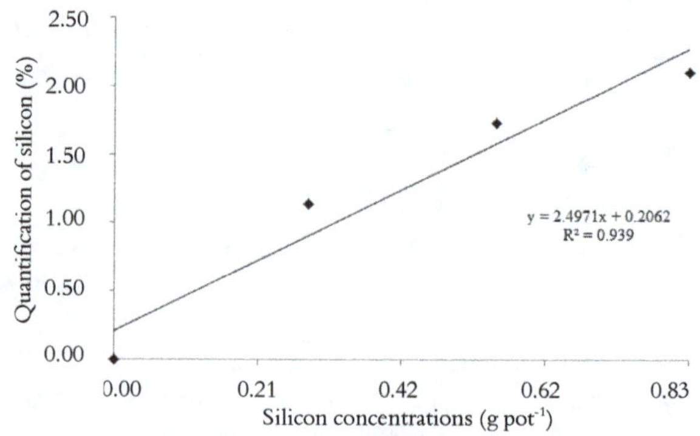

Figure 2. Quantification of silicon via X-ray microanalysis in passion fruit leaves (*Passiflora edulis* Sims f. *flavicarpa* Deg.) subjected to different concentrations of silicon.

In the silicon mapping of the abaxial epidermis of passion fruit leaves, the control did not show silicon, whereas the other treatments showed an increasing presence of silicon (Figure 3). The silicon polymerization on the lower surface of the leaf, a process called silicification, is common in grasses (Lux, Luxová, Hattori, Inanaga, & Sugimoto, 2002) and can occur in dicotyledonous plants, such as coffee (Pozza et al., 2004).

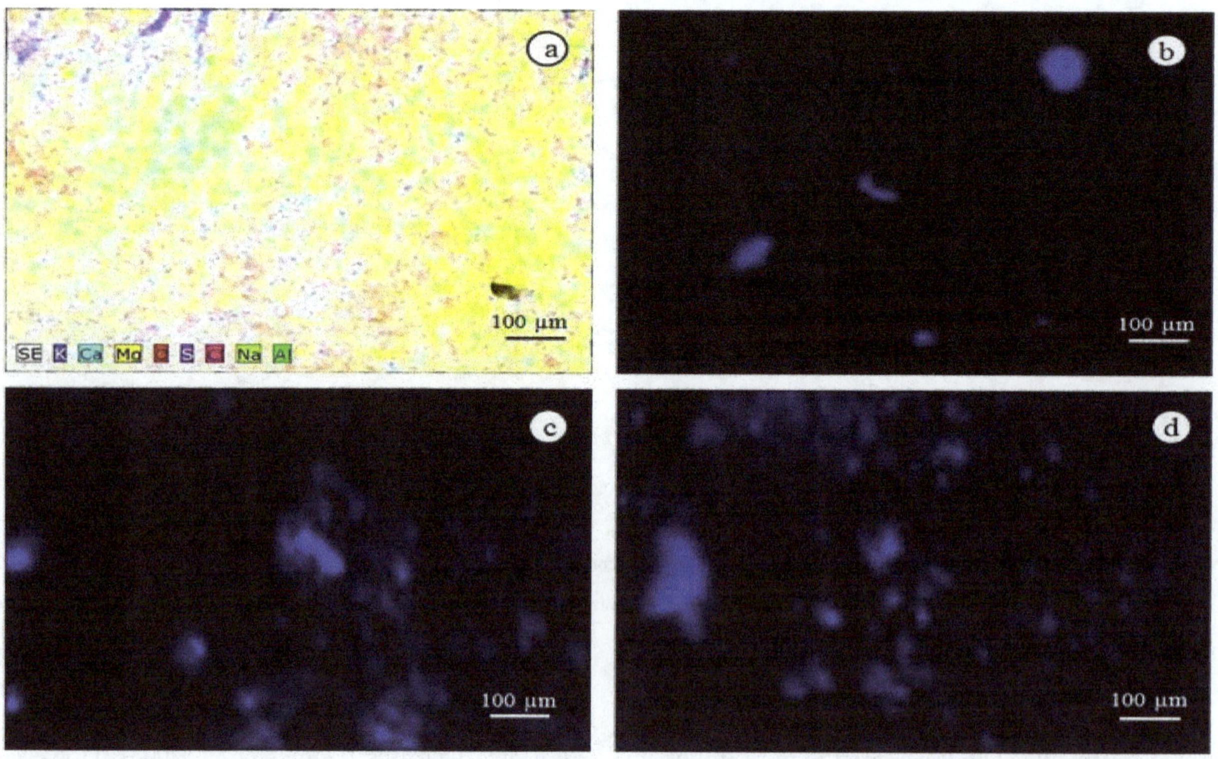

Figure 3. Silicon mapping on the abaxial epidermis of *Passiflora edulis* Sims f. *flavicarpa* Deg. leaves, showing the presence of silicon (in blue). a) Control showing mapping of various elements and absence of silicon; b) 0.28 g silicon pot^{-1}; c) 0.55 g pot^{-1}; d) 0.83 g pot^{-1}.

Silicon concentration

As shown in Table 2, no significant difference was observed among treatments for the silicon content in the stem; however, a significant difference did exist in the leaf and root. The concentration of 0.83 g silicon pot^{-1} represents a higher silicon content in the leaf compared to other treatments, and the concentrations of 0.28 and 0.55 g silicon pot^{-1} did not differ but showed higher silicon content in the leaf compared to the control. The concentrations of 0.55 and 0.83 g silicon pot^{-1} resulted in a higher silicon content in the root compared to the other treatments, and the 0.28 g silicon pot^{-1} treatment showed silicon content higher than observed in the control.

Tabela 2. Chemical analysis of silicon percentage in leaf, stem and root of passion fruit (*Passiflora edults* Sims f. *flavicarpa* Deg.) submitted to different silicon concentrations.

Silicon concentrations (g pot^{-1})	Leaf (%)	Stem (%)	Root (%)
Control	0.27 c	0.08 a	0.51 c
0.28	0.30 b	0.05 a	0.61 b
0.55	0.31 b	0.06 a	0.73 a
0.83	0.36 a	0.05 a	0.77 a
CV	8.60	38.60	10.18

*Means followed by the same lowercase letters in the same column belong to the same group, by Scott-Knott test ($p \leq 0.05$).

The presence of silicon in the control can be explained by the abundance of silicon in nature, where it is ubiquitous and is found even in water (Luz et al., 2006).

These results agree with those obtained by Ferreira, Botelho and Faria (2013), who studied the development of plum trees treated with silicon and found a positive linear effect for the leaf content due to the application of silicon doses.

Silicon accumulates in the leaf, so a lower amount of this element was found in the plant stem. However, higher levels were found in the root compared to the leaf, which has not been verified yet in other experiments of the application of silicon to passion fruit. According to Oliveira and Castro (2002), the average silicon content of the roots is lower compared with the stem and leaves, but in some cases, for example, in soybean, the silicon content in the root is greater than in the leaves. Fawe, Menzies, Cherif, and Belanger (2001) suggested that root silicon plays a role in the signaling network and can induce systemic resistance in other organs.

In the present study, the passion fruit was observed to be an intermediate accumulator of silicon in the roots and not an accumulator in the leaf and stem because, according to Ma, Miyake and Takahashi (2001), at relations above 1.0, plants are considered accumulators; between 1.0 and 0.5, they are considered intermediate; and under 0.5, they are nonaccumulators.

Flow cytometry

For a further elucidation of the effects of Si on this species, flow cytometry analysis was also performed.

These analyses have advantages over conventional genomics methods such as chromosome counts and other cytogenetic analysis because of the technical ease and speed (Chen, Hou, Zhang, Wang, & Tian, 2011; Nguyen et al., 2003).

No significant difference was observed in the DNA content among the treatments (Table 3); therefore, the application of the silicon did not modify the DNA content, which is important because the silicon maintains the genetic stability of the plants.

Tabela 3. Quantifications of DNA in passion fruit leaves (*Passiflora edults* Sims f. *flavicarpa* Deg.) analyzed by flow cytometry.

Silicon concentrations (g pot^{-1})	Leaf (%)	Stem (%)	Root (%)
Control	0.27 c	0.08 a	0.51 c
0.28	0.30 b	0.05 a	0.61 b
0.55	0.31 b	0.06 a	0.73 a
0.83	0.36 a	0.05 a	0.77 a
CV	8.60	38.60	10.18

*Means followed by the same lowercase letters in the same column belong to the same group, by Scott-Knott test ($p \leq 0.05$).

The flow cytometry technique is useful for distinguishing differences in the DNA content of plants maintained under stressful conditions, such as during micropropagation or when being subjected to chemical treatments. Many economically important species, such as *Vitis vinifera* (Yang et al., 2008; Prado et al., 2010), *Gossypium hirsutum*, (Jin et al., 2008), *Musa* spp. (Msogoya, Grout, & Roberts, 2011; Escobedo-GraciaMedrano, Maldonado-Borges, Burgos-Tan, Valadez-González, Ku-Cauich, 2014), *Passiflora* spp. (Silva et al., 2011), *Elaeis guineensis* (Madon, Heslop-Harrison, Schwarzacher, & Hashim, 2012), *Coffea arabica* (Clarindo, Carvalho, & Mendonça, 2012), *Prunus cerasus* (Vujović, Cerović, Ružić, 2012) and *Saccharum* spp (Nogueira, Pasqual, & Scherwinski-Pereira, 2013), have had alterations to their DNA content evaluated by flow cytometry.

The plants treated with silicon in this study showed no alteration in DNA content. This result is important because fertilization with silicon can occur at concentrations as high as 0.83 g per pot, which causes no harmful effect on the plant genome. However, due to the lack of studies on cytogenetics and silicon, more research in the area is needed.

In this work, the tomato (*Lycopersicon esculentum*) was used as the internal standard, having DNA content of 1.96 pg (Figure 4). This standard was chosen because it formed a peak in a channel near that of the sample, without overlapping that of the sample. The closer the two peaks, the lower the experimental error.

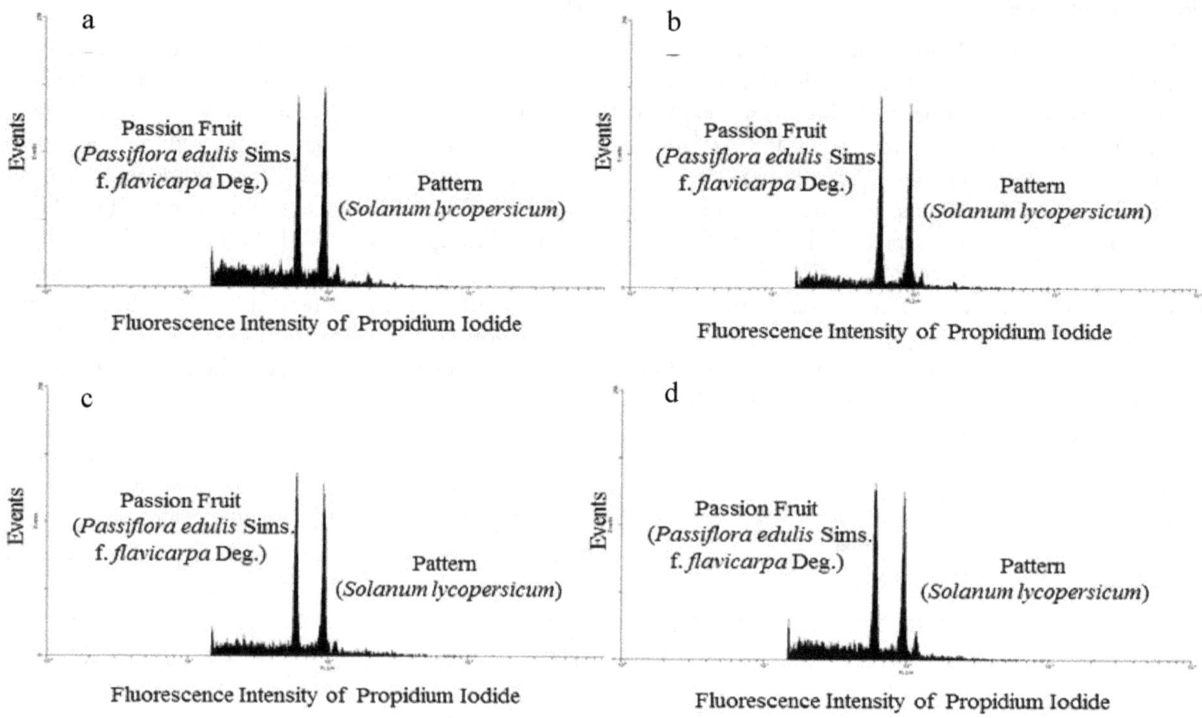

Figure 4. Histograms obtained by cytometric analysis in passion fruit (*Passiflora edulis* Sims f. *flavicarpa* Deg.). a) Control; b) 0.28 g silicon pot^{-1}; c) 0.55 g pot^{-1}; d) 0.83 g pot^{-1}.

Another positive point in this study was that the average coefficient of variation (CV) for the leaf samples ranged between 1.88 and 2.30. These values demonstrate the quality of the results obtained and the reliability of estimated amounts of DNA from the leaves of the passion fruit (Table 3), which can be seen by the thickness of the peaks in Figure 2.

Conclusion

The use of silicon at concentrations of 0.28 and 0.55 g pot^{-1} provides better passion fruit growth.

The absorption and deposition of silicon in the passion fruit leaves are proportional to the availability of this element in the plant.

The passion fruit is an intermediate root silicon accumulator plant.

Genetic stability is preserved in passion fruit treated with silicon.

Acknowledgements

We acknowledge the Conselho Nacional de Desenvolvimento Científico e Tecnológico (CNPq) for financial support and fellowships, and the laboratory for providing the electron microscopy.

References

Agrianual. (2014). *Anuário de Agricultura Brasileira*. São Paulo, SP: AgraFNP.

Almeida, E. L. P., Damatto Júnior, E. R., & Leonel, S. (2007). Nutrição mineral e adubação. In S. Leonel, & A. C. Sampaio (Eds.), *Maracujá doce: aspectos técnicos e econômicos* (p. 59-71). Jaboticabal, SP: Unesp.

Alves, E. (2004). *Introdução à microscopia eletrônica* (p. 88). Lavras, MG: UFLA/Faepe.

Bairu, M. W., Aremu, A. O., & Van staden, J. (2011). Somaclonal variation in plants: causes and detection methods. *Plant Growth Regulation, 63*(2), 147-173.

Cervi, A. C., Azevedo, M. A. M., & Bernacci, L. C. (2010). Passifloraceae. In R. F. Forzza, R. C., Leitman, P. M., Costa, A., Carvalho Junior, A. A. D., Peixoto, A. L., Walter, B. M. T., ... & Martinelli, G. (Ed.), *Catálogo de plantas e fungos do Brasil* (v. 2, p. 1432-1436). Rio de Janeiro, RJ: Jardim Botânico do Rio de Janeiro.

Chen, C., Hou, X., Zhang, H., Wang, G., & Tian, L. (2011). Induction of *Anthurium andraeanum* "Arizona" tetraploid by colchicine *in vitro*. *Euphytica, 181*(2), 137-145.

Clarindo, W. R., Carvalho, C. R., & Mendonça, M. A. C. (2012). Ploidy instability in long-term *in vitro* cultures of *Coffea arabica* L. monitored by flow cytometry. *Plant Growth Regulation, 68*(3), 533-538.

Dickinson, B. (1998). *Cell quest software: reference manual.* San Jose, CA: Becton Dickinson Immunocytometry Systems.

Dolezel, J., Binarova, P., & Lucretti, S. (1989). Analysis of nuclear DNA content in plant cells by flow cytometry. *Biologia Plantarum, 31*(2), 113-120.

Epstein, E. (1994). The anomaly of silicon in plant biology. *Proceedings National of Academy Science of the United State of America, 91*(1), 11-17.

Epstein, E., & Bloom, A. (2006). *Nutrição mineral de plantas: Princípios e perspectivas*. Londrina, PR: Planta.

Escobedo-GraciaMedrano, R. M., Maldonado-Borges, J. I., Burgos-Tan, M. J., Valadez-González, N., & Ku-Cauich, J. R. (2014). Using flow cytometry and cytological analyses to assess the genetic stability of somatic embryo-derived plantlets from embryogenic *Musa acuminata* Colla (AA) ssp. malaccensis cell suspension cultures. *Plant Cell, Tissue and Organ Culture, 116*(2), 175-185.

Espindula, M. C., Rocha, V. S., Souza, M. A., Grossi, J. A. S., & Souza, L. T. (2010). Doses e formas de aplicação de nitrogênio no desenvolvimento e produção da cultura do trigo. *Ciência Agrotécnica, 34*(6), 1404-1411.

Fawe, A., Menzies, J. G., Cherif, M., & Belanger, R. R. (2001). Silicon and disease resistance in dicotyledons. In L. E. Datnoff, G. H. Snyder, & G. H. Korndörfer (Eds.), Silicon in agriculture. *The Netherlands: Elsevier Science* (p. 159-169). Amsterdam, NL: Elsevier.

Ferreira, S. G. M., Botelho, R. V., Faria, C. M. D. R., Mateus, M. A. F., & Zaluski, W. L. (2013). Desenvolvimento e fitossanidade de ameixeiras tratadas com silício em sistema orgânico. *Revista Brasileira de Fruticultura, 35*(4), 1059-1065.

Ferreira, D. F. (2011). Sisvar: a computer statistical analysis system. *Ciência e Agroteclogia, 35*(6), 1039-1042.

Ferri, M. G. (1985). *Fisiologia vegetal* (2a ed., v. 1, p. 362, Revista Atual). São Paulo, SP: EPU.

Gomes, F. B., Moraes, J. C., & Assis, G. A. (2008). Silício e imidacloprid na colonização de plantas por *Myzus persicae* e no desenvolvimento vegetativo de batata inglesa. *Ciência Rural, 38*(5), 1209-1213.

Jarvis, S. C. (1987). The uptake and transport of silicon by perennial ryegrass and wheat. *Plant and Soil, 97*(2-3), 429-437.

Jin, S., Mushke, R., Zhu, H., Tu, L., Lin, Z., Zhang, Y., & Zhang, X. (2008). Detection of somaclonal variation of cotton (Gossypium hirsutum) using cytogenetics, flow cytometry and molecular markers. *Plant Cell Reports, 27*(8), 1303-1316.

Korndörfer, G. H., Pereira, H. S., & Nolla, A. (2004). *Análise de silício: solo, planta e fertilizante* (Boletim técnico 2, p. 34). Uberlândia, MG: GPSi/ICIAG/UFU.

Laviola, B. G., Martinez, H. E. P., Souza, R. B., & Alvarez, V. H. (2007). Dinâmica de cálcio e magnésio em folhas e frutos de *Coffea arabica*. *Revista Brasileira de Ciência do Solo, 31*(2), 319- 329.

Lux, A., Luxová, M., Hattori, T., Inanaga, S., & Sugimoto, Y. (2002). Silicification in sorghum (Sorghum bicolor) cultivars with different drought tolerance. *Physiologia Plantarum, 115*(1), 87-92.

Luz, J. M. Q., Guimarães, S. T. M. R., & Korndörfer, G. H. (2006). Produção hidropônica de alface em solução nutritiva com e sem silício. *Horticultura Brasileira, 24*(3), 295-300.

Ma, J. F., & Yamaji, N. (2008). Functions and transport of silicon in plants. *Cellular and Molecular Life Sciences, 65*(19), 3049-3057.

Ma, J. F., Miyake, Y., & Takahashi, E. (2001). Silicon as a beneficial element for crop plant. In L. E. Datnoff, G. H. Korndörfer, G. Snyder (Ed.), *Silicon in agriculture* (p. 17-39). New York, NY: *Elsevier Science*.

Madon, M., Heslop-Harrison, J. S., Schwarzacher, T., & Hashim, A. T. (2012). Analysis of oil palm calli and regenerants using flow and image cytometry and 18S-25S ribosomal DNA fluorescence in situ hybridisation (FISH). *Journal of Oil Palm Research, 24*, 1318-1329.

Marschner, H. (1995). *Mineral nutrition of higher plants* (p. 920). London, UK: Academic Press.

Msogoya, T. J., Grout, B. W., & Roberts, A. (2011). Reduction in genome size and DNA methylation alters plant and fruit development in tissue culture induced off-type banana (Musa spp.). *Journal of Animal and Plant Sciences, 11*(3), 1450-1456.

Nguyen, X. D., Eichler, H., Dugrillon, A., Piechaczek, C., Braun, M., & Klüter, H. (2003). Flow cytometric analysis of T cell proliferation in a mixed lymphocyte reaction with dendritic cells. *Journal of Immunological Methods, 275*(1/2), 57-68.

Nogueira, G. F., Pasqual, M., & Scherwinski-Pereira, J. E. (2013). Survival of sugarcane shoot tips after cryopreservation by droplet-vitrification. *Pesquisa Agropecuária Brasileira, 48*(11), 1524-527.

Oliveira, L. A., & Castro, N. M. (2002). Ocorrência de silica nas folhas de *Curatella americana* L. e de *Davilla elliptica* St. Hil. *Revista Horizonte Científico, 38*(1), 159-170.

Pereira, R. R. C., Moraes, J. C., Prado, E., & Dacosta, R. R. (2010). Resistance inducing agents on the biology and probing behaviour of the greenbug in wheat. *Scientia Agricola, 67*(4), 430-434.

Plucknett, D. L. (1971). The use soluble silicates in Hawaii agriculture. *University of Queensland, 1*(6), 203-223.

Pozza, A. A. A., Alves, E., Pozza, E. A., Carvalho, J. G., Montanari, M., Guimarães, P. T.G., ... Santos, D. M. (2004). Efeito do silício no controle da cercosporiose em três variedades de cafeeiro. *Fitopatologia Brasileira, 29*(2), 185-188.

Prado, R. M., & Natale, W. (2004). Efeitos da aplicação da escória de siderurgia ferrocromo no solo, no estado nutricional e na produção de matéria seca de mudas de maracujazeiro. *Revista Brasileira de Fruticultura, 26*(1), 140-144.

Prado, R. M., & Natale, W. (2005). Efeito da aplicação de silicato de cálcio no crescimento, no estado nutricional e na produção de matéria seca de mudas de maracujazeiro. *Revista Brasileira de Engenharia Agrícola e Ambiental, 9*(2), 185-190.

Prado, M. J., Rodriguez, E., Rey, L., González, M. V., Santos, C., & Rey, M. (2010). Detection of somaclonal variants in somatic embryogenesis regenerated plants of *Vitis vinifera* by flow cytometry and microsatellite markers. *Plant Cell, Tissue and Organ Culture, 103*(1), 49-59.

Raven, J. A. (1983). The transport and function of silicon in plants. *Biological Reviews of the Cambridge Philosophical Society*, *58*(2), 179-207.

Reis, T. H. P., Figueiredo, F. C., Guimarães, P. T. G., Botrel, P. P., & Rodrigues, C. R. (2008). Efeito da associação silício líquido solúvel com fungicida no controle fitossanitário do cafeeiro. *Coffee Science*, *3*(1), 76-80.

Ribeiro, R. V., Silva, L., Ramos, R. A., Andrade, C. A., Zambrosi, F. C. B., & Pereira, S. P. (2011). O alto teor de silício no solo inibe o crescimento radicular de cafeeiros sem afetar as trocas gasosas foliares. *Revista Brasileira de Ciências do Solo*, *35*(3), 939-948.

Richmond, K. E., & Sussman, M. (2003). Got silicon? The nonessential beneficial plant nutrient. Cur. Opin. *Plant Biology*, *6*(3), 268-272.

Ritchie, S. W., Hanway, J. J., Thompson, H. E., & Benson, G. O. (1994). *How a soybean plants develops* (Special Report 53, p. 20). Ames, IA: State University of Science and Techonology, Cooperative Extension Service.

Rodrigues, F. A., Mcnally, D. J., Datnoff, L. E., Jones, J. B., Labbé, C., Benhamou, N., ... Bélanger, R. R. (2004). Silicon enhances the accumulation of diterpenoid phytoalexins in rice: A potential mechanism for blast resistance. *Phytopathology*, *94*(2), 177-183.

Sávio, F. L., Silva, G. C., Teixeira, I. R., & Borém, A. (2011). Produção de biomassa e conteúdo de silício em gramíneas forrageiras sob diferentes fontes de silicato. *Semina: Ciências Agrárias*, *32*(1), 103-110.

Shibles, R. M., & Weber, C. R. (1966). Interception of solar radiation and dry matter production by various soybean planting patterns. *Crop Science*, *6*(1), 55-59.

Silva, C. V., Oliveira, L. S., Loriato, V. A. P., Silva, L. C., Campos, J. M. S., Viccini, L. F., ... & Otoni, W. C. (2012). Organogenesis from root explants of comercial populations of *Passiflora edulis* Sims and a wild passionfruit species, P. *cincinnata* Masters. *Plant Cell, Tissue and Organ Culture*, *107*(3), 407-416.

Smulders, M. J. M., & Klerk, G. J. (2011). Epigenetics in plant tissue culture. *Plant Growth Regulation*, *63*(2), 137-146.

Trotter, J. (2000). WinMDI©: version 2.8. La Jolla: The Scripps Research Institute, 2000. Retrieved on Jan 15, 2015 from <http://facs.scripps.edu/software.html.

Vujović, T., Cerović, R., & Ružić, D. (2012). Ploidy level stability of adventitious shoots of sour cherry 'Cacanski Rubin' and Gisela 5 cherry rootstock. *Plant Cell, Tissue and Organ Culture*, *111*(3), 323-333.

Werner, D. (1977). Silicate metabolism. In D. Werner (Ed.), *The biology of diatoms* (p. 110-149). Oxford, UK: Blackwells Scientific.

Yang, X. M., An, L. Z., Xiong, Y. C., Zhang, J. P., Li, Y., & Xu, S. J. (2008). Somatic embryogenesis from immature zygotic embryos and monitoring the genetic fidelity of regenerated plants in grapevine. *Biologia Plantarum*, *52*(2), 209-214.

9

Growth and physiology of jelly palm (*Butia capitata*) grown under colored shade nets

Irton de Jesus Silva Costa*, Bárbara Nogueira Souza Costa, Franscinely Aparecida de Assis, Adalvan Daniel Martins, Leila Aparecida Salles Pio and Moacir Pasqual

*Departamento de Agricultura, Universidade Federal de Lavras, Av. Doutor Sylvio Menicucci, 1001, Kennedy, 37200-000, Lavras, Minas Gerais, Brazil. *Author for correspondence. E-mail: jc.agro12@yahoo.com.br*

ABSTRACT. *Butia capitata* (Mart.) Becc., commonly known as jelly palm, is a palm tree native to the Cerrado biome. It is considered an endangered species owing to intensive extractive exploitation of its fruits, with slow and uneven seminiferous propagation as an aggravating factor. Thereby, there are some bottlenecks related to jelly palm seedling production that needs to be optimized. Light quality and intensity is very important to seedling production, as it controls the metabolism of plants and consequently their growth and development. Thus, the aim of this work is to evaluate the effect of shade nets on the growth and physiological aspects of jelly palm seedlings. The experimental design was completely randomized, with five treatments, ten replicates and eight plants per replicate, totaling 400 plants. Experimental treatments consisted of the use of shade nets with 50% shading and different radiation intensity: white (985 μmol m^{-2} s^{-1}), red (327 μmol m^{-2} s^{-1}), black (433 μmol m^{-2} s^{-1}), and silver (405 μmol m^{-2} s^{-1}). Full sun (1000 μmol m^{-2} s^{-1}) was used as the control treatment. Growth and physiological analyses of the plants were performed after six months of treatment. We concluded that the red shade net increases growth and biomass of *Butia capitata* plants. However, the use of shade nets had no effect on the tested physiological characteristics of this species. The red shade net is the most suitable for the production of *Butia capitata* seedlings in the southern region conditions of Minas Gerais State.

Keywords: shading; palm tree; colored nets; light.

Introduction

The genus *Butia* Becc. (Arecaceae) originated in South America, and the plants belonging to this genus are widely distributed in the south and southeast regions of Brazil (Lorenzi, Noblick, Kahn, & Ferreira, 2010). This species has significant ecological importance, especially in terms of providing food resources for native fauna (Sarmento & Villela, 2010).

Jelly palm (*Butia capitata* Mart. Becc.) is a palm tree native to the Cerrado and has great economic potential because its fruits are widely used by regional populations in juice, ice cream, and popsicle production (Martins, 2003; Moura, 2008).

Because of its wide use and the extractive production of this species, the commercial plantation of Jelly palm becomes important to extend the territorial limits. Commercial production is challenging, not only because the propagation of this palm tree is exclusively seminiferous (Lorenzi, 2004) but also because seed germination is low, slow, and uneven, requiring up to one year to complete the germination process (Aquino et al., 2007).

Another factor that must be taken into account during the production of plants is light, because it controls the metabolism of plants and consequently their growth and development. However, light intensity and spectral quality may cause morphophysiological changes in the plant (Meira, Martins, & Manganotti, 2012).

One of the tools used to manipulate the light spectrum transmitted to plants is the use of shade nets. Shade nets combine physical protection of the cultivation environment with differential solar radiation filtering to promote desirable physiological responses regulated by light (Henrique, Alves, Deuner, Goulart, & Livramento, 2011). Many desirable physiological responses have been documented for crop species, including Physalis (*Physalis* spp.; Silva et al., 2016), ipe (*Handroanthus* spp.; Sabino, Korpan, Ferneda, & Silva, 2016), rosemary (*Rosmarinus officinalis* L.; Souza, Silva, Oliveira, Santos Neto, & Santos, 2014), and yarrow (*Achillea millefolium*; Pinto et al., 2014).

Shade nets of different light spectra are commercially available. Black shade nets (sombrite nets) are considered neutral and help reduce the incidence of radiation on plants without affecting the spectral quality of light. Red shade nets reduce blue, green, and yellow waves and add waves in the red and far-red spectral range, with transmittance for wavelengths greater than 590 nm (Nomura et al., 2009; Costa, Chagas, Pinto, & Bertolucci, 2012). Silver shade nets increase light reflection and allow temperature control (Polysack Indústrias Ltda, 2011). Conversely, white shade nets do not affect the transmitted light spectrum (Henrique et al., 2011).

Costa, Silva, Sampaio, and Martins (2011) and Amarante, Steffens, Mota, and Santos (2007) report the importance of the transmitted light spectrum when using colored shade nets in fruit growing. They found that using different shade net colors affected a range of variables.

Given both the need to minimize Jelly palm propagation bottlenecks and the potential for optimizing the use of solar radiation for plant metabolism, the present study aimed to assess the effect of shade nets on Jelly palm growth and physiology.

Material and methods

The experiment was conducted in the Universidade Federal de Lavras (UFLA) Pomology Division's experimental area (21° 14′ S, 45° 00′ W, 918 m asl). According to the Köppen climate classification, the climate is Cwa (Dantas, Carvalho, & Ferreira, 2007).

Jelly palm (*Butia capitata*) fruits from the 2014/2015 harvest were picked, pulped by hand, and kept in the shade to dry for 4 d in the town of Mirabela, located in the northern region of the Minas Gerais State. The fruits were subsequently taken to the Tissue Culture Laboratory of the Department of Agriculture. The seeds were removed from the endocarp using a vise and were treated with carboxin-thiram (Vitavax-thiran® 200 SC). The seeds were immersed in full-strength fungicide and subsequently dried in the shade for 4h.

The seeds were then taken to the laminar flow hood, wherein the operculum of the embryonic cavity was removed from the seeds. The seeds were then sown in 50-seed-capacity plastic trays containing medium-texture vermiculite as substrate (60 mg + 300 mL distilled water). The trays were placed in a Biological Oxygen Demand incubator at 30°C, 80% humidity, and a 12-h photoperiod for a period of 30 days until full germination. Plants were transferred to pots of 1.1 kg containing soil substrate consisting of clayey dystroferric Red Latosol typical of the region, sand, and organic matter (2:1:1 v/v/v ratio) and were watered.

The treatments utilized the colored shade net ChromatiNET®, provided by the company *Polysack Plastic Industry* with 50% shade and different proportions of radiation: white (985 μmol m^{-2} s^{-1}), red (327 μmol m^{-2} s^{-1}), black (433 μmol m^{-2} s^{-1}) and silver (405 μmol m^{-2} s^{-1}). The net was stretched in a wooden structure of 2 x 2 x 1.5 m length, width and height, respectively. Control group plants were kept in full sun (1000 μmol m^{-2} s^{-1}). After six months, the plants were subjected to analysis, using 10 replicates per treatment.

Growth analysis – Plants were assessed for leaf number (units), root number (units), height (cm), stipe diameter (mm), root diameter (mm), leaf length (cm), taproot length (cm), fresh and dry shoot biomass (g) and fresh and dry root biomass (g). Dry mass of plant material was assessed after oven drying at 60°C for 72h.

Relative chlorophyll index (RCI) - RCI was determined using a portable chlorophyll meter (model SPAD-502; Soil and Plant Analysis Development), Minolta Co., Osaka, Japan. The reading was performed between 900 and 1000h. Before performing the readings, the device was calibrated using the reading checker, according to the recommendations of the manual (Minolta Camera, 1989). Readings were taken for two points in each plant, totaling 20 readings per treatment.

Chlorophyll content – Chlorophyll content was assessed using a simplified version of the Scopel, Barbosa, and Vieira (2011) protocol. Leaf tissue samples were collected and transported in a Styrofoam box with ice. Leaf discs with 1-cm diameter were transferred intact into test tubes with lids containing 10 ml of 80% acetone (v/v) and were stored for 24h in a cold room, protected from light. Extracts were then filtered and the resulting solution was placed in cuvettes. The reference sample (blank) consisted of 80% acetone (v/v) solution. The absorbance readings were performed in a spectrophotometer at 645, 652 and 663 nm and chlorophyll *a*, *b* and total contents were calculated using the obtained readings (Witham, Blaydes, & Devlin, 1971). The results were expressed as mg per gram fresh weight of leaf tissue (mg g^{-1}).

Gas exchange – Photosynthetic and transpiration rates of fully expanded leaves were evaluated using an infrared gas analyzer (IRGA) model LI-6400 equipped with a quantum sensor. Solar radiation was expressed as μmol m^{-2} s^{-1}. The photosynthetically active photon flux density was fixed in the device camera to 1,000 μmol m^{-2} s^{-1}. Solar radiation was measured at 405 μmol m^{-2} s^{-1} (40.5%) under the silver net, 327 μmol m^{-2} s^{-1} (32.7%) under the red net, 433 μmol m^{-2} s^{-1} (43.3%) under the black net, and 985 μmol m^{-2} s^{-1} (98.5%) under the white net. The control treatment was measured at 1,000 μmol m^{-2} s^{-1}, corresponding to 100% incident radiation.

Electrolyte leakage (protoplasmic tolerance) – Leaf protoplasmic tolerance was assessed by electrolyte release from leaf discs (Leopold, Musgrave, & Williams, 1981) immersed in 30 ml distilled water in test tubes. Ten leaf discs were collected in 10 replicates per treatment. First, electrical conductivity was measured with a conductivity meter (DIGIMED brand, CD 21A model), after the leaf discs had been immersed for 24h. The first reading was considered free conductivity (FC).

After this measurement, the test tubes were placed in a water bath at 100°C for 1h and then electrical conductivity was measured again, which was termed total conductivity (TC). The percentage of absolute integrity (PAI = 1 − FC/TC) was calculated from these data, according to the method described by Vasquez-Tello et al. (1990).

Relative water content (RWC) – Leaf discs from the youngest fully expanded leaf were removed from the center of the leaf blade of each plant and were massed to determine fresh mass (FM). Leaf discs were then placed in Petri dishes, where they were submerged in distilled water for a 24-h period to assess their turgid mass (TM). Leaf discs were then dried in a forced-air circulation oven at 70°C for 48h in order to assess their dry mass (DM). RWC was calculated using the following equation: RWC = {[(FM–DM) / (TM–DM)]} × 100, according to the method by Barrs and Weatherley (1962).

The experimental design used was completely randomized, with five treatments and 8 plants $plot^{-1}$, totaling 400 plants. All data obtained were subjected to analysis of variance, using the statistical software SISVAR (Ferreira, 2011). In the qualitative tests, the averages were compared using the Scott-Knott test at 5% probability.

Results and discussion

Shade nets positively influenced most of the assessed growth characteristics (Figure 1); shaded plants exhibited increased root growth and plant height compared to control plants (control).

The red, black, and silver net groups exhibited greater leaf length than the white net and control groups (Figure 1b). The red net group exhibited a larger stipe diameter than the other groups (Figure 1c).

The control and white net groups exhibited the lowest leaf length and plant height values. This suggests that Jelly palm plants are sensitive to light quality and that the 50% shading from the nets was beneficial.

This is consistent with the findings of Souza et al. (2011a), where the authors found that guaco (*Mikania glomerata*) plants grown under colored shade nets invested a greater quantity of photoassimilates in shoots than those grown in full sun, probably associated with the increased cell elongation of plants growing under shade.

Conversely, the low plant growth for the white net group most likely occurred because emitted light quality is the main factor controlling plant metabolism, and the white net failed to improve the transmitted light spectrum (Henrique et al., 2011).

The use of shade nets may cause changes in the radiation spectrum available to the plant, leading to metabolic adjustments in the photosynthetic system (Sabino et al., 2016), which may contribute to increasing growth characteristics such as those assessed in this study (i.e., taproot length, leaf length and plant height).

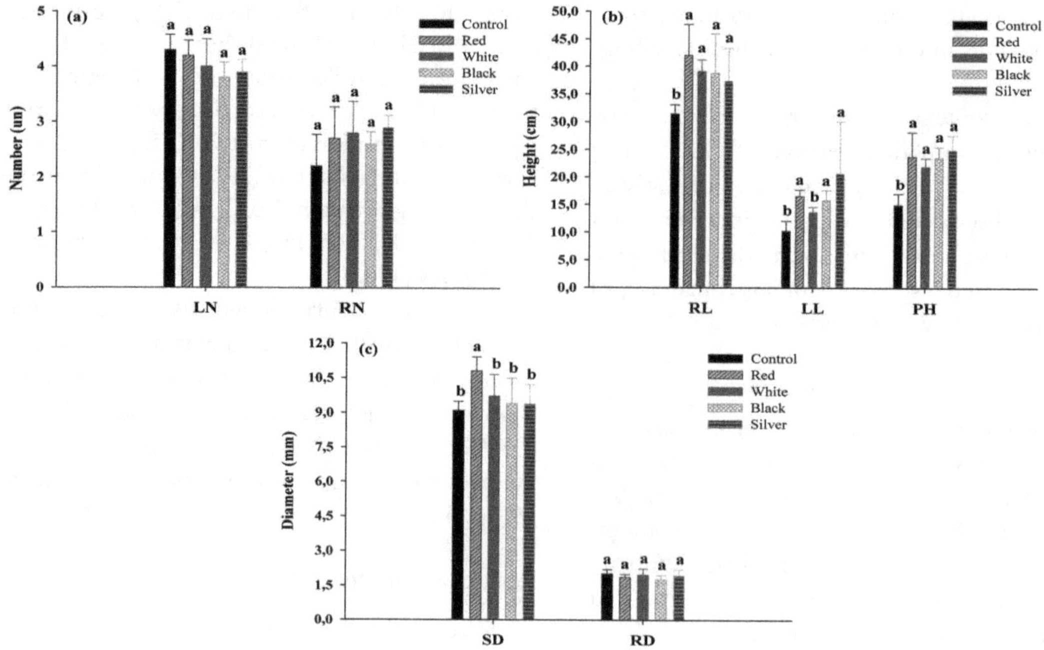

Figure 1. a) – Leaf number (LN), root number (RN), b) - taproot length (RL), leaf length (LL), plant height (PH), c) – stipe diameter (SD) and root diameter (RD) of Jelly palm (*Butia capitata*), under shade nets (Lavras, 2016).

Plants grown under red nets obtained improved incident and absorbed light quality. Therefore, they acquired increased stipe diameters. This phenotypic plasticity is favorable because it improves adaptation to the planting process in the field, thus avoiding possible damping off (Souza et al., 2014). Red nets increased stem diameter in yellow trumpet tree (*Handroanthus serratifolius*) and yellow trumpet tree of the Cerrado (*Handroanthus ochraceus*) plants (Sabino et al., 2016), which corroborates our results.

The use of colored shade nets aims to cause changes in the spectrum of solar radiation available to plants, producing metabolic adjustments in the photosynthetic system (Sabino et al., 2016). Henrique et al. (2011) report that using red light leads to positive responses in plant development.

No differences in leaf number, root number (Figure 1a) or root diameter (Figure 1c) occurred among treatments. Similarly, Henrique et al. (2011) found no effect of shade net color on leaf number in coffee (*Coffea arabica*) seedlings. This may have occurred because, although no differences were found in the number of leaves with respect to the treatments, the meshes provided the study species with the expansion of the organs that is confirmed by the biomass of the plant.

Fresh and dry biomass of roots and shoots (Figure 2) varied among treatments. Plants grown under red nets exhibited higher values of these growth parameters than the other treatments. Nonetheless, the plants grown under white, black, and silver nets exhibited higher values of fresh root biomass (FRB), dry root biomass (DRB), fresh shoot biomass (FSB) and dry shoot biomass (DSB) compared to the control group.

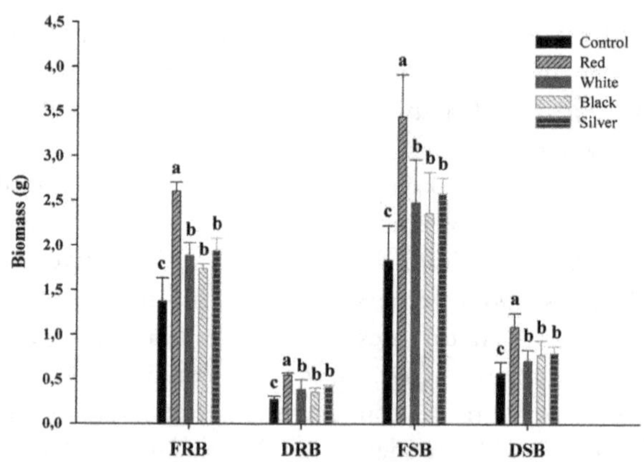

Figure 2. Fresh root biomass (FRB), dry root biomass (DRB), fresh shoot biomass (FSB) and dry shoot biomass (DSB) of Jelly palm (*Butia capitata*) under shade nets (Lavras, 2016).

In this context, the increased biomass accumulation of plants grown under red shade nets may be attributed to a stimulation of growth rate and plant vigor (Tinyane, Sivakumar, & Soundy, 2013).

The red shade nets allow plants to absorb more photons from the red wavelength, which benefit the activity of the photosynthetic pigments, leading to a higher production of photoassimilates (Taiz & Zeiger, 2013). Thus, plants cultivated under red shade nets usually show greater leaf and root mass (Nascimento, Pinto, Silva Jr, Castro, Santos, 2014; Oren-Shamir et al., 2001).

Henrique et al. (2011) reported that coffee plants grown under red nets exhibited higher dry and total leaf biomass than coffee plants grown under other types of shade nets. The same authors state that the red shade net was the most effective in promoting growth and development of coffee seedlings.

Similar to our study, lemon balm (*Melissa officinalis*) (Brant et al., 2009) and guaco (Souza et al., 2011a) plants grown under colored shade nets accumulated drier biomass than plants grown in full sun.

The colored shade nets represent an agricultural technology that combines the physical protection of the growing environment with the differential filtering of solar radiation (Henrique et al., 2011). The intensity and light quality are of considerable significance for the conversion of the chemical energy in the photosynthetic process, as well as for some morphogenic effects, which can be observed by the variations in the size of the leaves. In addition, the adaptive plasticity of the species associated with the accumulation of biomass depends on the adjustment of its photosynthetic apparatus and are subject to the different solar radiation conditions (Song, 1991).

For RCI, chlorophyll *a*, *b* and total chlorophyll, no difference was noted among the treatments, demonstrating that these characteristics were not affected by shade net color (Table 1).

Table 1. Relative chlorophyll index (RCI) and chlorophyll a, b and total content (mg·g^{-1} fresh matter) in Jelly palm (*Butia capitata*) plant leaves subjected to different shade nets (Lavras, 2016).

Treatments	RCI (%)	Chlorophyll a	Chlorophyll b	Chlorophyll total
Control	53.71 a	0.012 a	0.006 a	0.018 a
Red	53.39 a	0.010 a	0.007 a	0.014 a
White	53.31 a	0.011 a	0.005 a	0.016 a
Black	51.00 a	0.012 a	0.004 a	0.016 a
Silver	55.23 a	0.012 a	0.004 a	0.015 a
CV	12.16	39.81	49.33	35.60

*Averages followed by the same letters in columns belong to the same group according to the Scott-Knott test ($p \leq 0.05$).

Gonçalves et al., (2012) and Rego and Possamai (2006) report a strong correlation between the increase in chlorophyll content and lower intensity of radiation in mahogany and jequitibá-rosa, respectively. However, Sabino et al. (2016) and Henrique et al. (2011) analyzed ipe and coffee seedlings grown under colored shade nets, respectively, and found results that corroborate those found in this study.

For photosynthetic rate, internal carbon, transpiration rate, stomatal conductance and internal and external carbon ratio, no difference among treatments was found (Table 2). This implies that the varied spectra transmitted by the colored shade nets had no effect on the photosynthetic activity of the study species.

Table 2. Photosynthetic rate – A (μmol CO_2 m^{-2} s^{-1}), internal carbon – Ci (μmol CO_2 mol^{-1}), transpiration rate – E (mmol H_2O m^{-2} s^{-1}), stomatal conductance – Gs (mol H_2O m^{-2} s^{-1}) and internal and external carbon ratio - Ci/Ca of Jelly palm (*Butia capitata*) leaves, under shade nets (Lavras, 2016).

Treatments	A	Ci	E	Gs	Ci/Ca
Control	1.7658 a	284.6894 a	0.0005 a	0.0337 a	0.7551 a
Red	1.7149 a	290.3312 a	0.0006 a	0.0369 a	0.7779 a
White	1.7308 a	281.6456 a	0.0006 a	0.0311 a	0.7482 a
Black	1.7982 a	290.1410 a	0.0006 a	0.0375 a	0.7729 a
Silver	1.7235 a	281.2603 a	0.0005 a	0.0334 a	0.7495 a
CV	18.25	5.31	18.50	21.53	5.46

*Averages followed by the same letters in columns belong to the same group, according to the Scott-Knott test ($p \leq 0.05$).

Souza, Castro, Soares, Santos, and Alves (2011b) also found no differences in photosynthetic rate in guaco plants under colored shade nets. Additionally, stomatal conductance and photosynthesis were not changed in pitomba (*Talisia subalbens*) grown under different shade conditions, according to Nery et al. (2011).

Plants subjected to water stress conditions exhibit a reduction in stomatal opening and in the time during which the stomata remain open (Santos, Silva, & Silva Junior, 2014). Stomatal conductance values reflect the resistance associated with CO_2 diffusion through the stomatal opening (Silveira et al., 2013). We expected that plants in full sun would have a different photosynthesis rate when compared to other treatments because shading provided by the nets would minimize the loss of water. However, no difference was observed to this characteristic, which might be due to the mild temperature in the region when the experiment was conducted.

The different values of solar radiation in the treatments caused no change in stomatal conductance. Lobos (2012), also observed no changes in stomatal conductance, when studying blueberry (*Vaccinium corymbosum*) under different shade levels (40, 50, and 75% shade). In eucalyptus (*Eucalyptus citriodora*), the highest stomatal conductance values occur when solar radiation is maximal and the leaf water potential has not yet reached the minimum values capable of inducing stomatal closure, according to Ferreira, Davide, and Carvalho (1999). It appears that in our study, the irrigation of the pots was sufficient to maintain leaf water potential, even with the different solar radiation intensities and air temperatures experienced by the treated plants.

No difference in electrolyte leakage occurred among treatments. Electrolyte leakage is expressed as free and total conductivity and as a percentage of absolute integrity (Table 3).

Table 3. Electrolyte leakage of Jelly palm (*Butia capitata*) leaves under shade nets. Free (FC) and total (TC) conductivity and percentage of absolute integrity (PAI) (Lavras, 2016).

Treatments	FC (μs cm^{-1})	TC (μs cm^{-1})	PAI (%)
Control	33.91 a	135.77 a	0.74 a
Red	34.06 a	139.11 a	0.75 a
White	34.22 a	146.74 a	0.76 a
Black	34.55 a	145.73 a	0.76 a
Silver	35.07 a	130.87 a	0.73 a
CV	13.19	11.69	2.78

*Averages followed by the same letters in columns belong to the same group according to the Scott-Knott test ($p \leq 0.05$).

The release of electrolytes occurs at severe drought levels because of the increased quantity of reactive oxygen species (superoxides), free radicals, and "lysis" enzymes (Roy-Macauley, Zuily-Fodil, Kidric, Phan Thi, & Vieira da Silva, 1992), which can cause increased membrane permeability and irreversible damage to organelles and molecules present within the cells (Alonso, Queiroz, & Magalhães, 1997).

Thus, the present study aimed to observe through the analysis of electrolyte extravasation whether the plants in full sun would suffer some type of stress due to exposure to higher temperatures when compared to the plants that were under photoconverting mesh, since they had 50% shading.

Although plants went through a period of low temperatures, their water requirements were met adequately. This precluded stress, which was confirmed by electrolyte leakage analysis.

The increased concentration of electrolytes in leaf cells, measured by the total conductivity of the leaf discs, may be a mechanism that prevents tissue desiccation, given the decrease in the osmotic component of leaf water potential (James, Lawn, & Cooper, 2008).

Relative water content is a parameter that characterizes the degree of leaf hydration (Pimentel, Sarr, Diouf, Abboud, & Roy-Macauley, 2002). Relative water content is related to stress conditions; that is, in environments with high temperatures and drought, among others, and plants have decreased leaf water content. Because no significant difference in relative water content occurred among treatments Table 4 demonstrates that the different types of shade nets and the control treatment did not affect relative water content, indicating that the control plants did not experience this stress.

Table 4. Relative water content (RWC) of Jelly palm (*Butia capitata*) leaves under shade nets (Lavras, 2016).

Treatments	RWC (%)
Control	37.00 a
Red	39.10 a
White	36.86 a
Black	33.76 a
Silver	41.27 a
CV	17.43

*Averages followed by the same letters in columns belong to the same group, according to the Scott-Knott test ($p \leq 0.05$).

Conclusion

Red shade nets increased the growth and biomass of *Butia capitata* plants.

The use of shade nets had no effect on the physiological characteristics of *Butia capitata* plants.

The red shade net is the most suitable for the production of seedlings of *Butia capitata* in the southern conditions of Minas Gerais.

Acknowledgements

We thank the Coordenação de Aperfeiçoamento de Pessoal de Nível Superior (CAPES), Conselho Nacional de Desenvolvimento Científico e Tecnológico (CNPq) and Fundação de Amparo a Pesquisa do Estado de Minas Gerais (FAPEMIG) for funding this project.

References

Alonso, A., Queiroz, C. G. S., & Magalhães, A. C. (1997). Chilling stress leads to increased cell membrane rigidity in roots of coffee (*Coffea arabica* L.) seedlings. *Biochimica et Biophysica Acta, 1323*(1), 75-84. doi: 10.1016/S0005-2736(96)00177-0

Amarante, C. V. T., Steffens, C. A., Mota, C. S., & Santos, H. P. (2007). Radiação, fotossíntese, rendimento e qualidade de frutos em macieiras 'Royal Gala' cobertas com telas antigranizo. *Pesquisa Agropecuária Brasileira, 42*(7), 925-931. doi: 10.1590/s0100-204x2007000700003

Aquino, C. F., Lopes, P. S. N., da Silva, H. P., Neves, J. M. G., & de Moura, R. C. (2007). Resposta do coquinho-azedo à adubação mineral e orgânica em fase de desenvolvimento inicial. *Revista Brasileira de Agroecologia, 2*(2), 1374-1377.

Barrs, H. D., & Weatherley, P. E. A. (1962). Re-examination of the relative turgidity technique for estimating water deficit in leaves. *Australian Journal of Biological Science, 15*(3), 413-428. doi: 10.1071/BI9620413

Brant, R. S., Pinto, J. E. B. P., Rosa, L. F., Albuquerque, C. J. B., Ferri, P. H., & Corrêa, R. M. (2009). Crescimento, teor e composição do óleo essencial de melissa cultivada sob malhas fotoconversoras. *Ciência Rural, 39*(5), 1401-1407. doi: 10.1590/s0103-84782009005000083

Costa, A. G., Chagas, J. H., Pinto, J. E. B. P., & Bertolucci, S. K. V. (2012). Crescimento vegetativo e produção de óleo essencial de hortelã-pimenta cultivada sob malhas. *Pesquisa Agropecuária Brasileira, 47*(4), 534-540. doi: 10.1590/s0100-204x2012000400009

Costa, C. A., Silva, A. C., Sampaio, R. A., & Martins, E. R. (2011). Productivity of determinate growth tomato lines tolerant to heat under the organic system. *Horticultura Brasileira, 29*(4), 590-593. doi: 10.1590/S0102-05362011000400024

Dantas, A. A. A., Carvalho, L. G., & Ferreira, E. (2007). Classificação e tendência climática em Lavras, MG. *Ciência e Agrotecnologia, 31*(6), 1862-1866. doi: 10.1590/S1413-70542007000600039

Ferreira, C. A. G., Davide, A. C., & Carvalho, L. R. (1999). Relações hídricas em mudas de *Eucalyptus citriodora* Hook, em tubetes, aclimatadas por tratamentos hídricos. *Cerne, 5*(2), 95-104.

Ferreira, D. F. (2011). SISVAR: a computer statistical analysis system. *Ciência e Agroteclogia, 35*(6), 1039-1042. doi: 10.1590/S1413-70542011000600001

Gonçalves, J. F. C., Silva, C. E. M., Justino, G. C., & Nina Junior, A. R. (2012). Efeito do ambiente de luz no crescimento de plantas jovens de mogno (*Swietenia macrophylla* King). *Scientia Forestalis, 40*(95), 337-344.

Henrique, P. C., Alves, J. D., Deuner, S., Goulart, P. F. P., & Livramento, D. E. (2011). Aspectos fisiológicos do desenvolvimento de mudas de café cultivadas sob telas de diferentes colorações. *Pesquisa Agropecuária Brasileira, 46*(5), 458-465. doi: 10.1590/s0100-204x2011000500002

James, A. T., Lawn, R. J., & Cooper, M. (2008). Genotypic variation for drought stress response traits in soybean. II. Inter-relations between epidermal conductance, osmotic potential, relative water content, and plant survival. *Australian Journal of Agricultural Research, 59*(7), 670-678. doi: 10.1071/ar07160

Leopold, A. C., Musgrave, M. E., & Williams, K. M. (1981). Solute leakage resulting from leaf desiccation. *Plant Physiology, 68*(6), 1222-1225. doi: 10.1104/pp.68.6.1222

Lobos, G. A. (2012). Spectral irradiance, gas exchange characteristics and leaf traits of *Vaccinium corymbosum* L. 'Elliott' grown under photo-selective nets. *Environmental and Experimental Botany*, (75), 142-149. doi: 10.1016/j.envexpbot.2011.09.006

Lorenzi, H. (2004). *Palmeiras no Brasil: exóticas e nativas.* Nova Odessa, SP: Instituto Plantarum.

Lorenzi, H., Noblick, L. R., Kahn, F., & Ferreira, E. (2010). *Flora Brasileira Lorenzi: Arecaceae (Palmeiras).* Nova Odessa, SP: Instituto Plantarum.

Martins, E. R. (2003). *Projeto conservação de recursos genéticos de espécies frutíferas nativas do norte mineiro: coleta, ecogeografia e etnobotânica* (Relatório institucional). Montes Claros, MG: UFMG.

Meira, M. R., Martins, E. R., & Manganotti, A. S. (2012). Crescimento, produção de fitomassa e teor de óleo essencial de melissa (*Melissa officinalis*) sob diferentes níveis de sombreamento. *Revista Brasileira de Plantas Medicinais, 14*(2), 352-357. doi: 10.1590/s1516-05722012000200015

Minolta Camera. (1989). *Manual for chlorophyll meter SPAD-502.* Osaka, JN: Minolta Radiometric Instruments Division.

Nascimento, M. E., Pinto, J. E. B. P., Silva Jr, J. M., Castro, E. M., & Santos, F. M. (2014). Plasticidade foliar e produção de biomassa seca em *Copaifera langsdorffii* Desf. cultivada sob diferentes espectros de luz. *Revista de Ciências Agrárias: Amazonion Journal of Agricultural and Environmental Sciences, 57*(1), 41-48. doi: 10.4322/rca.2013.064

Nery, F. C., Oliveira, H. M. D., Alvarenga, A. A. D., Dousseau, S., Castro, E. M. D., & Campos, A. C. A. L. (2011). Initial development and gas exchange of *Talisia subalbens* (Mart.) Radlk. under different shading conditions. *Revista Árvore, 35*(1), 61-67. doi: 10.1590/s0100-67622011000100007

Nomura, E. S., Lima, J. D., Rodrigues, D. S., Garcia, V. A., Fuzitanil, E. J., & Silva, S. H. M. (2009). Crescimento e produção de antúrio cultivado sob diferentes malhas de sombreamento. *Ciência Rural, 39*(5), 1394-1400. doi: 10.1590/s0103-84782009000500014

Oren-Shamir, M., Gussakovsky, E. E., Eugene, E., Nissim-Levi, A., Ratner, K., Ovadia, R., Giller, Y., & Shahak, Y. (2001). Coloured shade nets can improve the yield and quality of green decorative branches of *Pittosporum variegatum*. *Journal of Horticultural Science and Biotechnology, 76*(3), 353-361. doi: 10.1080/14620316.2001.11511377

Pimentel, C., Sarr, B., Diouf, O., Abboud, A. C. S., & Roy-Macauley, H. (2002). Tolerância protoplasmática foliar à seca, em dois genótipos de caupi cultivados em campo. *Revista Universidade Rural Série Ciências da Vida, 22*(1), 7-14.

Pinto, J. E., Ferraz, E. O., Bertolucci, S. K., Silveira, H. R., Santos, A. R., & Silva, G. M. (2014). Produção de biomassa e óleo essencial em mil folhas cultivada sob telas coloridas. *Horticultura Brasileira, 32*(3), 321-326. doi: 10.1590/s0102-05362014000300014

Polysack Indústrias Ltda. (2011). *Malhas termorefletoras aluminizadas.* Retrieved on Oct. 27, 2016 from http://www.polysack.com.

Rego, G. M., & Possamai, E. (2006). Efeito do sombreamento sobre o teor de clorofila e crescimento inicial do Jequitibá-rosa. *Pesquisa Florestal Brasileira, 26*(53), 179-194.

Roy-Macauley, H., Zuily-Fodil, Y., Kidric, M., Phan Thi, A., & Vieira da Silva, J. (1992). Effect of drought stress on proteolytic activities in *Phaseolus* and *Vigna* leaves from sensitive and resistant plants. *Physiologia Plantarum*, *85*(1), 90-96. doi: 10.1034/j.1399-3054.1992.850114.x

Sabino, M., Korpan, C., Ferneda, B. G., & Silva, A. C. (2016). Crescimento de mudas de ipês em diferentes telas de sombreamento. *Nativa*, *4*(2), 61-65. doi: 10.14583/2318-7670.v04n02a01

Santos, L. M. J., Silva, E. C., & Silva Junior, C. D. (2014). Análise preliminar das trocas gasosas, fluorescência da clorofila e potencial hídrico de plantas de juazeiro em uma região semi-árida de Sergipe, Brasil. *Scientia Plena*, *10*(2), 1-10.

Sarmento, M. B., & Villela, F. A. (2010). Sementes de espécies florestais nativas do Sul do Brasil. *Informativo Abrates*, *20*(1-2), 39-44.

Scopel, W., Barbosa, J. Z., & Vieira, M. L. (2011). Extração de Pigmentos foliares em plantas de canola. *Unoesc & Ciência-ACET*, *2*(1), 87-94, 2011.

Silva, D. F. D., Pio, R., Soares, J. D. R., Nogueira, P. V., Peche, P. M., & Villa, F. (2016). The production of *Physalis* spp. seedlings grown under different-colored shade nets. *Acta Scientiarum. Agronomy*, *38*(2), 257-263. doi: 10.4025/actasciagron.v38i2.27893

Silveira, H. M., Ferreira, E. A., Silva, D. V., Castro Neto, M. D., Carvalho, F. P., Santos, J. B., & Silva, A. A. (2013). Características fisiológicas de cultivares de mandioca após aplicação do mesotrione. *Planta Daninha*, *31*(2), 403-409. doi: 10.1590/s0100-83582013000200018

Song, P.-S. (1991). Light and Plant Responses. Edited by T. H. Attridge. Edward Arnold. *Photochemistry and Photobiology*, *53*(5), 727. doi: 10.1111/j.1751-1097.1991.tb08503.x

Souza, G. S., Castro, E. M., Soares, A. M., Pinto, J. E. B. P., Resende, M. G., & Bertolucci, S. K. V. (2011a). Crescimento, teor de óleo essencial conteúdo de cumarina de plantas jovens de guaco (*Mikania glomerata* Sprengel) cultivadas sob malhas coloridas. *Revista Biotemas*, *24*(3), 1-11. doi: 10.5007/2175-7925.2011v24n3p1

Souza, G. S., Castro, E. M., Soares, Â. M., Santos, A. R., & Alves, E. (2011b). Teores de pigmentos fotossintéticos, taxa de fotossíntese e estrutura de cloroplastos de plantas jovens de *Mikania laevigata* Schultz Bip. ex Baker cultivadas sob malhas coloridas. *Semina: Ciências Agrárias*, *32*(Suppl1), 1843-1854. doi: 10.5433/1679-0359.2011v32suplp1843

Souza, G. S., Silva, J. S., Oliveira, U. C., Santos Neto, R. B., & Santos, A. R. (2014). Crescimento vegetativo e produção de óleo essencial de plantas de alecrim cultivadas sob telas coloridas. *Bioscience Journal*, *30*(1), 232-239. doi: 10.5007/2175-7925.2011v24n3p1

Taiz, L.; & Zeiger, E. (2013). *Plant Physiology* (5a ed.). Porto Alegre, RS: Artmed.

Tinyane, P. P., Sivakumar, D., & Soundy, P. (2013). Influence of photo-selective netting on fruit quality parameters and bioactive compounds in selected tomato cultivars. *Scientia Horticulturae*, (161), 340-349. doi: 10.1016/j.scienta.2013.06.024

Vasquez-Tello, A., Zuily-Fodil, Y., Pham Thi, A. T., & Vieira da Silva, J. (1990). Electrolyte and Pi leakages and soluble sugar content as physiological tests for screening resistance to water stress in *Phaseolus* and *Vigna* species. *Journal of Experimental Botany*, *41*(7), 827-32. doi: 10.1093/jxb/41.7.827

Witham, F. H., Blaydes, D. F., & Devlin, R. M. (1971). *Experiments in plant physiology*. New York, US: D. V. Nostrand.

10

Growth, development and content of flavonoids in calendula (*Calendula officinalis* L.)

Isabela Cristina Gomes Honório[1*], Filipe Pereira Giardini Bonfim[1], Sebastián Giraldo Montoya[2], Vicente Wagner Dias Casali[2], João Paulo Viana Leite[3] and Paulo Roberto Cecon[4]

*[1]Faculdade de Ciências Agronômicas, Universidade Estadual Paulista "Júlio de Mesquita Filho", Rua José Barbosa de Barros, 1780, 18610-307, Botucatu, São Paulo, Brazil. [2]Departamento de Fitotecnia, Universidade Federal de Viçosa, Viçosa, Minas Gerais, Brazil. [3]Departamento de Biologia Vegetal, Universidade Federal de Viçosa, Viçosa, Minas Gerais, Brazil. [4]Departamento de Estatística, Universidade Federal de Viçosa, Viçosa, Minas Gerais, Brazil. *Author for correspondence. E-mail: isabelagomeshonorio@gmail.com*

ABSTRACT. Flavonoids, the pharmaco-active compounds studied in this work, in general, are located in the shoots as natural pigments that protect the plants from damage by oxidizing agents. The experiment was conducted in the greenhouse of the Fitotecnia Department, Federal University of Viçosa, Minas Gerais Staty. The experimental design was completely randomized with five treatments and five replications. The treatments were the time of harvest: zero, three, six, nine, and twelve days after anthesis (DAA). The height of the plant (mm), stem diameter (mm), number of leaves and total flavonoid content (%, w w^{-1}) were determined. We developed growth curves for calendula with growth behavior assessed through time for height, stem diameter and number of leaves. The time of the harvest season positively influenced the total flavonoid contents and the average productivity of flowers, and the highest content of flavonoids was found three DAA. The average productivity of the flowers and the accumulation of flavonoids were heterogeneous among harvest times (0, 3, 6, 9 and 12 days after anthesis). The highest yield of flavonoids was obtained three days after anthesis; therefore, a harvest of the inflorescences is recommended every three days.

Keywords: secondary metabolism, phenolic compounds, harvest season, plant performance.

Introduction

Calendula (*Calendula officinalis* L.) is in the order Asterales and is a member of the family Asteraceae; the genus name *Calendula* is derived from the Latin word *Calendae*, which means 'first day of each month'. The genus has 29 species (Missouri Botanical Garden [Mobot], 2013), including *Calendula officinalis* L., which is herbaceous and widely cultivated in several countries for ornamental, medicinal and cosmetic purposes (Ramos et al., 1998). In Brazil, the plant is popular and is known as marigold and wonder-of-gardens. The origin of calendula is in the region of the Mediterranean Sea (Lorenzi & Matos, 2008; Silva Junior, 2006), and the plant is currently cosmopolitan and cultivated in many parts of the world, including Brazil (Bertoni et al., 2006).

Calendula is used for cosmetic purposes in moisturizing creams, for both pre- and postsun exposure, because of the amount of saponins contained in the plant and because the gums and mucilages have large wetting abilities (Nardi et al., 1991). Moreover, calendula has healing and antiseptic properties with actions that are sudorific and analgesic, affect the bile duct, are anti-inflammatory, antiviral and anti-emetic, and tone the skin with vasodilation (Martins, Castro, Castellani, & Dias, 2004). According Pagnano et al. (2008), a tincture of 5% calendula positively influences the generation of new cells involved in wound healing, the fibroblasts, and provides more satisfactory healing than other treatments applied to experimental wounds in rabbits.

The classic knowledge of social groups and communities on the use of plants is essential to discover active compounds and molecules that have some healing action. Data from the World Health Organization (WHO, 2003) show that approximately 80% of the world population uses some type of herb in search of relief from some type of unpleasant symptom. Of this total, at least 30% was of medical value (Martins et al., 2004). According to the WHO report released in the early 1990's, 65–80% of the population in developing countries depended on medicinal plants as the only form of access to basic health care (Martins et al., 2004). To target new pharmaceutical products against diseases that plague the population of the world, traditional knowledge about biodiversity is an important resource (World Health Organization [WHO], 2003).

In the production of quality medicinal plants, in addition to growing naturally without pesticides, the first aspect for consideration is undoubtedly planting and harvesting at the correct times.

In medicinal species, the variability is high in the production of substances with therapeutic activity. The ideal time to harvest varies according to the plant organ, the stage of development, the time of year and the time of day (Martins et al., 2004) and should be performed when the quality of plants is the highest. The ideal moment to harvest depends on three interrelated elements: the stage of development with the highest biomass, the stage of development with the highest biosynthesis of bioactive compounds, and finally, the seasonal variation in the content of active ingredients through the developmental stages of the plant (Corrêa Júnior & Scheffer, 2009).

Growth analysis describes the physical and physiological condition of the plant in a time interval, and therefore, follows the dynamics of productivity with the measurement of biochemical and physiological indices.

In this context, this study determined the optimum time to harvest the flower heads of marigold as determined by quality and total content of flavonoids. Additionally, the time of greatest production of flower heads of the species was determined.

Material and methods

Commercial marigold seeds, the Isla brand, cultivar 'Bonina Sortida', were acquired as defensive-free (naked seeds).

The treatment times were chosen based on the opening of the flower buds of calendula and the longevity of the inflorescences. As the inflorescences began to open, each chapter was labeled with a different colored ribbon for identification at harvest. The first day of the harvest of each inflorescence was documented.

The seeds were sown in July 2011 in a polystyrene tray with 128 cells (40 cm^3 per cell) filled with commercial substrate Tropstrato HA Vegetables®. Two marigold seeds per cell were planted at a depth of 1.0 cm. After ten days, 'the plants were thinned to one plant per cell.

One month after sowing, two seedlings were transplanted to pots with nine liters of a planting mixture. The potting mixture was composed of the soil mixture (Pure Land), cattle manure and sand.

The experiment was conducted from July to December 2011 in the greenhouse of the Plant Science Department, Federal University of Viçosa, Minas Gerais State, Brazil. The climate classification, according to Koeppen (1948), was type 'Cwa', and the geographic coordinates ranged from 42° 52' W to 42° 50' W longitude and from 20° 44' S to 20° 47' S latitude. The design was completely randomized with five treatments and five replications. Each experimental unit consisted of a pot, for a total of 25 experimental units. The treatments were as follows:

0–1 day after anthesis (DAA);
2–3 days after anthesis;
3–6 days after anthesis;
4–9 days after anthesis; and
5–12 days after anthesis.

A week after transplantation, the plants were examined weekly. One month after transplanting, the plants began to flower and were marked. The chapters were collected daily in the dry weather of morning from the beginning of flowering soon after the opening of the inflorescence for three months, as recommended by Pinto and Bertolucci (2002).

The control of weeds was periodic and manual to reduce the possible effects of weeds on the plants and to provide better sanitary quality of the harvested chapters, as lower soil contact with shoots minimizes microbiological contamination (Martins et al., 2004). The objective of the weekly and the final evaluation was to monitor the growth of plants according to the phase of development.

The variables that were evaluated are as follows: number of leaves, stem diameter (mm) and plant height (mm) measured with a digital caliper (Starrett brand 727), leaf area (cm^2, AF), weight of fresh inflorescences (g, PFFL), weight of dried inflorescences (g, PSFL), stem fresh weight (g, PFC), stem dry weight (g, PSC), fresh weight of leaves (g, PFF), weight of dried leaves (g, PSF), and total flavonoid content (%, TTF). The dry mass of stems and of leaves was obtained by drying in a forced-air oven at 75°C to constant weight. The dry mass of inflorescences was evaluated after drying in a forced-air oven at 40°C. The leaf area was estimated with a LI-Cor meter (model LI-3000). The data were subjected to regression analysis using the statistical program Sigma Plot.

The extraction of flavonoids was performed at the Laboratory of Biodiversity, Department of Biochemistry, Federal University of Viçosa (Minas Gerais State) in January 2012.

After drying and weighing, the inflorescences were manipulated and were separated according to the Farmacopéia Brasileira 4th Ed. (Farmacopéia Brasileira, 2002), which recommends for the drug of *Calendula officinalis*: 'The plant drug consists of whole or ground ligulate flowers, accompanied scarce flowers tubular, separated from the receptacle and involucral, dried bracts'. After separation, the plant material was ground in a Wiley mill; to obtain the plant drug, the material for extraction must have a particle size less than or equal to 0.8 mm.

The content of flavonoids was determined as described in the monograph of calendula in the Farmacopéia Brasileira (Farmacopéia Brasileira, 2002), with some modifications. The entire procedure was performed in triplicate, for a total of 75 samples.

Samples, 0.4 g of the powdered drug (crushed marigold flowers), were weighed and placed in 500 mL flat-bottomed flasks. One milliliter of aqueous 0.5% (w v^{-1}) hexamethylenetetramine was added to the flasks, with 20 ml of acetone and 2 mL of hydrochloric acid. The mixture was heated under reflux for 30 min. The sample was filtered through cotton into a 100 mL volumetric flask, and the residue of the drug with the cotton was returned to the same flat-bottomed flask. Another 20 mL of acetone was added, and the solution was brought to reflux for 10 min. This procedure was repeated twice. After cooling, the solution was filtered in the same 100 mL volumetric flask. After this first extraction, the volume of the flask was filled with acetone. Thus, we obtained the acetone extract of calendula.

The second step in the extraction of flavonoids used a separatory funnel with 20 ml of acetone extract and 20 mL of water and then 15 mL of ethyl acetate (per partition). This procedure was repeated three times, with 10 mL portions of ethyl acetate. Subsequently, the fraction in ethyl acetate was washed with two portions of distilled water, and this new solution of acetate was transferred to 50 mL volumetric flask with a volume of ethyl acetate (mother solution, SM). The marigold flavonoid extract was obtained from this new phase.

In the last stage of the extraction, 10 ml of the marigold flavonic solution was pipetted into 1.0 mL of an aluminum chloride reagent (1.0 g of aluminum chloride dissolved in a methanol solution of 5% acetic acid in a volumetric flask, v v^{-1}, 50 mL, and the volume was completed).

To read the spectrophotometer, the control was prepared by diluting 10 mL of the flavonoid extract in 25 mL in a volumetric flask with a methanol solution of 5% acetic acid (v v^{-1}). After 30 minutes, the absorbance of the solution was measured at 425 nm in a 1.0 cm cuvette using the zero adjustment control.

The percentage of flavonoids was calculated according to the following Equation 1:

$$TTF\ (\%; p/p) = \frac{A \times 62.500}{500 \times m \times (100 - PD)} \tag{1}$$

where:
TTF = Total flavonoids;
A = Absorbance;
m = Mass of the drug (g); and
PD = Loss on drying (%, w w^{-1}).

The loss on drying was determined with drying at 105°C. A 2.0 g sample of ground marigold in a petri dish, previously weighed and tared, was weighed and dried in a forced circulation oven at 105°C for four hours; in the greenhouse, the sample was weighed again when a constant weight was achieved.

The loss on drying was calculated with the following Equation 2:

$$PD = \frac{100 \times N}{P} \tag{2}$$

where:

PD = Loss on drying;
N = Moisture (mass loss in g);
P = Sample (g).

The result is expressed in percentage (w w^{-1}) of flavonoids calculated as hyperosides ($C_{21}H_{20}O_{12}$).

Results and discussion

Some variables were assessed only at the end of the experiment because they required destructive evaluations of plants. The descriptive statistics of the data are presented (Table 1).

Table 1. Descriptive Statistics of the Observed Data.

	Average	Deviation	CV (%)	Minimum value	Maximum value
AF (cm^2)	4466,7	1322,9	29,6	2059,5	7201,2
PFF (g)	184,7	74,1	40,1	52,3	378,7
PSF (g)	28,2	5,2	18,5	16,6	42,0
PFC (g)	187,0	38,1	20,4	110,4	258,1
PSC (g)	35,8	9,5	26,5	12,4	53,9
PFFL (g)	127,1	54,5	42,8	35,7	233,3
PSFL (g)	19,8	4,7	23,7	10,8	30,3

AF = Leaf Area; PFF = Weight of Fresh Leaf; PSF = Weight of Dry Leaf; PFC = Weight of Fresh Stem; PSC = Weight of Dry Stem; PFFL = Weight of Fresh Inflorescence; PSFL = Weight of Dry Inflorescence.

The overall performance of plants can be evaluated with these variables. Regarding the weights of dried leaves and stems, there was an accumulation of biomass during the study period, but the plant weights were largely composed of water. The leaf area values demonstrated the large photosynthetically active area of the plants. In addition to the availability of water, the soil might have positively influenced leaves, resulting in an increase in the number of leaves and leaf area.

The marigold plants grew exponentially with time (Figure 1), and the maximum height was reached (approximately 37 cm) 53 days after transplantation, stable growth then followed. According to Gomes, Vieira and Heredia (2007), the heights of marigold plants increased linearly during their study, with the average maximum height of 34.02 cm at 120 days after transplanting. The results were consistent with Vieira, Heredia and Ramos. (1999), who found average heights from 29.9 to 39.9 cm in marigold plants fertilized with phosphorus and bed-of-avian; however, linear growth was observed during the study period 90 day cycle. Furthermore, Moreira et al. (2005) found variable plant heights from 28.82 to 39.24 cm in plants treated with five levels of nitrogen and phosphorus.

The early flowering coincided with a period of growth (approximately 15 days after transplantation), and the growth in plant height was promoted because of the expression of a large number of inflorescences, which is characteristic of the species (Ferri, 1986). Because marigold plants continue to grow even after the onset of flowering, in this species, vegetative and reproductive phases occurred simultaneously. For the stem diameters, the largest diameter of 20.0 mm was observed 60 days after transplantation. According to Figure 2, the diameter of the stem continued to increase through time, and 60 days after transplantation, the diameter had not reached a maximum; more time would be required to detect the largest diameter. The diameter of the stem is an important indicator of plant growth because it should continue to increase throughout development of the plant. The radial growth of the stem is necessary to provide strong support for the shoot. In some cases, an accumulation of reserves develops in the stems, and the translocation of assimilates would occur steadily to maintain the vital activities of the plant. Additionally, via the stem, plants absorb water and nutrients from the soil for continued growth.

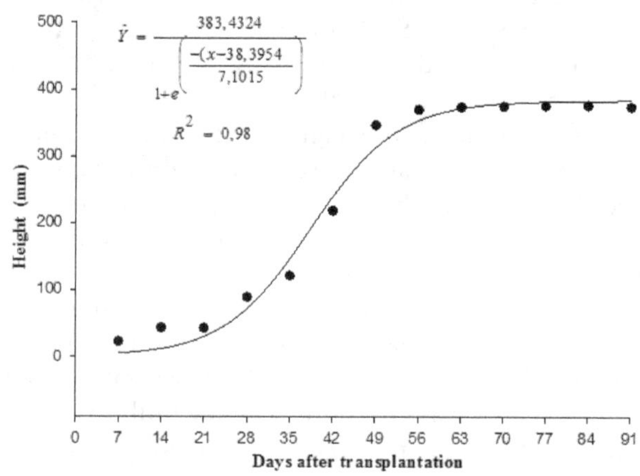

Figure 1. Plant height (mm) of marigolds (*Calendula officinalis* L.) during the growth period.

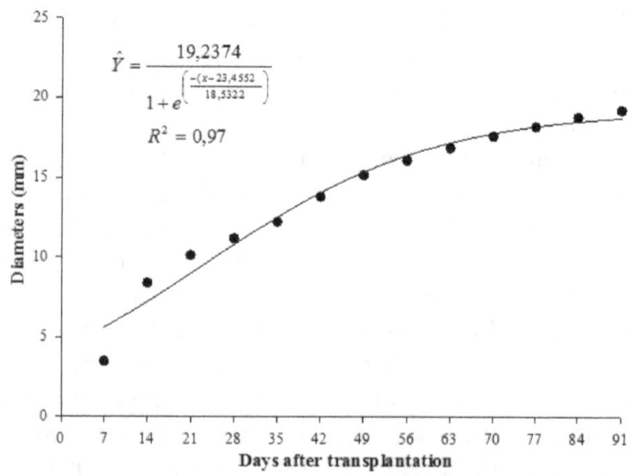

Figure 2. Stem diameter (mm) of marigolds (*Calendula officinalis* L.) during the growth period.

The number of marigold leaves increased exponentially (Figure 3) and had a similar growth curve to that of stem diameter, and after approximately 44 days, the increase in number of leaves stabilized. Leaves are an important indicator of production and determine the use of water by plants. Of major importance, the number of leaves is closely correlated with photosynthetic activity. During plant growth, the photosynthetic activity per leaf area increases with leaf age, until maximum expansion of the leaf, and then decreases thereafter until senescence (Taiz & Zeiger, 2006). If the plant is not in senescence and the soil is at field capacity, the normal supply of nitrogen to the leaves should occur, which increases the rate of photosynthesis and consequently the growth of the plant. Barman and Pal (1994) observed that the highest number of leaves per marigold plant was 146,10, which required 40 g m^{-2} mineral nitrogen; however, this relationship was not observed in this study because nitrogen was not applied to the soil and only organic fertilizer was used.

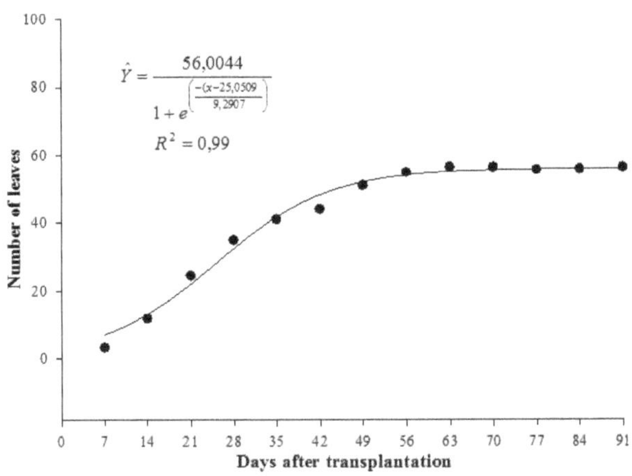

Figure 3. The number of leaves of marigold (*Calendula officinalis* L.) during the growth period.

The productivity of marigold chapters in grams per plant is shown in Figure 4. The highest productivity of inflorescences occurred zero days after anthesis (DAA), shortly after anthesis, resulting in a considerable reduction of the loss of important features of these flowers. Additionally, with time, productivity decreased because there was a great loss of parts or all of the flowers as a consequence of the initiation of flower senescence, which caused the gradual loss in characteristics of quality, such as color, appearance and integrity (Figure 4A). At the last harvest of the season (12 DAA), the drop of ligulate flowers and the general aging of the flower were evident, thus beginning the process of seed formation.

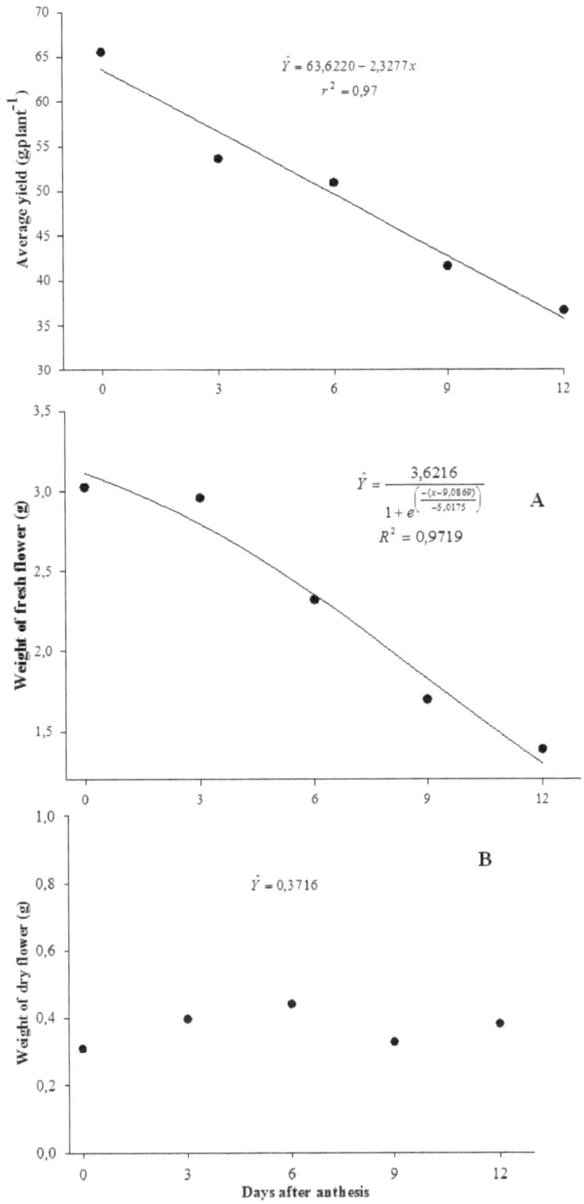

Figure 4. Average productivity of flowers, weights (g) of fresh (A) and dry (B) flowers of marigold (*Calendula officinalis* L.) in five harvest seasons for capitula.

Based on the results for the inflorescences, marigold has a type of short cycle and should be harvested at the right time, i.e., immediately after flower opening to avoid losses that would affect the quality of the raw material. Postharvest practices can increase the shelf life of the organoleptic properties of these flowers. For medicinal plants, postharvest practices are simple and were summarized in the drying marigold flowers. The aim of drying is to prevent spoilage by reducing the water content of the raw material, which has a negative effect on the action of enzymes; dehydration (Martins et al., 2004) can be performed naturally or artificially.

Moreira et al. (2005) found that the mass of dried flower heads of calendula varied according to the dose of P, and the maximum (4.70 g pot^{-1}) was obtained using 271.7 mg of P pot^{-1}. Because the mass of the dried flower heads also varied as a function of N application rates, the maximum (0.52 g pot^{-1}) was obtained with a dose of 199.2 mg N pot^{-1}.

The total flavonoid content was highest in marigolds 3.04 days after anthesis, accounting for approximately 0.70% of total flavonoids (Figure 5), calculated as hyperosides ($C_{21}H_{20}O_{12}$). According to the monograph of calendula described in the Farmacopéia Brasileira (Farmacopéia Brasileira, 2002), the drug in marigold plants is found in whole or ground ligulate flowers along with a few tubular flowers, separated from the receptacle and involucral dried bracts. The desiccated sample must contain no less than 0.4% of total flavonoids, calculated as hyperosides ($C_{21}H_{20}O_{12}$). Therefore, in this study, all marigold samples met the limit suggested by the pharmacopoeia.

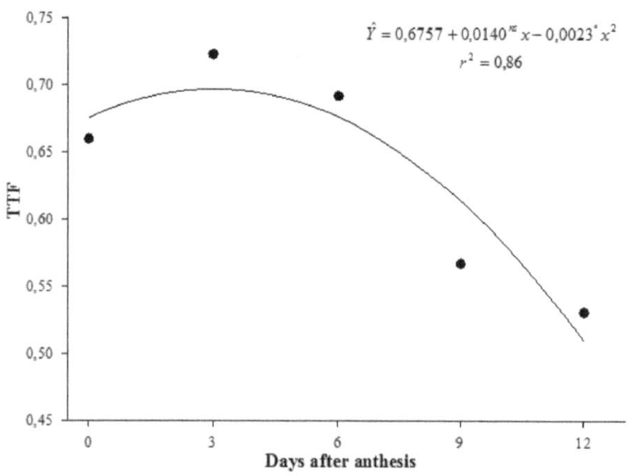

Figure 5. Content of total flavonoids (%) in flower heads of calendula (*Calendula officinalis* L.) at five harvest seasons. °Significant at 10% by t-test.

The quadratic curve for flavonoids indicated that soon after the flowers opened flavonoid content was not at a maximum. The following day we observed increased production of flavonoids by the plant, which could be explained by the high biosynthetic activity of younger tissues increasing the production of various compounds, including flavonoids. However, with the onset of flower senescence, the accumulation of flavonoids gradually diminished. Nevertheless, on the last date of assessment (12 DAA), sufficient amounts of flavonoids were found (approximately 0.53%) as hyperosides to classify the marigold plant as a drug.

Regarding marigold flavonoids, total flavonoids in ligulate flowers and receptacles are approximately 0.88 and 0.33%, respectively (Hadson, 1985). According Masterová, Grancaiová, Uhrinová, Ubik and Nagy (1991), the glycosidic flavonoids are major constituents of the anti-inflammatory effects of calendula.

Leite et al. (2005), in cultures with various doses of organic fertilizer, found that flavonoid production ranged from 0.52 to 0.60% flavonoids per t ha^{-1}; however, the authors emphasized that this ratio could increase with attack by aphids and other arthropods. This result, when compared with the present work, showed that with a large amount of organic fertilizer the content of flavonoids was close to the lowest level found in this study. Araújo et al. (2009) used mulch associated with increasing levels of organic fertilizer and obtained a value of 0.70% flavonoids with 59 t ha^{-1} of organic compost. Using chemical fertilization associated with mulch, Borella, Ribeiro, Freato, Mazzo and Barbosa (2011) observed a maximum yield of 0.79% of the total flavonoid content, and using organic fertilization associated with mulch, observed a yield of d 0.70%. The total flavonoid contents, after evaluation of their treatments, were within the range of those found in the present work.

In most studies that examined the content of flavonoids, some treatment was applied to increase the production of this metabolite. However, in this study, a treatment was not applied; for marigolds, the treatments were the time of harvest of the capitula. Nevertheless, high levels of total flavonoids were found, which demonstrated that it is not always necessary to apply large amounts of organic fertilizer for the production of flavonoids. To be classified as herbal drugs, good conduct and proper handling are required to produce large amounts of secondary metabolites in medicinal plants.

Andrade and Casali (1999) provided some recommendations to determine the ideal harvest times, which may vary because of the particular characteristics of each species. Thus, if the flowers are harvested in part, the ideal harvest time is at the beginning of flowering.

Because marigold inflorescences are widely required by the pharmaceutical industry, the optimum time to harvest the inflorescences is when the content of flavonoids is the highest. When the cultivated area is extensive, daily sampling of marigold inflorescences is impractical.

Therefore, based on the results of this study, the optimum time to harvest calendula flower heads for the maximum production of flavonoids was three days after anthesis. For chapters with the highest productivity, sampling should be avoided at bud opening. The optimum harvest season will depend on the purpose for growing calendula.

Conclusion

In the cultivation of plants to meet the requirements of producers, it is extremely important that the growth characteristics and the development of the species are understood to properly manage the species to maximize the intended purpose.

For marigolds, management is conducted with regular irrigation and monitoring of leaf development and stem thickening. Additionally, the correct timing of the harvest of capitula is essential because collection of flavonoids is the economic benefit of the plant.

Thus, the optimum time of harvest is every three days, which guarantees to the producer that flowers with maximum productivity, quality and total flavonoid contents will be collected in this period.

References

Andrade, F. M. C., & Casali, V. W. D. (1999). *Plantas medicinais e aromáticas: relação com o ambiente, colheita e metabolismo secundário*. Viçosa, MG: UFV.

Araújo, C. B. O., Santos, A. M., Fernandes, L. A., Martins, E. R., Sampaio, R. A., Costa, C. A., & Leite, G. L. D. (2009). Uso da adubação orgânica e cobertura morta na cultura da calêndula (*Calendulaofficinalis* L.). *Revista Brasileira de Plantas Medicinais, 11*(2), 117-123.

Barman, D., & Pal, P. (1994). Effect of nitrogen and phosphorus on seed yield in calendula (*Calendula officinalis* L.). *Orissa Journal of Agricultural Research,* 7(1), 17-21.

Bertoni, B. W., Damião Filho, C. F., Moro, J. R., França, S. C., & Pereira, A. M. S. (2006). Micropropagação de *Calendula officinalis* L. *Revista Brasileira de Plantas Medicinais, 8*(2), 48-54.

Borella, J. C., Ribeiro, N. S., Freato, A. M. R., Mazzo, K. F., & Barbosa, D. M. (2011). Influência da adubação e da cobertura morta na produtividade e no teor de flavonoides de *Calendula officinalis* L. (Asteraceae). *Revista Brasileira de Plantas Medicinais, 13*(2), 235-239.

Corrêa Júnior, C., & Scheffer, M. C. (2009). *Boas práticas agrícolas (BPA) de plantas medicinais, aromáticas e condimentares* (2a ed.). Curitiba, PR: Emater.

Farmacopéia Brasileira. (2002). *Comissão permanente de revisão da farmacopéia brasileira* (4a ed.). São Paulo, SP: Atheneu.

Ferri, M. G. (1986). *Fisiologia Vegetal*. São Paulo, SP: Pedagógica e Universitária Ltda.

Gomes, H. E., Vieira, M. C., Heredia, Z. N. A. (2007). Density and plant arrangement on *Calendula officinalis* L. yield. *Revista Brasileira de Plantas Medicinais, 9*(3), 117-123.

Hadson, V. (1985). Qualitative and quantitative analysis of flavonoids in plant products of cosmetic interest. *Clujul Medical, 58*(4), 378-381.

Koeppen, W. (1948). *Climatologia*. Buenos Aires, ARG: Gráfica Panamericana.

Leite, G. L. D., Araújo, C. B. O., Amorim, C. A. D., Pêgo, K. P., Martins, E. R., & Santos, E. A. M. (2005). Níveis de adubação orgânica na produção de calêndula e artrópodes associados. *Arquivos do Instituto Biológico, 72*(2), 227-233.

Lorenzi, H., & Matos, F. J. A. (2008). *Plantas medicinais no brasil: nativas e exóticas cultivadas* (2a ed.). São Paulo, SP: Instituto Plantarum.

Martins, E. R., Castro, D. M., Castellani, D. C., & Dias, J. E. (2004). *Plantas medicinais*. Viçosa, MG: UFV.

Masterová, L., Grancaiová, S., Uhrinová, V., Ubik, K., & Nagy, M. (1991). Flavonoids in flowers of *Calendula officinalis* L. *Chemical Papers, 45*(1), 105-108.

Missouri Botanical Garden. (2013). *W3 tropics*. Retrevide from http://www.tropicos.org/Name/2709695.

Moreira, P. A., Marchetti, M. E., Vieira, M. C., Novelino, J. O., Gonçalves, M. C., & Robaina, A. D. (2005). Desenvolvimento vegetativo e teor foliar de macronutrientes da calêndula (*Calendula officinalis* L.) adubada com nitrogênio e fósforo. *Revista Brasileira de Plantas Medicinais, 8*(1), 18-23.

Nardi, U.; Cagliardi, L.; Prampolini, F. (1991). La Calendula. *Erboristeria Domani,* (3), 62-67.

Pagnano, L. O., Baraldi-Artoni, S. M., Pacheco, M. R., Santos, E., Oliveira, D., & Lui, J. F. (2008). Morfometria de fibroblastos e fibrócitos durante o processo cicatricial na pele de coelhos da raça Nova Zelândia Branco tratados com calêndula. *Ciência Rural, 38*(6), 1662-1666.

Pinto, J. E. B. P., & Bertolucci, S. K. V. (2002). *Cultivo e processamento de plantas medicinais*. Lavras, MG: UFLA.

Ramos, A., Edreira, A., Vizoso, A., Betancourt, J., López, M., & Décalo, M. (1998). Genotoxicityof na extract of *Calendula officinalis* L. *Journal of Ethnopharmacology, 61*(1), 49-55.

Silva Junior, A. A. (2006). *Essentiaherba*. Florianópolis, SC: Epagri.

Taiz, L., & Zeiger, E. (2006). *Fisiologia Vegetal* (3a. ed.). Porto Alegre, RS: Artmed.

Vieira, M. C., Heredia, N. A., & Ramos, M. B. M. (1999). Crescimento e produção de capítulos de (*Calendula officinalis* L.), em função de cama-de-aviário semi-decomposta e de fósforo. *Revista Brasileira de Plantas Medicinais*, 1(2), 45-52.

World Health Organization (WHO). (2003). *Draft WHO guidelines on safety monitoring and pharmaco vigilance of herbal medicines*. Amsterdan, NLD: WHO.

11

Influence of bio-regulators on the seed germination and seedling growth of onion cultivars

Raquel Leszczynski[1], Alessandro de Lucca e Braccini[2*], Leandro Paiola Albrecht[3], Carlos Alberto Scapim[2], Gleberson Guillen Piccinin[4] and Lilian Gomes de Moraes Dan[4]

*[1]Programa de Pós-graduação em Genética e Melhoramento, Universidade Estadual de Maringá, Maringá, Paraná, Brazil. [2]Departamento de Agronomia, Universidade Estadual de Maringá, Av. Colombo, 5790, 87020-900, Maringá, Paraná, Brazil. [3]Universidade Federal do Paraná, Palotina, Paraná, Brazil. [4]Programa de Pós-graduação em Agronomia, Universidade Estadual de Maringá, Maringá, Paraná, Brazil. *Author for correspondence. E-mail: albraccini@uol.com.br*

ABSTRACT. Application of bio-regulators improves the productivity of cash crops. Current experiment evaluates the effects of different doses of a bio-regulator on the seed quality of three onion cultivars ('Baia Periforme', 'Bola Precoce' and 'Híbrida Superex'). Six doses of bio-regulator at 0, 2.5, 5.0, 10.0, 20.0 and 30.0 mL kg^{-1} were applied and results were compared to those of three controls. The experimental design was entirely randomized under the factorial arrangement 3 x 6 + 3, with four replications. Data were subjected to analysis of variance ($p < 0.05$), and all seed treatments were compared with controls by Dunnett's test. The analysis of regression described the effects of the increasing doses of bio-regulator on germination and vigor responses which were evaluated by the first count of the germination test, the percentage of germination and the length of the roots and shoots of the normal seedlings. The application of the bio-regulator did not improve the percentage of seed germination and no positive response of the seedlings was reported from the cultivars 'Baia Periforme' and 'Bola Precoce'. However, seedling vigor of 'Híbrida Superex' was positively affected by the application of the bio-regulator.

Keywords: *Allium cepa* L., bio-regulator, seed quality.

Introduction

Onion (*Allium cepa* L.) is a condimental plant cultivated under different weather conditions since ancient times. It is the third horticultural product worldwide and the second source of net income in agriculture, with China currently ranked as the biggest onion producer (BOITEUX; MELO, 2004; CHEMELLO, 2005).

In Brazil, tomato, potato and onion are the third most important economic produces with the southern, southwestern and northeastern regions featuring the highest acreages in onion planting. Onion growers in all these regions are farmer families whose members participate in the entire production system, aiming at supplying the Brazilian market with this fresh produce (BOEING, 2002; RESENDE; COSTA, 2005; OLIVEIRA; BOITEUX,

2004). However, the constraints in onion cultivation require new approaches so that the usual levels of crop productivity may be exceeded.

In plant growth and development, the seed germination, vegetative growth, flowering, fruit growth and maturation are influenced by several endogenous and exogenous factors. Plant development is regulated by six types of hormones: auxin, gibberellin, cytokinin, ethylene, abscisic acid and brassinosteroid. When applied to plants at low concentration (10^{-4} M), phytohormones promote, inhibit or modify physiological and morphological processes (CASTRO; VIEIRA, 2001). Gibberellins are important to promote seed germination because they induce the synthesis of enzymes through genetic mechanisms that support the breakdown and the mobilization of seed reserves from the endosperm into the embryo axis. The cytokinins interacting with the auxins promote cell division due to their role in cell cycle and participation in the mechanisms of cell lengthen and differentiation. Further, auxins affect cell length through ATPase activation which is necessary to promote the cell wall acidity and therefore make possible the action of gibberellin-synthesized hydrolytic enzymes (CASTRO; VIEIRA, 2001; RAVEN et al., 2007; TAIZ; ZEIGER, 2009).

Metabolism, morphogenesis and plant growth depend on the signalling transduction pathway for plant hormones. These signalling phytohormones are important to maintain the regulatory mechanisms during the entire plant development (TAIZ; ZEIGER, 2009). In fact, the pathways transmit information on the development and physiological status of plant cells and tissues (CASTRO; VIEIRA, 2001). The origin, biosynthesis, chemical structure, mechanisms of transport, action and physiological effects of these substances should be currently known to enhance the plants' physiological responses to these chemicals or their counterparts (VIEIRA; CASTRO, 2004). Thus, plant hormones are of paramount importance in controlling the mechanisms of plant productivity.

The application of bio-regulators has been recently recommended to increase the productivity of several crops, such as soybean, beans, maize and rice (KLAHOLD et al., 2006; ALBRECHT et al., 2009, 2010). Bio-regulators are natural or synthetic substances applied to seeds and other parts of plant to produce the plant's hormone effects. In general, they increase plant productivity although they have not been applied to agriculture systems in which plants are not propped by high technology (VIEIRA; CASTRO, 2001; VIEIRA; SANTOS, 2005). Since several benefits have been obtained through the use of these bio-regulators, the mixture of one or more of these substances or their mixture with amino acids, nutrients and vitamins is usually called bio-stimulants (CASTRO; VIEIRA, 2001). Bio-regulators have increased the percentage of seed germination and plant height of *Strelitzia reginae* (GARCIA et al., 2006), the percentage of seed germination and the germination rate of *Annona cherimola* Mill. x *Annona squamosa* L. cv. Gefrer, the percentage of emergence and plant development of *Passiflora edulis* Sims. f. *flavicarpa* Deg. (FERREIRA et al., 2007), and the development and crop yield of *Phaseolus vulgaris* (ALLEONI et al., 2000). The dry weight of the seedling at the three trifoliate stages and the dry weight of 1,000 grain were increased.

Bio-regulators with a wide range of concentrations were applied on the seeds or sprayed on the leaves of soybean and common bean crops (ALLEONI et al., 2000; MOTERLE et al., 2008) or only on the seeds of *Passiflora edulis* Sims. f. *flavicarpa* Deg. and *Strelitzia reginae* (FERREIRA et al., 2007; GARCIA et al., 2006). Whereas bio-regulator at 250 mL ha^{-1} was applied on the seeds and 750 mL ha^{-1} was sprayed on the leaves of common beans (ALLEONI et al., 2000), 12 and 16 mL kg^{-1} were applied on the seeds of *Passiflora edulis* Sims. f. *flavicarpa* Deg. (FERREIRA et al., 2007), and 4.07 mL kg^{-1} increased by 14.2% the percentage of normal seedlings of *Annona cherimola* Mill. x *Annona squamosa* L. cv. Gefrer. On the other hand, high doses of bio-stimulants applied on cotton seeds affected negatively seedling emergence (VIEIRA; SANTOS, 2005). No difference either in plant height or in pod numbers per plant was detected in soybean (MOTERLE et al., 2008).

The presence of inconsistent reports in the literature may be due to the few species investigated, environmental conditions and crop management. Furthermore, industrialized phytohormones hardly ever work alone because the target tissue has endogenous hormones that may influence responses (VIEIRA; CASTRO, 2004). Thus, the efficacy of bio-regulators coupled to different concentrations should be determined (VIEIRA; CASTRO, 2004). The products' composition, concentration and proportion in the chemical solution may increase the plants' growth and development by stimulating the cell length and division and the differentiation of plant tissues, with the consequent increase in the absorption nutrient and efficient use of water by the plants (STOLLER DO BRASIL, 1998).

Onion seeds, which need more time to germinate than other kitchen garden species, constitute a constraint in onion production. Seedling growth is also slow even when the photoperiod and temperature are satisfactory for the plant's development and the bulbs' formation (OLIVEIRA; BOITEUX, 2004; OLIVEIRA et al., 2004). Moreover, intrinsic factors may also influence the other stages of the crop's development (OLIVEIRA; BOITEUX, 2004). Current experiment evaluates the percentage of seed germination and the growth of seedlings from three cultivars of onion (*Allium cepa* L.) after the application of several concentrations of the bio-regulator on the seeds.

Material and methods

The experiment was conducted in the Laboratory of Seed Technology at the Applied Agriculture Research Nucleus (NUPAGRI), Center of Agrarian Sciences of the State University of Maringá (UEM), Maringá, Paraná State, Brazil, during the growth season 2009-2010.

Cultivars, treatments and experimental design

Current experiment comprised three onion genotypes ('Baia Periforme', 'Bola Precoce' and 'Híbrida Superex'), six doses of bio-regulator at 0, 2.5, 5.0, 10, 20 and 30 mL kg^{-1} of seeds and three controls. The bio-regulator used in this experiment was the Stimulate®, a liquid product with 50 mg L^{-1} (0.005%) indole-3-butyric acid (auxin), 50 mg L^{-1} (0.005%) gibberellin acid, and 90 mg L^{-1} (0.009%) kinetin (cytokinin) (STOLLER DO BRASIL, 1998). Treatments were completely randomized under the factorial arrangements of 3 onion genotypes and six doses of bio-regulator, with four replications. The treatments were compared with three controls (without seed imbibition in distilled water or bio-regulator), one for each genotype (3 x 6) + 3.

All the laboratory equipments used in this experiment were cleansed with alcohol 70% and commercial sodium hypochloride. Laboratory instruments were treated under ultraviolet light for one hour; the water and the blotting paper were disinfected with alcohol 70%. The seeds were weighed with an analytical scale with a 0.001 g readability of. Micro-pipettes were used to apply the solution on the seeds in graduate 50 mL beakers. The container was then stirred for 1 min. so that the solution could be distributed on all the seeds.

Seed quality

Seed germination

Four replications or sub-samples with 25 seeds for each treatment were used. The seeds were sown in plastic gerboxes on three sheets of towel paper moistened with distilled water, the amount of which was approximately 2.5 times the mass of dry paper. After the treatments, the gerboxes were transferred to a B.O.D. germination chamber, regulated to maintain a constant temperature of 20°C during 12 days, in the absence of light. The first germinated (normal seedlings) seed count occurred on the 6th day and the final count on the 12th day. Seedlings with primary root longer than or equal to 2 mm were considered germinated (HADAS, 1976). Results were expressed as a percentage of seed germination (BRASIL, 2009; MARCOS FILHO et al., 1987).

First count of seed germination

Seed germination was determined by normal seedling percentage obtained in the first count of the germination test, or rather, those that germinated faster. The test was based on the principle that the high percentage of samples that produced normal seedlings in the germination test's first count was more vigorous and thus determined the relative vigor of the seed lot (BRASIL, 2009; MARCOS FILHO et al., 1987; NAKAGAWA, 1999).

Seedlings root and shoot length

Procedures were undertaken following Nakagawa (1999), in which onion seedling lengths was measured. Samples that demonstrate the highest average values were considered the most vigorous.

Four replications of 10 seeds sown in plastic gerboxes on three sheets of paper towel on hanging distilled-water moistened towels were used. The amount of water was equivalent to 2.5 times the mass of the dry paper. Two lines were drawn lengthwise in the upper third of the germination paper where the seeds were placed, at an interleaved manner, between the papers. The plastic gerboxes were positioned at an angle of 45° in a B.O.D germination chamber for 12 days at 20°C, in the dark. At the end of this period, the sections of the germinated normal seedlings (primary root and shoot) were measured by a ruler and average results given in centimetres.

Statistical analysis

After testing data normality, the analysis of variance was calculated to determine initial treatment effects on seed quality as measured by each test. Significant effects were described by analysis of regression. Dunnett's test compared treatments and control at level of significance $p < 0.05$ by statistical package SISVAR (BANZATO; KRONKA, 2008).

Results and discussion

The application of the product showed significant difference on cultivar responses ($p < 0.05$). The first count of normal seedlings, for example, was significantly different only for seeds of the cultivar 'Híbrida Superex' (Figure 1). The cubic model showed the highest percentage of normal seedlings (77%) at dose 24.18 mL kg^{-1} (Figure 1) unlike the application of dose 8.31 mL kg^{-1} (66%). Whereas the 'Baia Periforme' had 6.8% of normal seedlings, the 'Bola Precoce' had 5.8%.

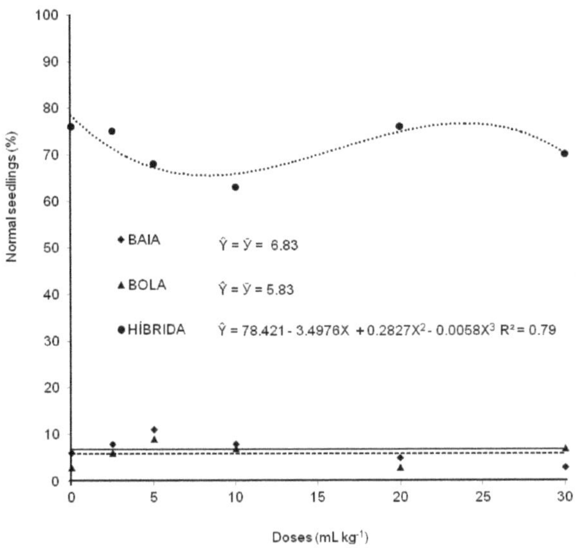

Figure 1. Polynomial curve shows the effects of doses of bio-regulator on the percentage of normal seedlings evaluated for different onion cultivars using the first count of the germination test (Maringá, Paraná State, Brazil, 2009).

Dose increase of the bio-regulator applied on the seeds of 'Baia Periforme' caused a significant fall on seed germination percentage ($p < 0.05$). In fact, the estimated control (dose 0) had 22% of normal seedlings, whereas dose 30 mL kg^{-1} had only 7% (Figure 2). Thus, decrease in the percentage of normal seedlings was 0.49% for every 1 mL of bio-regulator kg^{-1} of seed. No significant differences ($p > 0.05$) occurred with the other two cultivars. Average normal seedlings was 21% for 'Bola Precoce' and 77% for 'Híbrida Superex'. Further, quadratic responses were reported for other species and indicated losses in the percentage of normal seedlings at the highest doses of bio-regulators (ALBRECHT et al., 2010; ÁVILA et al., 2008). High physiological quality of seeds from the cultivar 'Híbrida Superex' enhanced the importance of the genotype in maintaining seed vigour and the germination of horticulture plants (MARCOS FILHO, 2005; PRETE; GUERRA, 1999).

Only the 'Híbrida Superex' had significant responses to treatments. Length of primary roots was significantly different from that in 'Híbrida Superex'. The quadratic model showed 0.73 cm as the highest root length at 18 mL kg^{-1} (Figure 3). On average, the other cultivars showed 0.26 cm and indicated that seed treatment had no influence on root length.

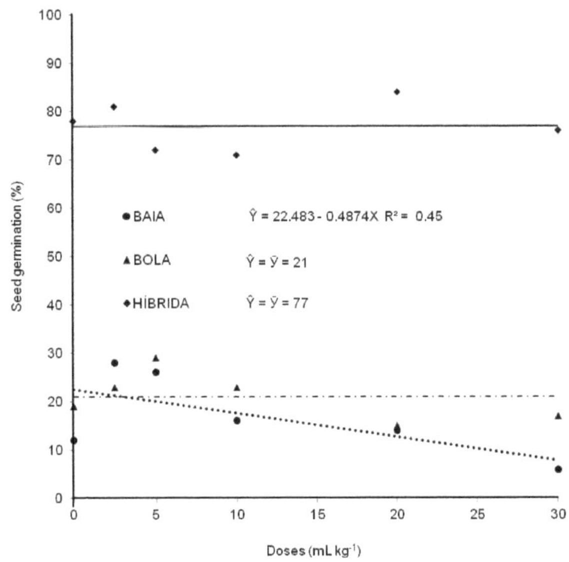

Figure 2. Linear curves describe the effects of bio-regulator doses on the percentage of normal seedlings evaluated for different onion cultivars using the germination test (Maringá, Paraná State, 2009).

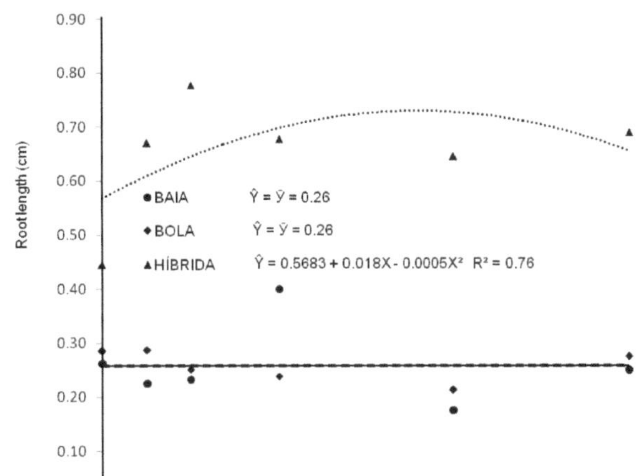

Figure 3. Polynomial curves describe the effects of bio-regulator doses on seedling root length of different onion cultivars (Maringá, Paraná State, 2009).

Responses from the shoot length were similar to the root length (Figure 4), but the maximum shoot length 2.34 cm was obtained at 15.24 mL kg^{-1}. Cultivars 'Baia Periforme' and 'Bola Precoce' were 0.10 and 0.17 cm long, respectively. Responses from cultivar 'Híbrida Superex' indicate that the product had an effect similar to auxin application. High root length in jenipapo was found after applying bio-regulator at 10 mL L^{-1} (PRADO NETO et al., 2007). This treatment was 84.31% higher than that of control. In all the cases, the dose to improve root growth was lower than that applied to enhance shoot growth. High doses inhibit the root and cause shoot growth (RAVEN et al., 2007; TAIZ; ZAIGER, 2009). Similar to what was reported for 'Híbrida Superex', positive effects have been reported for other species as, for example, increases in the physiological and sanitary quality of soybean (*Glycine max*) seeds (ALBRECHT et al., 2010; ÁVILA et al., 2008) and increases in soybean yield (MOTERLE et al., 2008). In fact, the application of bio-regulator has improved the seed quality of some kitchen garden plants (PRADO NETO et al., 2007).

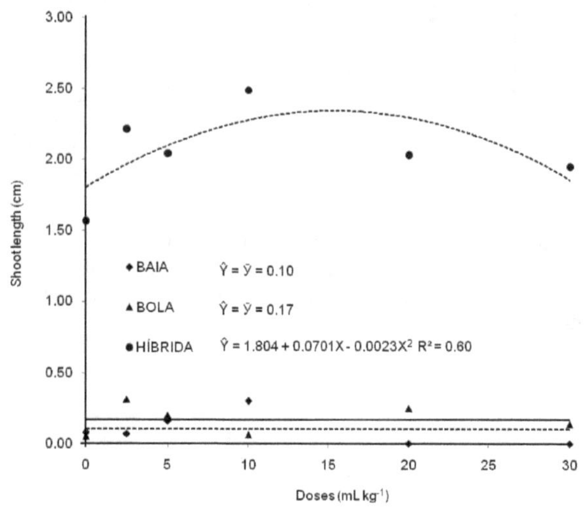

Figure 4. Polynomial curves describe the effects of bio-regulator doses on seedling shoot length of different onion cultivars (Maringá, Paraná State, 2009).

With the exception of 'Hibrida Superex', only the percentage of normal seedlings was different from that of the treatment with bio-regulator and control (Table 1 and 2), discriminated by Dunnett's test ($p < 0.05$). All doses, except doses 2.5 and 5.0 mL kg^{-1}, applied on the seeds of 'Baia Periforme', reduced the percentage of seed germination by 66.7, 55.6, 38.9 and 83.3%, respectively. Seed germination percentage of cultivar 'Bola Precoce' was reduced by 72.72% using bio-regulator at 20 mL kg^{-1} of seeds (Table 1). Comparison between growth and seedling components by Dunnett's test (Table 2) showed no difference between the treatments and control ($p > 0.05$). Current responses corroborate the literature in which few reports have been produced on kitchen garden products such as the onion, for example. Therefore, the treatment of onion seeds with bio-regulator depended on the dose and the plant cultivar. The cultivar 'Híbrida Superex' produced positive responses to the application of this bio-regulator.

Table 1. Percentage means of normal onion seedlings at the time of the first and final counts of the germination test (Maringá, Paraná State, 2009).

Doses (mL kg^{-1})	Normal seedlings[1]					
	First count (%)			Percentage of seed germination (%)		
	Baia	Bola	Híbrida	Baia	Bola	Híbrida
0	6.00 ^{ns}	3.00 ^{ns}	76.00 ^{ns}	12.00 -	19.00 -	78.00 ^{ns}
2.5	8.00 ^{ns}	6.00 ^{ns}	75.00 ^{ns}	28.00 ^{ns}	23.00 -	81.00 ^{ns}
5	11.00 ^{ns}	9.00 ^{ns}	68.00 ^{ns}	26.00 ^{ns}	29.00 -	72.00 ^{ns}
10	8.00 ^{ns}	7.00 ^{ns}	63.00 ^{ns}	16.00 -	23.00 -	71.00 ^{ns}
20	5.00 ^{ns}	3.00 ^{ns}	76.00 ^{ns}	14.00 -	15.00 -	84.00 ^{ns}
30	3.00 ^{ns}	7.00 ^{ns}	70.00 ^{ns}	6.00 -	17.00 -	76.00 ^{ns}
Control	11.00	15.00	81.00	36.00	55.00	82.00

[1]Superior (+), inferior (-) ($p < 0.05$), or non-significant differences (^{ns}) ($p > 0.05$) when compared to control, by Dunnett's test, with every cultivar.

Table 2. Means of primary root and shoot length of onion seedlings (Maringá, Paraná State, 2009).

Doses (mL kg^{-1})	Length (cm)[1]					
	Primary root			Seedling shoot		
	Baia	Bola	Híbrida	Baia	Bola	Híbrida
0	0.2625 ^{ns}	0.2850 ^{ns}	0.4450 ^{ns}	0.0750 ^{ns}	0.0500 ^{ns}	1.5675 ^{ns}
2.5	0.2250 ^{ns}	'0.2875 ^{ns}	0.6700 ^{ns}	0.0675 ^{ns}	0.3125 ^{ns}	2.2150 ^{ns}
5	0.2325 ^{ns}	0.2500 ^{ns}	0.7775 ^{ns}	0.1575 ^{ns}	0.2000 ^{ns}	2.0425 ^{ns}
10	0.4000 ^{ns}	0.2375 ^{ns}	0.6775 ^{ns}	0.3000 ^{ns}	0.0625 ^{ns}	2.4875 ^{ns}
20	0.1750 ^{ns}	0.2125 ^{ns}	0.6450 ^{ns}	0.0000 ^{ns}	0.2500 ^{ns}	2.0325 ^{ns}
30	0.2500 ^{ns}	0.2750 ^{ns}	0.6900 ^{ns}	0.0000 ^{ns}	0.1375 ^{ns}	1.9500 ^{ns}
Control	0.3300	0.2625	0.6525	0.3450	0.0325	1.6275

[1] Superior (+), inferior (-) ($p < 0.05$), or non-significant differences (^{ns}) ($p > 0.05$) when compared to control, by Dunnett's test, with every cultivar.

Conclusion

The application of bio-regulator did not improve the percentage of seed germination and no positive seedling response was observed from cultivars 'Baia Periforme' and 'Bola Precoce'. However, the seedling vigor of 'Híbrida Superex' was improved by the application of the bio-regulator.

Acknowledgements

The authors would like to thank the graduate programme in plant breeding at the State University of Maringá, the Stoller do Brasil Ltda. for technical support, CAPES for funding the first author and for the partial granting of the experiment. A.L. Braccini and C.A. Scapim are fellow researchers of the Conselho Nacional de Desenvolvimento Científico e Tecnológico (CNPq).

References

ALBRECHT, L. P.; BRACCINI, A. L.; ÁVILA, M. R.; BARBOSA, M. C.; RICCI, T. T.; ALBRECHT, A. J. P. Aplicação de biorregulador na produtividade do algodoeiro e qualidade de fibra. **Scientia Agraria**, v. 10, n. 3, p. 191-198, 2009.

ALBRECHT, L. P.; BRACCINI, A. L.; SCAPIM, C. A.; ÁVILA, M. R.; ALBRECHT, A. J. P.; BARBOSA, M. C. Qualidade das sementes de soja produzidas sob manejo com biorregulador. **Revista Brasileira de Sementes**, v. 32, n. 4, p. 39-48, 2010.

ALLEONI, B.; BOSQUEIRO, M.; ROSSI, M. Efeito de reguladores vegetais de Stimulate® no desenvolvimento e produtividade do feijoeiro (*Phaseolus vulgaris* L.). **Publicatio UEPG**, v. 1, n. 6, p. 23-35, 2000.

ÁVILA, M. R.; BRACCINI, A. L.; SCAPIM, C. A.; ALBRECHT, L. P.; TONIN, T. A.; STÜLP, M. Bio-regulator application, agronomic efficiency, and quality of soybean seeds. **Scientia Agricola**, v. 65, n. 6, p. 604-612, 2008.

BANZATO, D. A.; KRONKA, S. N. **Experimentação agrícola**. 3. ed. Jaboticabal: FUNEP, 2008.

BOEING, G. **Fatores que afetam a qualidade da cebola na agricultura familiar catarinense**. Florianópolis: Instituto CEPA/SC, 2002.

BOITEUX, L. S.; MELO, P. C. T. Taxonomia e origem. In: OLIVEIRA, V. R.; BOITEUX, L. S. (Ed.). **Cultivo da cebola**. Brasília: Embrapa Hortaliças, 2004. (Sistema de cultivos). Available from: <http://www.cnph.embrapa.br/paginas/sistemas_producao/cultivo_da_cebola.htm>. Access on: Dec. 14, 2009.

BRASIL. Ministério da Agricultura, Pecuária e Abastecimento. **Regras para análise de sementes**. Brasília, 2009.

CASTRO, P. R. C.; VIEIRA, E. L. **Aplicações de reguladores vegetais na agricultura tropical**. Guaíba: Livraria e Editora Agropecuária, 2001.

CHEMELLO, E. **A química na cozinha apresenta**: As cebolas. São Paulo: Editora Cia da Escola, 2005. ano 6, n. 2. Available from: http://www.ciadaescola.com.br/zoom/materia.asp?materia=263>. Access on: Dec. 15, 2011.

FERREIRA, G.; COSTA, P. N.; FERRARI, T. B.; RODRIGUES, J. D.; BRAGA, J. F.; JESUS, F. A. Emergência e desenvolvimento de plântulas de maracujazeiro azedo oriundas de sementes tratadas com bioestimulante. **Revista Brasileira de Fruticultura**, v. 29, n. 3, p. 595-599, 2007.

GARCIA, A. S.; BRANQUINHO, E. G. A.; MENUCHI, A. C. T. P.; ERLACHER, K. C.; DOMINGUES, M. C. S. Efeitos de reguladores vegetais na germinação e desenvolvimento da semente *Strelitzia reginae*. **Thesis**, v. 5, ano III, p. 161-176, 2006.

HADAS, A. Water uptake and germination of leguminous seeds under changing external water potential in osmotic solution. **Journal of Experimental Botany**, v. 27, n. 3, p. 480-489, 1976.

KLAHOLD, C. A.; GUIMARÃES, V. F.; ECHER, M. M.; KLAHOLD, A.; CONTIERO, R. L.; BECKER, A. Resposta da soja (*Glycine max* L. Merrill) à ação de bioestimulante. **Acta Scientiarum. Agronomy**, v. 28, n. 2, p. 179-185, 2006.

MARCOS FILHO, J. **Fisiologia de sementes de plantas cultivadas**. Piracicaba: Fealq, 2005.

MARCOS FILHO, J.; CÍCERO, S. M.; SILVA, W. R. **Avaliação da qualidade fisiológica das sementes**. Piracicaba: Fealq, 1987.

MOTERLE, L. M.; SANTOS, R. F.; BRACCINI, A. L.; SCAPIM, C. A.; BARBOSA, M. C. Efeito da aplicação de biorregulador no desempenho agronômico e produtividade da soja. **Acta Scientiarum. Agronomy**, v. 30, n. 5, p. 701-709, 2008.

NAKAGAWA, J. Testes de vigor baseados no desempenho das plântulas. In: KRZYZANOWSKI, F. C.; VIEIRA, R. D.; FANÇA NETO, J. B. (Ed.). **Vigor de sementes**: conceitos e testes. Londrina: Abrates, 1999. Cap. 2, p. 1-24.

OLIVEIRA, V. R.; BOITEUX, L. S. **Cultivo da cebola**. Brasília: Embrapa Hortaliças, 2004. (Sistema de cultivos). Available from: <http://www.cnph.embrapa.br/paginas/sistemas _producao/cultivo_da_cebola.htm>. Access on: Dec. 14, 2009.

OLIVEIRA, V. R.; LEITE, D. L.; SANTOS, C. A. F.; COSTA, N. D.; MELO, P. C. T. Cultivares. In: OLIVEIRA, V. R.; BOITEUX, L. S. (Ed.). **Cultivo da cebola**. Brasília: Embrapa Hortaliças, 2004. (Sistema de cultivos). Available from: <http://www.cnph.embrapa.br/paginas/sistemas _producao/cultivo_da_cebola.htm>. Access on: Dec. 14, 2009.

PRADO NETO, M.; DANTAS, A. C. V. L.; VIEIRA, E. L.; ALMEIDA, V. O. Germinação de sementes de jenipapeiro submetidas à pré-embebição em regulador e estimulante vegetal. **Ciência e Agrotecnologia**, v. 31, n. 3, p. 693-698, 2007.

PRETE, C. E. C.; GUERRA, E. P. Qualidade fisiológica das sementes. In: DESTRO, D.; MONTALVÁN, R. (Ed.). **Melhoramento genético de plantas**. Londrina: UEL, 1999. Cap. 34, p. 661-676.

RAVEN, P. H; EVERT, R. F; EICHHORN, S. E. **Biologia Vegetal**. 7. ed. Rio de Janeiro: Ed. Guanabara Koogan, 2007.

RESENDE, G. M.; COSTA, N. D. Características produtivas e conservação pós-colheita de cebola em diferentes espaçamentos de plantio. **Horticultura Brasileira**, v. 23, n. 3, p. 707-711, 2005.

STOLLER DO BRASIL. **Stimulate Mo em hortaliças**. Cosmópolis: Stoller do Brasil. Divisão Arbore, 1998. v. 1. (Informativo técnico).

TAIZ, L.; ZEIGER, E. **Fisiologia vegetal**. 4. ed. Porto Alegre: Artmed, 2009.

VIEIRA, E. L.; CASTRO, P. R. C. Ação de bioestimulante na germinação de sementes, vigor das plântulas, crescimento radicular e produtividade de soja. **Revista Brasileira de Sementes**, v. 23, n. 2, p. 222-228, 2001.

VIEIRA, E. L.; CASTRO, P. R. C. **Ação de bioestimulante na cultura da soja (*Glycine max* (L.) Merrill)**. Cosmópolis: Stoller do Brasil, 2004.

VIEIRA, E. L.; SANTOS, C. M. G. **Stimulate na germinação de sementes, vigor de plântulas e crescimento inicial do algodoeiro**. In: CONGRESSO BRASILEIRO DE ALGODÃO, 5., 2005. Available from: <http//www.cnpa.embrapa.br/produtos/algodão/publicações/trabalhos_ba5/390.pdf>. Access on: Feb. 5, 2010.

12

Primary leaf growth and coleoptile length in maize seedlings

Camila Aparecida Lopes*, **Maria Laene Moreira de Carvalho, Michelle Conceição Vasconcelos, Ana Maria Pereira Ribeiro and Heloísa Oliveira dos Santos**

*Laboratório de Análise de Sementes, Departamento de Agricultura, Universidade Federal de Lavras, 37200-000, Lavras, Minas Gerais, Brazil. *Author for correspondence. E-mail: camilalopes_sjc@hotmail.com*

ABSTRACT. Evaluation of the physiological potential of maize seeds during the certification process is performed by a germination test. However, some tests show a delay in the growth of aerial seedling organs that does not reflect seed lot behavior in the field. This study aims to investigate the effects of genotype, water, light and temperature on differential primary leaf growth compared to coleoptile length in maize seedlings. Seeds from two simple hybrids and their reciprocal crosses that showed slow primary leaf growth were evaluated according to physiological quality, morphology, histochemistry, and enzymatic expression. There were differences among maize genotypes regarding initial primary leaf growth when compared to the coleoptile, especially when germination was tested without light. Water availability and optimal temperature range did not affect the differential growth of the primary leaf in the hybrids. Pericarp thickness, lignin content and differences in enzyme expression in different genotypes were also not related to changes in the relationship of primary leaf growth with coleoptile length during the maize seed germination test.

Keywords: *Zea mays* L.; germination test; lignin; enzymes.

Introduction

The germination test is the typical way to evaluate the physiological potential of maize seeds during the certification process (minimum 85% germination according to the commercialization standards established in normative instruction 45 and law 10 711). The germination percentage is determined by the number of normal seedlings at the end of the test, which is carried out under optimal conditions for germination. At the time of the evaluation, seedlings should contain all the essential structures, allowing the analyst to judge the possibility of that seedling to continue its development and become a productive plant.

One of the reasons that a maize seedling may be considered abnormal is the size of the primary leaf compared to the coleoptile length. If the primary leaf is less than half the coleoptile length by the time of the assessment, the seedling is considered abnormal (Brasil, 2009). In some genetic backgrounds, it has been observed that seedlings exhibit delayed growth of aerial structures, specifically the primary leaf, which leads them to be classified as abnormal seedlings and adversely affects germination test results. These results, however, do not always reflect the performance of seed lots in the field.

The growth rates of primary leaf and coleoptile may vary in some cases; they are both genetically controlled and influenced by environmental conditions during germination (Kurtyka, Burdach, & Karez, 2011). This study aims to investigate the effects of genotype, water, light, temperature and chemical treatment on the differential growth of primary leaf in relation to coleoptile length in maize seedlings.

Material and methods

Seeds from two simple hybrids and their reciprocal crosses were used. All demonstrated slow primary leaf growth. Simple hybrid 1 was the product of Line A x Line B, and hybrid 4 was its reciprocal cross (Line B x Line A); simple hybrid 2 was the product of Line A' x Line C, and hybrid 3 was its reciprocal cross (Line C x Line A'). Lines A and A' were from the same genetic background but carried different trans genes. The seeds to be evaluated (classified in sieve 22) were obtained from the four hybrids at the same time and under the same climate conditions.

Assay 1: A germination test was used to assess the effect of water, light, temperature and hybrid origin. The seeds were placed on paper towel substrate (Germitest) moistened with a volume of distilled water in an amount equal to 2.0, 2.5, and 3.0 times the paper weight at temperatures of 20°C, 25°C, and 30°C and in a 20-30°C range. The seeds were subjected to two treatments, one dark and the other with 12 hours of dark and 12 hours light. The factorial design used was 4 x 3 x 4 x 2 (4 hybrids, 3 volumes of water, 4 temperatures, and 2 light conditions) in a completely randomized design with four replicates of 25 seeds per treatment. Evaluations were performed four, five, six and seven days after sowing. The data collected were the number of normal seedlings (at four and seven days), the number of seedlings with primary leaf growth less than 50% of coleoptile length, and the number of seedlings with primary leaf growth between 50% and 75% of coleoptile length. The data were evaluated for data normality and homogeneity of variance among treatments. A normal data distribution was found only for the normal seedlings. Then, a constant value equal to 3 was added (Lucio, Couto, Lopes, & Storck, 2011; Yakamura, 1999), and the Box-Cox transformation was performed (Box & Cox, 1964). After processing, the results were subjected to analysis of variance and means compared with a significance level of 5% by the Scott-Knott test using SISVAR® software (Ferreira, 2011).

Assay 2: A germination test was performed in sand (TGA). The test was carried out in germination chambers with 25°C temperature and a photoperiod of 12 hours light and 12 hours lightless. Seeds of the four hybrids were sown in trays with sand that had been sieved, washed and sterilized (autoclaved to one atmosphere and 120°C for sixty minutes) with water equivalent to 60% of field capacity in a completely randomized design with four replicates of 25 seeds per treatment. Evaluations were performed at three, five and seven days. The data collected were the number of normal seedlings and the number of seedlings with differential primary leaf growth in relation to coleoptile length. The data were evaluated for data normality and homogeneity of variance among treatments and then subjected to analysis of variance and means compared with a significance level of 5% by the Scott-Knott test using SISVAR® software (Ferreira, 2011).

Assay 3: Seed pericarp thickness was assessed. Sections were made in the middle region of the seeds of each hybrid and reciprocal cross using a microtome (Olympus CBL, model BX60). Five replicate sections of approximately 20 microns were obtained for each material. The sections were dyed for 60 seconds with three drops per replicate of1% Saflablau (0.9% Astra + 0.1% Safran). Images of the pericarp were obtained using a stereoscopic microscope adapted with a camera. The pericarp thickness was measured with Image J® software (Rasband, 2012). The measurements obtained were submitted to analysis of variance and means compared with a significance level of 5% by the Scott-Knott test using SISVAR® software (Ferreira, 2011).

Assay 4: Histochemical analysis was performed to evaluate lignin deposition on the coleoptile wall. Seedlings with four days of growth were sectioned transversely at the transition region between the coleoptile and first leaf in a table microtome and stained with phloroglucinol solution (1 g 100 ml^{-1}) for 5 minutes. An unstained section was used as control. The slides were evaluated in a stereoscopic microscope. Evaluation of lignin deposition in the coleoptile wall was visually performed in all five replicates.

Assay 5: The expression of α-amylase and β-amylase enzymes was evaluated using the following samples for all hybrids: ten seeds soaked for 72 hours, ten dry seeds and ten seedlings with four days of growth. The seeds and seedlings were homogenized with liquid nitrogen and stored in a freezer at -81°C until analysis. Extraction of β- and α-amylase and gel development were performed according to Alfenas (2006).

Results and discussion

The classes of maize seedlings observed in the germination assessments are shown in Figure 1. Slow development of the primary leaf in relation to the coleoptile is shown in Figure 1A, with primary leaf growth less than 50% of coleoptile length.

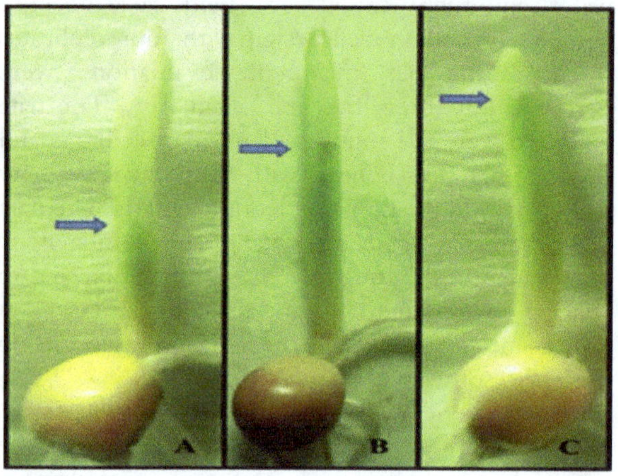

Figure 1. Maize seedlings after five-day germination test. A) abnormal seedling with primary leaf growth less than 50% of coleoptile length; B) normal seedling with primary leaf growth between 50 and 75% of coleoptile length; C) normal seedling.

Assay 1: The 'normal seedling' group showed significant interaction effects of hybrid identity and temperature in the germination test at the fourth day. Lower germination rates were observed at 20°C for hybrids 2 and 3 (Table 1). According to Zucarelli, Cavariani, Oliveira, and Nakagawa (2011), temperatures below the optimum recommended for maize seed germination slow down the imbibition process. Lower temperatures also enable difficulties in the reorganization of cell membranes and consequently cause delays in germination (Carvalho, Sediyama, Reis, Dias, & Moreira, 2009), influenced by physiological quality and the genotype (Grzybowski, Vieira, & Panobianco, 2015).

Table 1. Percentage of normal maize seedlings assessed at the fourth day after sowing.

Hybrid	Temperature			
	20-30°C	20°C	25°C	30°C
1	98 aA	95 aA	99 aA	99 aA
2	96 aA	87 bB	96 aA	98 aA
3	96 aA	85 bB	96 aA	97 aA
4	98 aA	92 aB	98 aA	97 aA
	CV(%) = 4.45			

Means followed by the same lowercase letter (in columns) or the same capital letter (in rows) do not differ by the Scott-Knott test at 5% of probability.

There was also a triple interaction among hybrid identity, temperature and light conditions for seedlings with primary leaf growth less than 50% of coleoptile length. In Table 2, the seeds of hybrid 4 showed more seedlings with primary leaf growth less than 50% of coleoptile length when subjected to temperatures of 30°C or a range from20-30°C under lightless conditions. The results contradict the findings of Deng, Yang, Zhang, and Li (2015) and Zucarelli, Cavariani, Portugal, and Nakagawa (2008) that the temperature of 30°C was more suitable for germination and growth of maize seedlings.

Table 2. Percentages of maize seedlings with primary leaf growth less than 50% of coleoptile length, assessed at the fourth day after sowing.

Hybrid	Light	Temperature			
		20-30°C	20°C	25°C	30°C
1	Light	0 aA	0 aA	0 aA	0 aA
2	Light	0 aA	0 aA	0 aA	0 aA
3	Light	0 aA	0 aA	0 aA	0 aA
4	Light	0 aA	0 aA	0 aA	0 aA
1	Lightless	0 bA	0 aA	0 aA	0 bA
2	Lightless	0 bA	0 aA	0 aA	0 bA
3	Lightless	0 bA	0 aA	0 aA	0 bA
4	Lightless	4 aA	0 aB	0 aB	4 aA
		CV(%) = 1.78			

Means followed by the same lowercase letter (in columns) or the same capital letter (in rows) do not differ by the Scott-Knott test at 5% of probability.

In the 'seedlings with primary leaf growth between 50 and 75% of coleoptile length' group, there was a significant interaction between hybrid identity and light conditions. In lightless conditions, hybrid 3 presented the lowest percentage of seedlings with primary leaf growth between 50% and 75% of coleoptile length. Under light, only hybrid 4 showed any seedlings with this characteristic (Table 3). Lima, Santos, and Melo (2010) found that maize plants grown in light less conditions developed an elongated mesocotyl and coleoptile and had primary leaves rolled inside the coleoptile.

Table 3. Percentages of maize seedlings with primary leaf growth between 50 and 75% of coleoptile length, assessed at the fourth day after sowing.

Hybrid	Light	Lightless
1	0 bB	5 aA
2	0 bB	4 aA
3	0 bA	0 bA
4	4 aA	4 aA
	CV (%) = 2.87	

Means followed by the same lowercase letter (in columns) or the same capital letter (in rows) do not differ by the Scott-Knott test at 5% of probability.

In the 'seedlings with primary leaf growth less than 50% of coleoptile length' group after five days, only hybrid identity was a significant factor, indicating the influence of genotype on uneven growth. The highest percentage of seedlings with primary leaf growth less than 50% of coleoptile length at five days was observed for hybrid 4 (Table 4). According to Mertz et al. (2009) and Nerling, Coelho, and Nodari (2013), the physiological qualities of seeds are genetically inherited from their parents, so different hybrids of the same species may show varying vigor, germination and emergence in the same production conditions.

Table 4. Percentages of maize seedlings with primary leaf growth less than 50% of coleoptile length, assessed at the fifth day after sowing.

Hybrid	Seedling
1	0 b
2	0 b
3	0 b
4	2 a
CV (%) = 1.51	

Means followed by the same letter do not differ by the Scott-Knott test at 5% probability.

In the seedlings with primary leaf growth between 50% and 75% of coleoptile length' group after five days, there was a significant quadruple interaction among hybrid identity, water, temperature and light conditions. A higher percentage of seedlings with primary leaf growth between 50% and 75% of coleoptile length was observed for hybrid 4 in lightless conditions (Table 5). Because the percentage of seedlings with primary leaf growth between 50% and 75% of coleoptile length varied on the fifth day after sowing, it was not possible to determine what temperature or volume of water had the best effect on germination.

Table 5. Percentages of maize seedlings with primary leaf growth between 50% and 75% of coleoptile length, assessed at the fifth day after sowing.

			Temperature			
Hybrid	Water	Light conditions	20-30°C	20°C	25°C	30°C
1	2,0	Lightless	0 aB	4 aA	5 aA	3 aA
2	2,0	Lightless	0 aB	0 bB	3 bA	4 aA
3	2,0	Lightless	2 aA	2 bA	2 bA	0 bA
4	2,0	Lightless	2 aB	7 aA	7 aA	0 bB
1	2,0	Light	2 aA	0 bA	2 bA	0 aA
2	2,0	Light	0 aB	3 aA	0 bB	0 aB
3	2,0	Light	0 aA	0 bA	0 bA	0 aA
4	2,0	Light	2 aB	0 bB	4 aA	0 aB
1	2,5	Lightless	2 bA	3 aA	2 b A	3 aA
2	2,5	Lightless	6 aA	4 aA	4 aA	0 bB
3	2,5	Lightless	2 bA	2 b A	0 b A	0 bA
4	2,5	Lightless	5 aA	3 aA	5 aA	0 bB
1	2,5	Light	0 aB	0 bB	3 aA	0 aB
2	2,5	Light	0 aB	2 bB	4 aA	0 aB
3	2,5	Light	0 aA	0 bA	2 bA	0 aA
4	2,5	Light	0 aB	3 aA	0 bB	0 aB
1	3,0	Lightless	2 bB	0 bB	4 bA	2 aB
2	3,0	Lightless	0 bB	4 aA	3 bA	0 aB
3	3,0	Lightless	2 bA	3 aA	2 cA	2 aA
4	3,0	Lightless	3 aB	3 aB	8 aA	2 aB
1	3,0	Light	0 aA	0 bA	0 bA	0 aA
2	3,0	Light	0 aA	2 bA	2 bA	0 aA
3	3,0	Light	0 aA	2 bA	0 bA	0 aA
4	3,0	Light	0 aB	3 aA	3 aA	0 aB
CV (%) = 1.43						

Means followed by the same lowercase letter (in columns) or the same capital letter (in rows) do not differ by the Scott-Knott test at 5% of probability.

In the 'seedlings with primary leaf growth less than 50% of coleoptile length' group after six days, there was a significant triple interaction among hybrid identity, temperature and water. Hybrid 4 presented the highest percentage of seedlings with primary leaf growth less than 50% of coleoptile length at a temperature of 25°C or range of 20°C to 30°C and water volume 3.0 times the paper weight (Table 6). This result differs from the findings of Lipchinsky, Sharova, and Medvedev (2013), who reported that decreased temperature caused an instant stretching of the coleoptile's cell wall.

Table 6. Percentages of maize plants with primary leaf growth less than 50% of coleoptile length, assessed at the sixth day after sowing.

		Temperature			
Hybrid	Water	20-30°C	20°C	25°C	30°C
1	2,0	0 aA	0 aA	0 aA	0 aA
2	2,0	0 aA	0 aA	0 aA	0 aA
3	2,0	0 aA	0 aA	0 aA	0 aA
4	2,0	0 aA	0 aA	0 aA	0 aA
1	2,5	0 aA	0 aA	0 aA	0 aA
2	2,5	0 aA	0 aA	0 aA	0 aA
3	2,5	0 aA	0 aA	0 aA	0 aA
4	2,5	0 aA	0 aA	0 aA	0 aA
1	3,0	0 bA	0 aA	0 bA	0 aA
2	3,0	0 bA	0 aA	0 bA	0 aA
3	3,0	3 aA	0 aB	0 bB	0 aB
4	3,0	3 aA	0 aB	2 aA	0 aB
CV(%) = 2.33					

Means followed by the same lowercase letter (in columns) or the same capital letter (in rows) do not differ by the Scott-Knott test at 5% of probability.

In the 'seedlings with primary leaf growth between 50% and 75% of coleoptile length' variable was a significant interaction between temperature and light. The highest percentage of seedlings with primary leaf growth between 50% and 75% of coleoptile length occurred at 20°C in the absence of light, six days after sowing (Table 7). Sbrussi and Zucarelli (2014) reported that temperatures below 20°C caused the absence of normal seedlings across different lots of maize seeds due to delayed germination. Zucarrelli et al. (2008) observed that a temperature of 20°C reduced the development of the aerial structures of maize plants when compared with seedling development at 30°C.

Table 7. Percentages of maize plants with primary leaf growth between 50% and 75% of coleoptile length, assessed at the sixth day after sowing.

	Temperature			
Light conditions	20-30°C	20°C	25°C	30°C
Lightless	0 aB	3 aA	0 aB	0 aB
Light	0 aA	1 bA	0 aA	0 aA
CV(%) = 1.96				

Means followed by the same lowercase letter (in columns) or the same capital letter (in rows) do not differ by the Scott-Knott test at 5% of probability.

In the 'normal seedling' group after seven days, hybrid identity showed a significant effect. The percentage of normal seedlings was lower for hybrid 2 (Table 8).

Table 8. Percentage of normal maize seedlings, assessed at the seventh day after sowing.

Hybrid	Average
1	100 a
2	97 b
3	99 a
4	98 a
CV (%) = 2.29	

Means followed by the same letter do not differ by the Scott-Knott test at 5% probability.

At seven days after sowing, there were no seed lings with primary leaf growth less than 50% of coleoptile length.

For the 'seedlings with primary leaf growth between 50% and 75% of coleoptile length' group there was a significant quadruple interaction among hybrid, temperature, water and light. The highest percentage of seedlings with primary leaf growth between 50% and 75% of coleoptile length was observed in hybrid 4,followed by hybrids 2 and 3, which had a higher percentage of this seedling type than hybrid 1 (Table 9) under light less conditions with water quantity 2.0 to 3.0 times paper weight. This was also observed by Ávila, Braccini, and Scapim (2007), who reported a reduction in the length of maize plants when water availability was reduced. However, according to Yap (1981), an excess of water is also detrimental to the development of maize seedlings. It was not possible to determine which temperature showed the largest percentage of seedlings with this feature, since such seedlings were observed at all temperatures.

According to the Brasil (2009) germination test for maize seeds, normal seedling evaluation is performed between the fourth and the seventh day after sowing. Due to the results observed in this test, it is recommended that an evaluation at seven days be performed for hybrids with differential growth, in order to prevent inaccurate reports. It is noteworthy that most seed companies perform only one count at five days which could affect the test results.

Assay 2: In the sand germination test (TGA) three days after sowing, the lowest percentage of normal seedlings was observed for hybrid 3. However, at the fifth and the seventh day there was no difference in the percentage of normal seedlings among all hybrids (Table 10). In sand substrate, 100% of seedlings had emerged at seven days, and in this test no seedlings with abnormal primary leaf growth in relation to the coleoptile were observed. This fact shows that although these seeds presented delays in aerial organ growth, those results do not reflect the performance of the hybrids in the field, where despite the delay they produce healthy plants. Conceição et al. (2012) also failed to differentiate maize seed lots through the sand germination test, emphasizing the need for adequate choice of substrate in the germination test, especially for differential growth materials.

Table 9. Percentage of maize seedlings with primary leaf growth between 50% and 75% of coleoptile length, assessed at the seventh day after sowing.

			Temperature			
Hybrid	Water	Light conditions	20-30°C	20°C	25°C	30°C
1	2,0	Lightless	0 aA	0 bA	0 aA	2 cA
2	2,0	Lightless	0 aA	0 bA	0 aA	2 cA
3	2,0	Lightless	0 aB	0 bB	1 aB	8 aA
4	2,0	Lightless	0 aB	5 aA	0 aB	4 bA
1	2,0	Light	0 aA	0 b A	0 bA	0 aA
2	2,0	Light	0 aB	3 aA	0 bB	0 aB
3	2,0	Light	0 aA	0 bA	0 bA	0 aA
4	2,0	Light	0 aB	0 bB	3 aA	0 aB
1	2,5	Lightless	0 aB	3 aA	0 bB	0 aB
2	2,5	Lightless	0 aB	0 bB	3 aA	0 aB
3	2,5	Lightless	0 aA	0 bA	0 bA	0 aA
4	2,5	Lightless	1 aB	0 bB	5 aA	0 aB
1	2,5	Light	0 aA	0 bA	0 aA	0 aA
2	2,5	Light	0 aA	0 bA	0 aA	0 aA
3	2,5	Light	0 aA	1 bA	1 aA	0 aA
4	2,5	Light	0 aB	3 aA	0 aB	0 aB
1	3,0	Lightless	0 aA	0 aA	0 cA	0 aA
2	3,0	Lightless	0 aA	0 aA	2 bA	0 aA
3	3,0	Lightless	0 aB	0 aB	5 aA	1 aB
4	3,0	Lightless	1 aB	1 aB	8 aA	1 aB
1	3,0	Light	0 aA	0 bA	0 bA	0 aA
2	3,0	Light	0 aA	1 bA	0 bA	0 aA
3	3,0	Light	0 aA	2 aA	0 bA	0 aA
4	3,0	Light	0 aB	1 bB	3 aA	0 aB
CV(%) = 1.07						

Means followed by the same lowercase letter (in columns) or the same capital letter (in rows) do not differ by the Scott-Knott test at 5% of probability.

Table 10. Percentage of normal maize seedlings in the sand germination test, assessed at three to seven days after sowing.

Hybrid	3 days	5 days	7 days
1	16 a	100 a	100 a
2	12 a	96 a	100 a
3	4 b	100 a	100 a
4	16 a	100 a	100 a
CV(%)	38.50	5.12	2.25

Means followed by the same letter in columns do not differ by the Scott-Knott test at 5% probability.

Assay 3: In the histological analysis, differences were observed in the pericarp thickness of hybrid seeds. Hybrid 1 showed greater thickness (Table 11). When working with reciprocal maize crosses, José, Von Pinho, Von Pinho, Ramalho, and Silva (2005) also reported differences among the pericarp structures, corroborating the results obtained in this experiment. According to Tracy and Galinat (1987), the thickness and density of the cell walls in the pericarp vary among maize genotypes, affecting seed quality and germination.

However, the hybrid with the highest pericarp thickness was not the hybrid that showed slow primary leaf growth. There was no difference in the pericarp thickness of the hybrid 4 that could explain its greater susceptibility to differential primary leaf growth compared to the coleoptile. Thus, it was found that the pericarp did not present resistance to the germination of hybrid maize seed and did not influence differential primary leaf growth compared to the coleoptile.

Table 11. Pericarp thickness of hybrid maize seeds and their reciprocal crosses.

Hybrid	Pericarp thickness (μm)
1	203.12 a
2	132.79 b
3	138.13 b
4	148.62 b
	CV (%) = 10.65

Means followed by the same letter do not differ significantly by the Scott-Knott test at 5% probability.

Assay 4: In the histochemical test, it was observed that the four hybrids showed a small amount of lignin in the cell wall of the coleoptile (Figure 2).

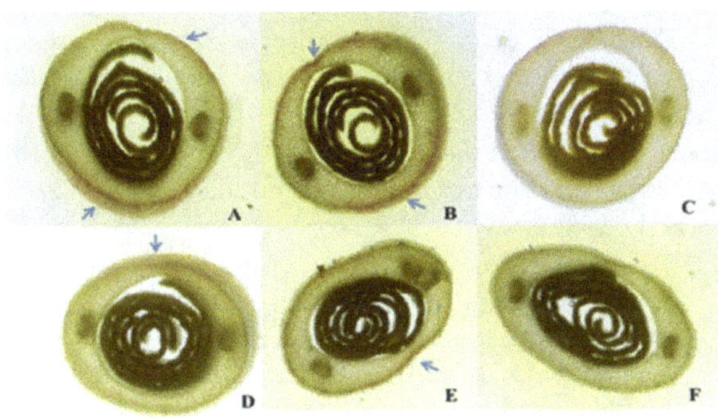

Figure 2. Histochemical test in maize seedlings showing the presence of lignin. A) Hybrid 1; B) hybrid 4; C) unstained control (Hybrid 1 and 4); D) hybrid 2; E) hybrid 3; F) unstained control (hybrids 2 and 3). Magnified 200x.

Musel et al. (1997) also observed the presence of lignin in the coleoptile cell wall of maize seedlings and considered its presence a potential regulator of coleoptile growth. However, Hohl, Greiner, and Schopfer (1995) observed that the elongation of the coleoptile was not affected by lignification of the wall. In this work, despite the finding of lignin in the hybrid seedlings' coleoptile walls, it was not possible to say that the presence of lignin was responsible for differential growth of primary leaf in relation to the coleoptile of maize seedlings, because lignin was present in all hybrids.

Assay 5: As shown in Figure 3, dry seeds possessed low expression of α-amylase in all hybrids. However, the expression of this enzyme was observed in soaked seeds. According to Silva-Neta et al. (2015), α-amylase activity is higher after root protrusion, because it increases the available reserves to promote seedling growth. Soaked seeds from hybrid 1 showed the lowest expression of α-amylase and soaked seeds from hybrid 3 and 4 showed the highest expression of this enzyme. According to Oliveira et al. (2013), genotypes with greater expression of α-amylase enzyme can easily provide carbohydrates to the embryo, resulting in higher germination and vigor. However, in this study, hybrid 4, despite having the highest activity of α-amylase enzyme, was also the hybrid with the highest percentage of seedlings with slow growth of aerial organs. In seedlings with four days of growth, the expression of α-amylase enzyme was lower than in soaked seeds. Seedlings from hybrid 1 had the highest expression of this enzyme, and seedlings from hybrid 4 had the lowest expression.

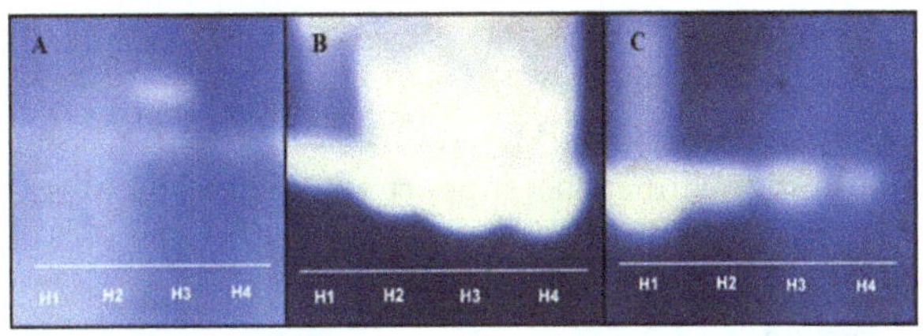

Figure 3. Enzymatic profiles for α-amylase enzyme (E.C. 3.2.1.1) from maize simple hybrids and their reciprocal crosses. A) Dry seeds; B) soaked seeds; C) seedlings with four days of growth.

Enhanced expression of β-amylase enzyme occurs in dry seeds (Figure 4), where this enzyme is preexisting. The expression of this enzyme was greater in dry seeds of hybrid 3 than in the remaining hybrids. Seeds from hybrid 4 showed very little expression of β-amylase, even when dry. In soaked seeds, the expression of β-amylase was lower than in dry seeds, but the enzyme was still present. In soaked seeds from hybrids 1, 2, and 3 the expression of this enzyme was higher than in hybrid 4. As expected, there was no expression of β-amylase enzyme in seedlings with four days of growth; the β-amylase enzyme is preexisting in dry maize seed and assists in immediate starch hydrolysis after soaking, until α-amylase is synthesized again.

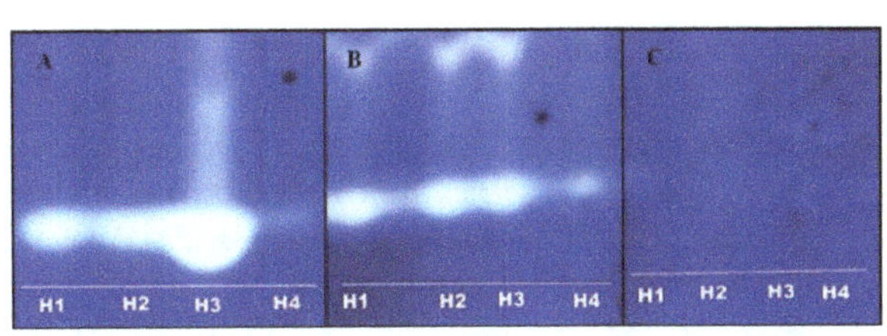

Figure 4. Enzymatic profiles for β-amylase (E.C. 3.2.1.2) enzyme from maize simple hybrids and their reciprocal crosses. A) Dry seeds; B) soaked seeds; C) seedlings with four days of growth.

Conclusion

There are differences among hybrid maize genotypes regarding the initial growth of primary leaf in relation to coleoptile, especially under lightless germination conditions.

Water availability and temperatures within the optimal range (20° to 30°C) do not affect the response of these hybrids in relation to differential growth of the primary leaf.

Pericarp thickness, lignin content and differences in enzyme expression among different genotypes are not related to changes in the relationship of primary leaf growth to coleoptile length during the maize seed germination test.

Acknowledgements

To the Coordination for Improvement of Higher Education Personnel (Capes) and the National Council for Scientific and Technological Development (CNPq) for granting scholarship funding and financial support for the development of research.

References

Alfenas, A. C. (2006). *Eletroforese de iso enzimas e proteínas afins: fundamentos e aplicações em plantas e microorganismos*. Viçosa, MG: UFV, 2006.

Ávila, M. R., Braccini, A. L., & Scapim, C. A. (2007). Teste de comprimento de plântulas sob estresse hídrico na avaliação do potencial fisiológico das sementes de milho. *Revista Brasileira de Sementes, 29*(2), 117-124. doi: 10.1590/S0101-31222007000200016.

Box, G. E. P., & Cox, D. R. (1964). An analysis of transformation. *Journal of the Royal Statistical Society, 26*(2), 211-252.

Brasil. (2009). Ministério da Agricultura, Pecuária e Abastecimento *Regras para análise de sementes*. Brasília, DF: MAPA.

Brasil. (2003). *Lei n° 10.711, de 5 de agosto de 2003. Dispõe sobre o Sistema Nacional de Sementes e Mudas e dá outras providências*. Brasília, DF: MAPA.

Brasil. (2013). *Instrução Normativa n° 45, de 17 de setembro de 2013. Padrões para produção e comercialização de sementes de milho (Zea mays L.)*. Brasília, DF: MAPA.

Carvalho, L. F., Sediyama, C. S., Reis, M. S., Dias, D. C. F. S., & Moreira, M. A. (2009). Influência da temperatura de embebição da semente de soja no teste de condutividade elétrica para avaliação da qualidade fisiológica. *Revista Brasileira de Sementes, 31*(1), 9-17. doi: 10.1590/S0101-31222009000100001.

Conceição, P. M., Sediyama, C. A. Z., Vieira, R. F., Galvão, J. C. C., Corrêa, M. L. P., & Conceição, P. S. (2012). Estimativa do vigor de sementes de milho através da avaliação do sistema radicular de plântulas. *Ciência Rural, 42*(4), 600-606. doi: 10.1590/S0103-84782012000400004.

Deng, B., Yang, K., Zhang, Y., & Li, Z. (2015). The effects of temperature on the germination behavior of white, yellow, red and purple maize plant seeds. *Acta Phisiologiae Plantarum, 37*(174), 1-11. doi: 10.1007/s11738-015-1937-1.

Ferreira, D. F. (2011). Sisvar: a computer statistical analysis system. *Ciência e Agrotecnologia, 35*(6), 1039-1042. doi: 10.1590/S1413-70542011000600001.

Grzybowski, C. R. S., Vieira, R. D., & Panobianco, M. (2015). Teste de estresse na avaliação do vigor de sementes de milho. *Revista Ciência Agronômica, 46*(3), 590-596. doi: 10.5935/1806-6690.20150042.

Hohl, M., Greiner, H., & Schopfer, P. (1995). The cryptic-growth response of maize coleoptiles and its relationship to H202- dependent cell wall stiffening. *Physiology Plant, 94*(3), 491-498. doi: 10.1111/j.1399-3054.1995.tb00959.x.

José, S. C. B. R., Von Pinho, E. V. R., Von Pinho, R. G., Ramalho, M. A. P., & Silva Filho, J. L. (2005). Características físicas do pericarpo de sementes de milho associadas com tolerância à alta temperatura de secagem. *Revista Brasileira de Sementes, 27*(1), 125-131. doi: 10.1590/S0101-31222005000100015.

Kurtyka, R., Burdach, Z., & Karez, W. (2011). Temperature dependence of growth in maize seedlings and excised coleoptile segments. *Polish Journal of Environmental Studies, 20*(5), 1221-1225.

Lima, O. R. F., Santos, G. G., & Melo, M. B. (2010). Influência da luz na germinação do milho. *Arte e Ciência, 1*(1), 1-8.

Lipchinsky, A., Sharova, E. I., & Medvedev, S. S. (2013). Elastic properties of the growth-controlling outer cell walls of maize coleoptile epidermis. *Acta Physiologiae Plantarum, 35*(7), 2183-2191. doi: 10.1007/s11738-013-1255-4.

Lucio, A. D., Couto, M. R. M., Lopes, S. J., & Storck, L. (2011). Transformação box-cox em experimentos com pimentão em ambiente protegido. *Horticultura Brasileira, 29*(1), 38-42. doi: 10.1590/S0102-05362011000100007.

Mertz, L. M., Henning, F. A., Cruz, H. L., Meneghello, G. E., Ferrari, C. S., & Zimmer, P. D. (2009). Diferenças estruturais entre tegumentos de sementes de soja com permeabilidade contrastante. *Revista Brasileira de Sementes, 31*(1), 23-29. doi: 10.1590/S0101-31222009000100003.

Musel, G., Schindler, T., Bergfeld, R., Ruel, K., Jacquet, G., Lapierre, C., ... Schopfer, P. (1997). Struture and distribution of lignin in primary and secondary cell walls of maize coleoptiles analyzed by chemical and immunological probes. *Planta, 201*(2), 146-156.

Nerling, D., Coelho, C. M. M., & Nodari, R. O. (2013). Genetic diversity for physiological quality of seeds from corn (*Zea mays L.*) intervarietal cross breeds. *Journal of Seed Science, 35*(4), 449-456. doi: 10.1590/S2317-15372013000400006.

Oliveira, G. E., Von Pinho, R. G., Andrade, T., Von Pinho, E. V. R., Santos, C. D., & Veiga, A. D. (2013). Physiological quality and amylase enzyme expression in maize seeds.

Ciência e Agrotecnologia, *37*(1), 40-48. doi: 10.1590/S1413-70542013000100005.

Rasband, S. W. (2012). ImageJ - Image Processing and Analysis in Java. *Astrophysics Source Code Library, 1*(1), 6013.

Sbrussi, C. A. G., & Zucarelli, C. (2014). Germinação de sementes de milho com diferentes níveis de vigor em resposta à diferentes temperaturas. *Semina: Ciências Agrárias*, *35*(1), 215-226. doi: 10.5433/1679-0359.2014v35n1p215.

Silva-Neta, I. C., Von Pinho, E. V. R., Veiga, A. D., Von Pinho, R. G., Guimarães, R. M., Caixeta, F., Santos, H. O., & Marques, T. L. (2015). Expression of genes related to tolerance to low temperature for maize seed germination. *Genetics and Molecular Research*, *14*(1), 2674-2690. doi: 10.4238/2015.March.30.28.

Tracy, W. F., & Galinat, W. C. (1987). Thickness and cell layer number of the pericarp of sweet corn and some of its relatives. *Horticultural Science*, *22*(4), 645-647.

Yakamura, K. (1999). Transformation using (x + 0.5) to stabilize the variance of populations. *Journal Researches on Population Ecology*, *41*(3), 229-234. doi: 10.1007/s101440050026.

Yap, S. K. (1981). Collection, germination and storage of dipterocarp seeds. *Malasyan Forester*, *44*(2/3), 281-300.

Zucarelli, C., Cavariani, C., Portugal, G., & Nakagawa, J. (2008). Potencial fisiológico de sementes de milho hidratadas pelo método do substrato de papel toalha. *Revista Brasileira de Sementes*, *30*(3), 122-129. doi: 10.1590/S0101-31222008000300016.

Zucarelli, C., Cavariani, C., Oliveira, E. A. P., & Nakagawa, J. (2011). Métodos e temperaturas de hidratação na qualidade fisiológica de sementes de milho. *Revista Ciência Agronômica*, *42*(3), 684-692. doi: 10.1590/S1806-66902011000300015.

13

Growth, grain yield and calcium, potassium and magnesium accumulation in common bean plants as related to calcium nutrition

Lucas da Silva Domingues[1], Nerinéia Dalfollo Ribeiro[2*], Jerônimo Luiz Andriolo[2], Micheli Thaise Della Flora Possobom[2] and Allan Emanoel Mezzomo Zemolin[2]

*[1]Curso de Agronomia, Universidade Tecnológica Federal do Paraná, Dois Vizinhos, Paraná, Brazil. [2]Departamento de Fitotecnia, Centro de Ciências Rurais, Universidade Federal de Santa Maria, Avenida Roraima, 1000, 97105-900, Santa Maria, Rio Grande do Sul, Brazil. *Author for correspondence. E-mail: nerineia@hotmail.com*

ABSTRACT. The objective of this work was to evaluate the plant growth and grain yield characteristics and the accumulation of calcium, potassium and magnesium in the leaves and grains of common bean cultivars grown with different calcium concentrations in the nutrient solution. Two experiments were conducted with nutrient solutions in a soilless system in a greenhouse. In the first experiment, five calcium concentrations (1.10, 1.65, 2.20, 2.75 and 3.30 mmol L^{-1}) and three common bean cultivars (Iraí, BRS Expedito and Carioca) were evaluated. Higher calcium concentrations (2.20, 2.75, 3.30, 3.85, 4.40 and 4.95 mmol L^{-1}) and two common bean cultivars (BRS Expedito and Carioca) were tested in the second experiment. Dry mass of the shoot and root, grain yield and calcium accumulation in the leaves and grains increased linearly in common bean plants supplied with calcium concentrations between 2.20 and 4.95 mmol L^{-1}. The high calcium concentration did not reduce the accumulation of calcium, potassium and magnesium in the leaves and grains. Common bean plants grown with high calcium concentrations present higher dry mass of the shoot and root, high grain yield and high calcium concentration in the leaves and grains.

Keywords: dry mass, nutritional effect of calcium, *Phaseolus vulgaris*.

Introduction

The availability of calcium to bean plants depends on its concentration in the soil solution and also on chemical soil properties, such as acidity and aluminum concentration (Quintana, Harrison, Palta, Nienhuis, & Kmiecik, 1999; Fageria, Stone, & Moreira, 2008; Ouertatani, Regaya, Ryan, & Gharbi, 2011). With low calcium availability, a reduction in plant height, leaf area and shoot and root growth in bean plants has been reported (Leal & Prado, 2008). Bean plants grown in a nutrient solution with depleted calcium concentrations presented signs of deficiency, such as necrosis of the root tips after 28 days and necrosis of the shoot apices after 30 days of growth, and these symptoms apparently caused a halt in growth, as indicated by the observed absence of xylem sap during harvesting (Schmitt et al., 2013). In fact, calcium serves vital physiological functions in plants, such as the organization of cell walls by calcium pectates, stabilization of membranes, pollen germination, pollen tube growth and elongation of the roots (Ge, Tian, & Russel, 2007; Gilliham et al., 2011).

Liming has been the main management practice used to neutralize soil acidity and aluminum and indirectly supplies calcium to crops. Liming effectively increased grain yield, the dry mass of the shoot and the number of pods and grains per plant in common beans (Fageria et al., 2008) and fava beans (Ouertatani et al., 2011). However, calcium is involved in complex reactions with other chemical elements present in the soil solution, and the soluble quantities available for plant uptake are difficult to estimate. The majority of research evaluating the response of bean plants to calcium was conducted in soil trials under field conditions, and the results have been contradictory for snap beans (Quintana et al., 1999) and common beans (Fageria et al., 2008). In fact, such results cannot separate the role of calcium as a soil neutralizer and its relations with other cations, such as potassium and magnesium, from its role as a nutrient for plant growth.

However, in a soilless system using a nutrient solution in a greenhouse it is possible to grow plants with the highest degree of control over factors such as nutrient uptake, water absorption, pH and salinity and also to isolate the effect of variations of other nutrients on plant growth and grain yield. Common bean plants supplied with high calcium concentrations in the nutrient solution showed an increase in the dry mass of the shoot and roots and higher calcium accumulation in the shoot and roots (Souza Junior, Nascimento, & Martinez, 1998; Silva, Moraes, & Souza, 2011). In these experiments, the plants were grown under controlled conditions and fed with nutrient solutions for periods ranging from 30 to 45 days.

Snap bean plants supplied with high calcium concentrations in the nutrient solution presented higher dry mass of the stem and leaves and high calcium accumulation in the leaves, followed by reproductive organs and stem (Pomper & Grusak, 2004; Schmitt et al., 2013). Previous studies also showed that an increase in calcium concentration in the nutrient solution resulted in an increase in fresh weigh, number of pods per plant, calcium accumulation in the pods and quality of snap beans (Favaro, Braga Neto, Takahashi, Miglioranza, & Ida, 2007). Cabot, Sibole, Barceló, and Poschenrieder (2009) observed that an increase in calcium in the nutrient solution decreased the potassium and magnesium concentrations in the leaves of snap bean plants. Snap beans are consumed in the form of green pods. In Brazil, common beans are preferred for consumption in the form of dry grains.

Common bean plants supplied with nutrient solutions with high calcium concentrations during the entire developmental cycle may have higher dry mass of the shoot and roots in the different phenological stages, high grain yield and high calcium, potassium and magnesium concentrations in the leaves and grains. These results were not found in the literature for dry beans. Furthermore, the portioning of calcium, potassium and magnesium in parts of the common bean plant needs to be further studied to be able to enhance the calcium concentration in the plant's edible parts by breeding. Therefore, the objective of this work was to evaluate the plant growth and grain yield characteristics and the accumulation of calcium, potassium and magnesium in the leaves and grains of common bean cultivars grown with different calcium concentrations in the nutrient solution.

Material and methods

Two experiments were performed in a closed soilless system inside a greenhouse at the Plant Science Department of the Federal University of Santa Maria, Santa Maria, Rio Grande do Sul State, Brazil (lat 29° 42' S, long 53° 49' W and 95 m altitude). The growing device used was described by Domingues, Ribeiro, Andriolo, Possobom, and Zemolin (2014). The plants were grown in 4 dm^3 black polypropylene pots filled with sand (1.2 to 2.4 mm gauge) previously washed with a 1% sodium hypochlorite solution.

The nutrient solutions were prepared according to the recommendations of Domingues et al. (2014). The electrical conductivity (EC) was adjusted so that a deviation of more than 10% from the initial value was recorded. For this, water or aliquots of new nutrient solution were added as required. The addition of KOH or 1 N H_2SO_4 into the nutrient solution was performed whenever necessary to maintain the pH between 5.5 and 6.5.

Experiment 1 was conducted from March to June 2011 in a completely randomized design with split plots and six repetitions. The main plots consisted of five calcium concentrations in the nutrient solution: 1.10, 1.65, 2.20, 2.75 and 3.30 mmol L^{-1}. The concentrations of other nutrients in each nutrient solution were adjusted to maintain the ionic equilibrium as follows, for total N, K^+, $H_2PO_4^-$, Ca^{++} and Mg^{++}, respectively, in mmol L^{-1}: 6.76, 4.50, 1.50, 1.00 and 1.00 in T1; 8.15, 5.00, 1.00, 1.00 and 1.00 in T2; 8.03, 5.00, 1.00, 1.00 and 1.75 in T3; 8.61, 5.70, 2.50, 1.00 and 1.75 in T4; 9.54, 5.25, 1.50, 1.50 and 2.00 in T5. The micronutrient quantities, in mg L^{-1}, were: 0.03 Mo, 0.42 B, 0.06 Cu, 0.50 Mn, 0.22 Zn and 1.00 Fe in all treatments. Three common bean cultivars, Iraí, BRS Expedito and Carioca, were grown in sub-plots. The Iraí cultivar comes from the Andean gene pool; it has cranberry-type grains (beige tegument with red stripes), determinate growth habit (type I) and early cycle. The BRS Expedito comes from the Middle American gene pool and has black commercial group grains, indeterminate growth habit with short guides (type II) and an intermediate cycle. The Carioca comes from the Middle American gene pool and has carioca-type grains (beige tegument with brown streaks), indeterminate growth habit with long guides (type III) and intermediate cycle. Twelve plants per treatment were used as the experimental units.

Three seeds per pot were sown on March 18, 2011, but only one plant per pot was kept after thinning at the first trifoliate leaf stage (V3). At the flowering stage (R6), six plants were collected, and the dry mass of the stems, leaves and shoot (stems + leaves) were determined after drying in an oven with forced air circulation (65 - 70°C) until a constant mass was reached. At the maturation stage (R9), another six plants of each treatment were collected, and the number of pods and grains per plant, number of grains per pod and grain yield (g plant^{-1} at 13% average moisture) were determined.

To determine the calcium concentration, 5 g samples of the dry mass of the stem and leaves at the R6 stage and the grains at the R9 stage were finely ground in an analytical knife mill to obtain particles smaller than 1 mm. The resulting flour was exposed to nitric-perchloric digestion, and the reading was performed using an atomic absorption spectrophotometer according to the method described by Domingues et al. (2014).

The second experiment was conducted from September to December 2011 to evaluate the effect of higher calcium concentrations than those used in Experiment 1 on plant growth, grain yield and the partitioning of calcium, potassium and magnesium. The same experimental design as the previous experiment was used. The main plots consisted of six calcium concentrations in the nutrient solution: 2.20 (T1), 2.75 (T2), 3.30 (T3), 3.85 (T4), 4.40 (T5) and 4.95 (T6) mmol L^{-1}, and two common bean cultivars, BRS Expedito and Carioca, were grown in sub-plots. The nutrient concentrations in T1, T2 and T3 were the same as in Experiment 1, and those of T4, T5 and T6 were as follows, for total N, K^+, $H_2PO_4^-$, Ca^{++} and Mg^{++}, respectively, in mmol L^{-1}: 9.67, 2.00, 1.00, 1.50 and 2.00 in T4; 11.06, 2.00, 1.00, 2.00 and 2.50 in T5; 12.44, 2.00, 1.00, 2.00 and 2.50 in T6. Micronutrients were supplied as in Experiment 1. Twenty-four plants per treatment were used as the experimental units.

Three seeds per pot were sown on September 15, 2011, but only one plant per pot was kept after thinning at the first trifoliate leaf stage (V3). In each developmental stage, third leaf trifoliate (V4), flowering (R6), pod filling (R8) and maturation (R9), six plants were collected from each treatment for the determination of the dry mass of the plant. The roots were washed in running water to remove the sand. The dry mass of the shoot and roots were determined after drying in an oven with forced air circulation (65 - 70°C) until a constant mass was reached. At pod filling, the number of pods per plant and the dry mass of the pods were also evaluated. At maturation, the number of grains per plant was also counted, and the grain yield at 13% of average moisture was determined.

Samples of 5.0 g of the leaves at the V4, R6 and R8 stages and of the grains at the R9 stage of each treatment were finely ground to determine the nutrient concentration in the tissues, as described for Experiment 1. The determination of the calcium, potassium and magnesium concentrations was performed with an atomic absorption spectrophotometer using a wavelength of 422.70 nm for calcium, 766.50 nm for potassium and 202.50 nm for magnesium.

The data were subjected to variance analysis considering the completely randomized design with split plots and six repetitions, except for the nutrient concentration in the tissues, for which three repetitions were used. The F test (p value < 0.05) was used to test the hypotheses of the main effects and for the calcium concentration x cultivar interaction, considering all effects as fixed. When the effect of the calcium concentration and the calcium concentration x cultivar interaction were significant, regression analysis was done by the method of orthogonal polynomials, and the equation at the higher significant grade was retained. In cases in which the second grade equation fitted by regression analysis was also significant, the point of maximum (or minimum) technical efficiency was estimated by the equation ($X = -\hat{b}_1 / 2\hat{b}_2$), $\hat{b}_1$ and $\hat{b}_2$ being the coefficients of the first and second grade, respectively, obtained from the estimated equation of regression analysis. When the effect of the cultivars was significant, the comparison of means was carried out using the Tukey test (p = 0.05) in the first experiment and by the F test in the second experiment (p = 0.05). Statistical analyses were performed using the Microsoft® Office Excel spreadsheet and the Sigma Plot and Genes softwares.

Results and discussion

Experiment 1

In the variance analysis, a significant calcium concentration x cultivar interaction was observed only for the number of grains per pod ($p < 0.05$) (Table 1). However, the main effects (calcium concentration and cultivar) were significant for the dry mass of the stems, leaves and shoot, the number of pods and grains per plant, the grain yield and the calcium concentration in the grains. For the calcium concentration in the stems and leaves, a significant effect was observed only for the calcium concentration in the nutrient solution. A significant effect for the calcium concentration was previously described for the dry mass of the shoot (Favaro et al., 2007; Silva et al., 2011) and the number of pods per plant (Favaro et al., 2007) in bean plants supplied with calcium in the nutrient solution. Significant differences between snap bean cultivars for the calcium accumulation in the pods were previously reported by Pomper and Grusak (2004). Therefore, the increase in calcium concentration in the rooting medium was efficient to increase plant growth and grain yield in the bean plants, and the genetic variability among bean cultivars can be observed.

For the dry mass of the stems, leaves and shoot and the calcium concentration in the stems, significant effects were found for the calcium concentration in the nutrient solution, but the regressions of the 1st, 2nd and 3rd grades were not significant (Table 1). For these characteristics, there was no response, linear, quadratic or cubic, to the calcium concentration in the rooting medium. Other models showed no biological explanation and therefore were not presented. However, common bean plants grown with different calcium concentrations in the nutrient solution over a period of 45 days showed a quadratic response for the dry mass of the shoot (Silva et al., 2011). In the present work, the common bean plants were grown until the end of their cycle using different calcium concentrations, and this can justify the differences observed.

The number of pods and grains per plant and the grain yield increased due to the effect of the calcium concentration in the nutrient solution in the range between 1.1 and 3.3 mmol L^{-1} (Figures 1a, b and c). In snap beans, Favaro, Braga Neto, Takahashi, Miglioranza, and Ida (2007) reported that the number of green pods harvested 10-12 days after anthesis also increased with the calcium concentration in the nutrient solution. Similar effects were recorded by Domingues et al. (2014) for the number of pods and grains per plant, the number of grains per pod and the grain yield of common bean plants grown with a calcium concentration of 3.85 mmol L^{-1} in the nutrient solution.

Table 1. Variance and regression analysis for the dry mass of: stems (DMST, g), leaves (DML, g) and shoot (DMS, g); number of: pods per plant (NP), grains per plant (NG) and grains per pod (NGP); grain yield (yield, g plant^{-1}); calcium concentration in: stems (Ca in stems, g kg^{-1} of dry matter - DM), leaves (Ca in leaves, g kg^{-1} DM) and grains (Ca in grains, g kg^{-1} DM) evaluated in three common bean cultivars grown with five calcium concentrations in the nutrient solution in the first experiment.

Source of variation	Mean square						
	DF	DMST	DML		DMS	NP	NG
Concentration (A)	4	2.20*	3.46*		10.79*	47.28*	1318.76*
Error A	25	0.79	0.84		2.85	7.51	190.36
Cultivar (B)	2	14.61*	19.8*		60.72*	180.11*	13149.9*
A X B	8	1.39^{ns}	0.77^{ns}		3.77^{ns}	10.81^{ns}	463.31^{ns}
Error B	50	0.70	1.57		3.87	8.58	238.07
Grade 1	1	2.93^{ns}	4.60^{ns}		14.36^{ns}	63.04*	1743.55*
Grade 2	1	1.99^{ns}	3.23^{ns}		10.39^{ns}	57.05^{ns}	1624.90^{ns}
Grade 3	1	0.79^{ns}	1.08^{ns}		3.97^{ns}	5.60^{ns}	116.96^{ns}
Deviations	1	3.08	4.93		14.35	63.45	1719.57
Mean		4.25	4.99		9.25	10.47	46.76
C.V. (%)		19.62^{4}- 15.68^{5}	18.37^{4}-8.12^{5}		18.27^{4}-16.85^{5}	16.17^{4}-17.96^{5}	19.50^{4}-18.95^{5}
Source of variation	DF	NGP	Yield	DF	Ca in stems	Ca in leaves	Ca in grains
Concentration (A)	4	0.25^{ns}	174.54*	4	11.18*	43.37*	0.08*
Error A	25	0.21	18.03	10	1.17	12.19	0.43
Cultivar (B)	2	43.94*	278.00*	2	10.58^{ns}	52.22^{ns}	0.57*
A X B	8	0.60*	17.93^{ns}	8	0.97^{ns}	26.6^{ns}	0.53^{ns}
Error B	50	0.23	28.97	20	0.69	15.09	0.5
Grade 1	1	0.28^{1}-0.16^{2}-0.03^{3ns}	241.27*	1	22.93^{ns}	66.45^{ns}	0.05*
Grade 2	1	0.25^{1}-0.05^{2}-0.02^{3ns}	227.92^{ns}	1	9.47^{ns}	34.11*	0.04^{ns}
Grade 3	1	0.02^{1}-0.05^{2}-0.01^{3ns}	2.95^{ns}	1	3.46^{ns}	21.99^{ns}	0.03^{ns}
Deviations	1	0.01^{1}-0.06^{2}-0.01^{3}	226.02	1	8.84	50.70	1.98
Mean		4.40	14.52		5.60	16.16	0.95
C.V. (%)		10.43^{4}-8.26^{5}	12.65^{4}-17.06^{5}		19.33^{4}-14.85^{5}	11.60^{4}-14.03^{5}	14.36^{4}-11.75^{5}

*Significant at 5% probability by F test. ^{ns}not significant. ^{1}Value corresponding to BRS Expedito cultivar. ^{2}Value corresponding to Carioca cultivar. ^{3}Value corresponding to Iraí cultivar. ^{4}Coefficient of variation of the main plot. ^{5}Coefficient of variation of the subplot.

These results might be due to the effect of calcium on pollen germination and the growth of the pollen tube, leading to higher fertility (Ge et al., 2007) and thereby leading to an increased number of pods and grains per plant and higher grain yield (Figure 1a, b and c). However, although a linear relationship was fitted for these characteristics, there should be a maximum calcium concentration above which this effect will no longer be recorded.

The highest calcium concentration in the leaves was reached at a calcium concentration of 2.64 mmol L^{-1} in the nutrient solution (point of maximum technical efficiency) (Figure 1d). Snap bean plants were responsive to increases in the calcium concentration in the nutrient solution, and the highest calcium accumulation was observed in the leaves (Favaro et al., 2007; Cabot, Sibole, Barceló, & Poschenrieder, 2009; Schmitt et al., 2013).

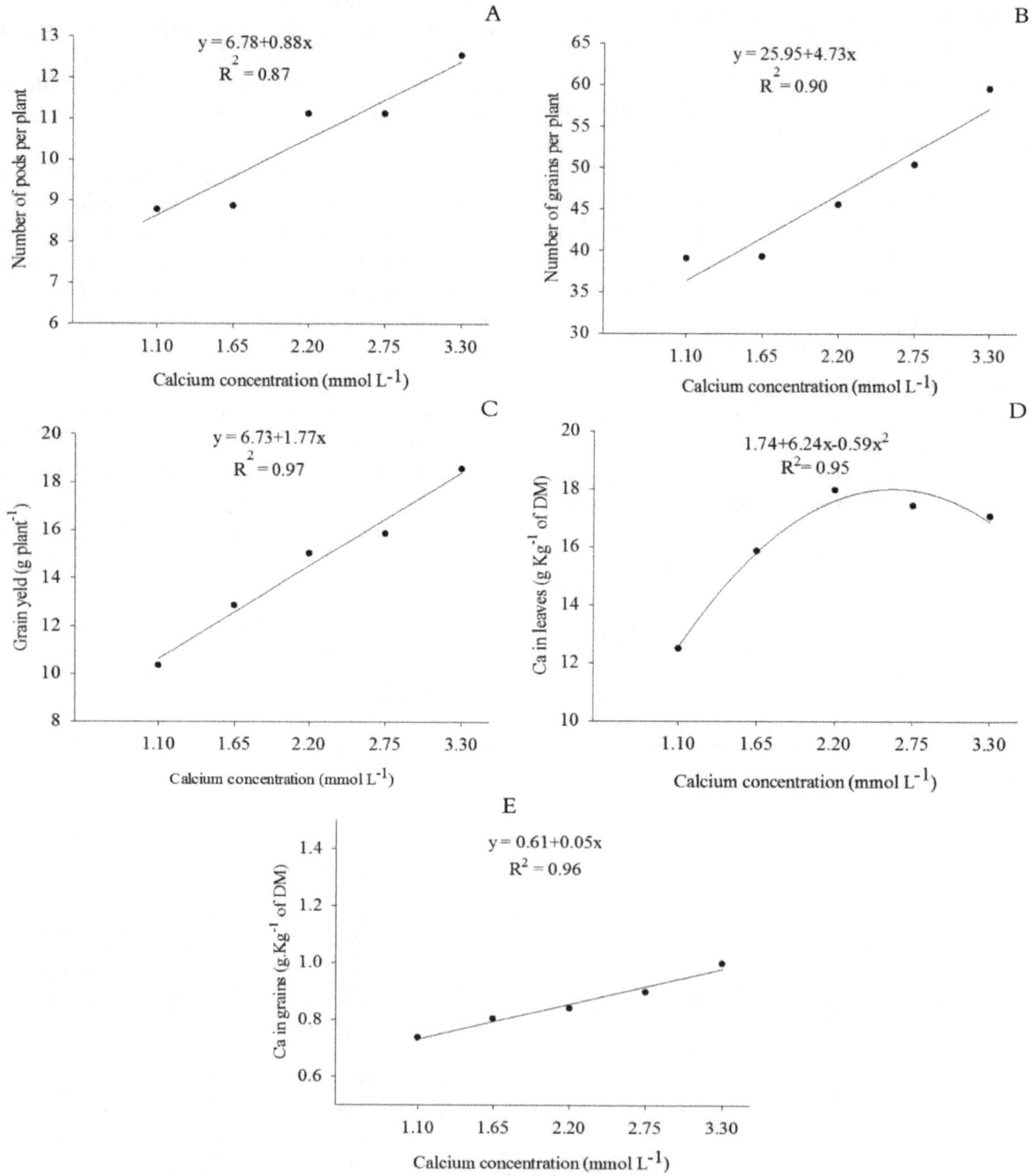

Figure 1. Number of pods per plant; a), number of grains per plant; b), grain yield (g $plant^{-1}$); c), calcium concentration in leaves (Ca in leaves, g kg^{-1} of dry matter - DM); d) and calcium concentration in grains (Ca in grains, g kg^{-1} DM); e), of three common bean cultivars grown with five calcium concentrations in the nutrient solution in the first experiment.

In the present work, the decrease at the upper calcium concentrations in the nutrient solution may have been a simultaneous effect of the reduction in the water flow in the plant and the higher electrical conductivity of the nutrient solutions. This experiment was conducted in autumn, and the plants reached full leaf area growth in later autumn, when the atmospheric water demand and transpiration were lower. When plant water demand is low, the plants might be more sensitive to salinity. This hypothesis is supported by the results in the literature reporting that the highest calcium accumulation in the shoot of common bean plants was obtained at a calcium concentration of 1.40 mmol L^{-1} in the nutrient solution (Souza Junior et al., 1998).

The calcium concentration in the grains increased 35% and fitted a linear regression in the range of treatments (Figure 1e). In snap beans grown in soil, it has been reported that calcium fertilization was not effective to increase the calcium concentration in immature pods (Quintana et al., 1999). These differences may also be related to the water flow in the plant during its growing period. In indeterminate growth snap bean plants, the leaves and pods grow simultaneously and may compete for water during periods of high transpiration demand by the plants. At the final developmental stages of determinate growth in plants, nutrients are remobilized from the vegetative organs to the grains, and the nutrient concentration in the grains can increase. In the present experiment, the calcium uptake, transport and accumulation in the plant organs might also be enhanced because this ion was in a readily available form in the nutrient solution. This is not the case in plants grown in soil, where the uptake of calcium might be depressed by interactions with other nutrients and/or chemical elements dissolved in the soil solution.

The common bean cultivars differed in their plant growth and grain yield characteristics (Table 2). The BRS Expedito and Carioca cultivars showed higher dry mass of the stems, leaves and shoot and higher number of pods and grains per plant and number of grains per pod than the Iraí cultivar. This is explained by the fact that the BRS Expedito and Carioca cultivars, which are of indeterminate growth habit, had vigorous shoot growth, resulting in shading of the Iraí plants. Therefore, the Iraí cultivar showed the lowest values in dry mass of the plant and grain yield components. The highest grain yield was obtained by the Carioca cultivar.

Table 2. Mean values for the dry mass of stems (DMST, g), dry mass of leaves (DML, g), dry mass of shoot (DMS, g), number of pods per plant (NP), number of grains per plant (NG), number of grains per pod (NGP), grain yield (yield, g $plant^{-1}$) and calcium concentration in grains (Ca in grains, g kg^{-1} of dry matter - DM) evaluated in three common bean cultivars grown with five calcium concentrations in the nutrient solution in the first experiment.

Cultivar	DMST	DML	DMS	NP	NG	NGP	Yield	Ca in grãos
BRS Expedito	4.84 a*	5.11 a	9.94 a	10.88 a	54.57 a	5.12 a	14.27 b	1.27 a
Carioca	4.45 a	5.75 a	10.19 a	12.70 a	62.68 a	5.10 a	17.69 a	0.93 ab
Iraí	348 b	4.13 b	7.62 b	7.85 b	23.05 b	3.01 b	11.62 b	0.63 b
Mean	4.25	4.99	9.25	10.47	46.76	4.40	14.52	0.95
C.V.%	15.68	8.12	16.85	17.96	18.95	8.26	17.06	11.75

*Means followed by the same letter in the column do not differ significantly ($p = 0.05$) by Tukey's test. C.V.%: Coefficient of variation.

The calcium accumulation in the grains differed among the cultivars, with BRS Expedito differing significantly from Iraí (Table 2). BRS Expedito is a cultivar in use in Brazil featuring black grains with a high calcium concentration in the grains (Jost et al., 2013). This cultivar showed high calcium accumulation in the grains when grown with low and high calcium availability in the nutrient solution (Domingues, Ribeiro, Andriolo, Possobom, & Zemolin, 2014). These results suggest that the common bean cultivars with black grains such as BRS Expedito might be higher accumulators of calcium in the grains. Differences in calcium accumulation in the pods of snap bean cultivars have also been reported in the literature (Pomper & Grusak, 2004; Quintana et al., 1999). According to Pomper and Grusak (2004), snap bean plants of the Labrador cultivar, which exhibit low pod calcium concentration, showed higher whole-plant water uptake, and this led to a dilution of the calcium concentration in the xylem stream, and thus less total calcium was transported to the developing pods.

Experiment 2

A significant calcium concentration x cultivar interaction ($p < 0.05$) was observed for the potassium concentration in the leaves at R8 and for the magnesium concentration in the grains at R9 (Table 3). For these characteristics, the response of the common bean cultivars differed depending on the calcium concentration supplied in the nutrient solution.

Significant effects of the calcium concentration were observed on the dry mass of the shoot at R6, R8 and R9, dry mass of the roots at R6 and R9, dry mass of the pods at R8, grain yield, calcium concentration in the leaves at R8 and in the grains at R9, potassium concentration in the grains at R9 and magnesium concentration in the leaves at R8. Thus, the calcium availability in the rooting medium affected these characteristics. A significant effect of the cultivars ($p < 0.05$) was observed for the dry mass of the roots at R6 and R9, number of pods per plant at R8, dry mass of the pods at R8, potassium concentration in the grains at R9 and magnesium concentration in the leaves at R8. The existence of significant differences among the cultivars for these characteristics allows the selection of superior common bean cultivars.

In the second experiment the dry mass of the shoot at R6 and R8 and the dry mass of the root at R6 showed significant response to the increase in the calcium concentration in the nutrient solution (Figures 2a, b and c). Shoot growth increased at calcium concentrations in the nutrient solution above 3.30 mmol L^{-1} in both plant phenological stages (Figures 2a and c), while root growth increased linearly at R6 in the range between 2.20 and 4.95 mmol L^{-1} (Figure 2b). Silva, Moraes, and Souza (2011) observed that the growth of common bean plants at 45 days after planting was reduced by a calcium concentration of 0.46 mmol L^{-1} in the nutrient solution. In later plant stages, the calcium requirements may be higher, and reduction in growth at higher calcium concentrations could not be excluded. High root growth under high calcium availability has also been reported in the literature for bean plants (Souza Junior et al., 1998; Favaro et al., 2007; Silva et al., 2011). In the present work, the quite large range of concentrations used as treatments produced a surprisingly linear response without a saturation point and without any signs of toxicity observed in the plants.

Table 3. Variance and regression analysis for the dry mass of the: shoot at third trifoliate leaf (DMS V4, g), roots at third trifoliate leaf (DMR V4, g), shoot at flowering (DMS R6, g), roots at flowering (DMR R6, g), shoot at pod filling (DMS R8, g) and roots at pod filling (DMR R8, g); number of pods at pod filling (NP R8); dry mass of: pods at pod filling (DMP R8, g), shoot at maturity (DMS R9, g) and roots at maturity (DMR R9, g); number of grains per plant at maturity (NG R9), grain yield (yield, g plant^{-1}); calcium in the: leaves at third trifoliate leaf (Ca L V4, g kg^{-1} of dry matter - DM), leaves at flowering (Ca L R6, g kg^{-1} DM), leaves at pod filling (Ca L R8, g kg^{-1} DM) and grain at maturity (Ca G R9, g kg^{-1} DM); potassium in the: leaves at pod filling (K L R8, g kg^{-1} DM) and grains at maturity (K G R9, g kg^{-1} DM); magnesium in the: leaves at pod filling (Mg L R8, g kg^{-1} DM) and grains at maturity (Mg G R9, g kg^{-1} DM) evaluated in two common bean cultivars grown with six calcium concentrations in the nutrient solution in the second experiment.

Source of variation	Mean square							
	DF	DMS V4	DMR V4	DMS R6	DMR R6	DMS R8	DMR R8	NP R8
Concentration (A)	5	0.38^{ns}	0.03^{ns}	27.92^{*}	1.99^{*}	56.04^{*}	0.54^{ns}	19.55^{ns}
Error A[1]	30	0.63	0.02	5.36	0.59	10.93	0.49	9.92
Cultivar (B)	1	0.45^{ns}	0.02^{ns}	8.00^{ns}	21.23^{*}	0.11^{ns}	0.10^{ns}	84.5^{*}
A X B	5	0.39^{ns}	0.005^{ns}	6.16^{ns}	0.47^{ns}	5.13^{ns}	0.21^{ns}	2.26^{ns}
Error B[2]	30	0.21	0.009	4.77	0.43	8.67	0.54	8.70
Grade 1	1	0.85^{ns}	0.006^{ns}	41.99^{*}	4.05^{*}	112.47^{*}	0.69^{ns}	28.00^{*}
Grade 2	1	0.94^{ns}	0.002^{ns}	25.45^{*}	0.77^{ns}	23.73^{ns}	0.59^{ns}	34.83^{*}
Grade 3	1	0.10^{ns}	0.001^{ns}	2.40^{ns}	0.18^{ns}	3.84^{ns}	0.08^{ns}	6.03^{ns}
Deviations	1	0.83	0.14	69.80	4.99	140.11	1.35	28.89
Mean		1.30	0.14	16.85	2.79	21.03	4.33	19.13
C.V. (%)		19.31[3]-11.31[4]	22.05[3]-21.50[4]	17.18[3]-12.96[4]	13.74[3]-23.64[4]	15.72[3]-14.00[4]	16.20[3]-17.06[4]	16.46[3]-15.41[4]
Source of variation	DF	DMP R8	DMS R9	DMR R9	NG R9	Yield	DF	Ca L V4
Concentration (A)	5	30.09^{*}	66.41^{*}	1.28^{*}	287.42^{ns}	39.38^{*}	5	10.54^{ns}
Error A[1]	30	9.94	22.21	0.53	138.62	8.46	12	5.40^{*}
Cultivar (B)	1	93.16^{*}	34.03^{ns}	34.72^{*}	320.88^{ns}	6.90^{ns}	1	17.05^{ns}
A X B	5	7.12^{ns}	77.09^{ns}	3.10^{ns}	306.02^{ns}	11.06^{ns}	5	1.57^{ns}
Error B[2]	30	8.12	25.67	0.71	230.16	10.61	12	6.01
Grade 1	1	39.08^{*}	80.21^{ns}	2.14^{ns}	366.42^{ns}	55.70^{*}	1	18.69^{*}
Grade 2	1	51.62^{*}	66.42^{ns}	1.05^{ns}	128.51^{ns}	14.33^{ns}	1	9.77^{*}
Grade 3	1	4.54^{ns}	22.97^{ns}	0.83^{ns}	223.68^{ns}	7.70^{ns}	1	8.92^{ns}
Deviations	1	55.24	106.04	2.40	718.56	98.46	1	15.32
Mean		17.92	30.69	5.18	28.27	24.51		15.66
C.V. (%)[3]		17.59[3]-15.90[4]	15.35[3]-16.51[4]	14.05[3]-16.29[4]	13.01[3]-16.77[4]	11.87[3]-13.29[4]		14.84[3]-15.64[4]
Source of variation	DF	Ca L R6	Ca L R8	Ca G R9	K L R8	K G R9	Mg L R8	Mg G R9
Concentration (A)	5	18.67^{ns}	96.71^{*}	0.56^{*}	959.95^{*}	21.91^{*}	3.86^{*}	1.14^{*}
Error A[1]	12	7.18	3.92	0.07	13.72	2.13	0.27	0.41
Cultivar (B)	1	3.82^{ns}	16.88^{ns}	0.21^{ns}	1887.7^{*}	136.11^{*}	14.68^{*}	2.68^{*}
A X B	5	12.41^{ns}	9.25^{ns}	0.02^{ns}	150.09^{*}	5.41^{ns}	0.62^{ns}	0.47^{*}
Error B[2]	12	4.13	4.12	0.01	8.52	8.79	0.23	0.48
Grade 1	1	14.32^{*}	175.07^{*}	0.15^{*}	410.5[6]-81.6^{7ns}	75.30^{ns}	5.65^{ns}	0.57- 0.58^{*5}
Grade 2	1	54.45^{ns}	63.52^{ns}	0.54^{ns}	245.9[6]-590.7^{7ns}	23.40^{ns}	2.93^{ns}	0.64- 0.50^{ns}
Grade 3	1	5.29^{ns}	86.62^{ns}	0.01^{ns}	222.8[6]-$161.3^{7n.s.}$	0.38^{ns}	2.47^{ns}	0.42- 0.72^{ns}
Deviations	1	1.34	158.41	0.09	151.7[6]-221.7[7]	10.47	8.28	1.22- 1.46
Mean		19.58	32.39	1.51	50.83	31.44	9.42	2.10
C.V. (%)[3]		13.68[3]-10.38[4]	6.11[3]-6.27[4]	5.76[3]-6.81[4]	7.28[3]-5.74[4]	4.64[3]-9.43[4]	5.55[3]-5.16[4]	9.62[3]-10.42[4]

*Significant at 5% probability by F test. ns not significant. [1]Error A: main plot error. [2]Error B = subplot error. [3]Variation coefficient of main plot. [4]Coefficient of variation of the subplot. [5]Significant only for BRS Expedito cultivar. [6]Value for BRS Expedito cultivar. [7]Value for Carioca cultivar.

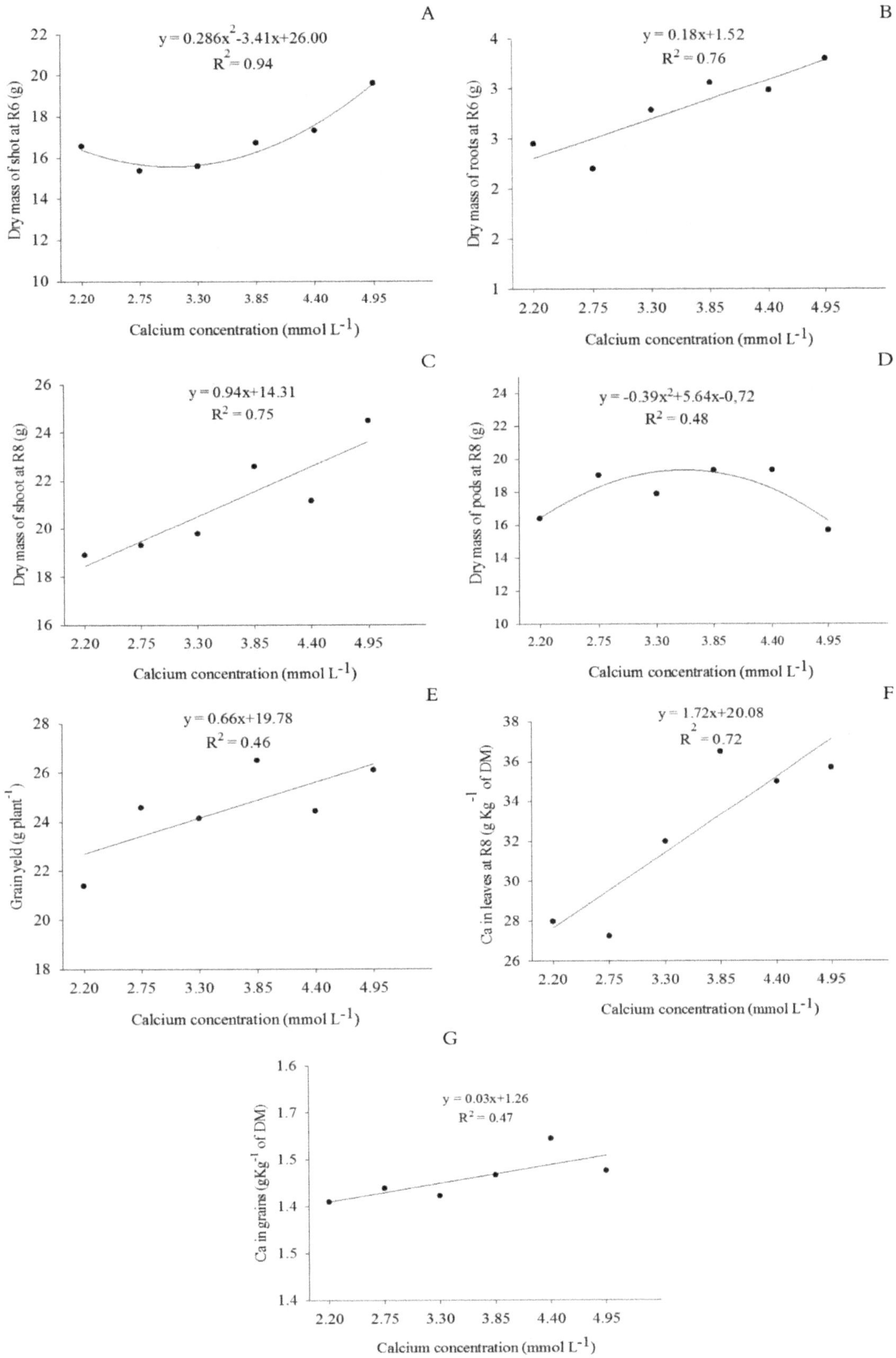

Figure 2. Dry mass of: shoot at flowering (R6) (g); a), roots at flowering (R6) (g); b), shoot at pod filling (R8) (g); c), and pods at pod filling (R8) (g); d), grain yield (g plant^{-1}); e), calcium concentration in the leaves at pod filling (Ca in leaves at R8, g kg^{-1} of dry matter - DM); f) and in the grains (Ca in grains, g kg^{-1} DM); g), of two common bean cultivars grown with six calcium concentrations in the nutrient solution in the second experiment.

In Experiment 1, the number of pods per plant increased with the increasing calcium concentration in the rooting medium (Figure 1a), but this effect was not significant in Experiment 2 (Table 3). Perhaps the genetically determined potential number was reached. The dry mass of the pods at R8 increased until a calcium concentration of 3.55 mmol L^{-1} in the nutrient solution (point of maximum technical efficiency) (Figure 2d). When supplying snap bean plants with calcium in the rooting medium, Favaro et al. (2007) also observed an improvement in the dry mass of the green pods. Nevertheless, the grain yield increased linearly (Figure 2e), confirming the tendency fitted in the first experiment (Figure 1c). In bean plant growth in soil, it has been reported that calcium fertilization was effective to increase the grain yield, the number of pods and the number of grains per plant in common beans (Fageria et al., 2008) and fava beans (Ouertatani et al., 2011). The increase in grain yield with calcium fertilization was mainly associated with an increase in the dry mass of the shoot (Fageria et al., 2008)

The leaves of the common bean plants were large accumulators of calcium at the phenological stages of third leaf trifoliate (15.66 g kg^{-1} of DM), flowering (19.58 g kg^{-1} of DM) and pod filling (32.39 g kg^{-1} of DM), followed by the grains at maturity (1.51 g kg^{-1} of DM) (Table 3). In snap bean cultivars, Pomper and Grusak (2004) also observed that the calcium concentration in the leaves differed according to the phenological stage, and the maximum value was found at the pod filling stage. Schmitt et al. (2013) supplied snap bean plants with calcium in the nutrient solution for six weeks and verified that the calcium accumulation was higher in the leaves (64%), followed by the reproductive organs (16%), stem (15%) and roots (5%). Understanding the partitioning of calcium in common bean plants in different phenological stages is very important to increase the calcium concentration in the edible parts of the plant.

The calcium concentration increased linearly in the leaves at R8 (Figure 2f) and also in the grains at R9 (Figure 2g). In the leaves, the calcium concentration averaged 28%, ranging from 27.98 g kg^{-1} of DM at the lowest to 35.68 g kg^{-1} DM at the highest calcium concentration in the nutrient solution. Jáuregui-Zùniga, Ferrer, Calderón, Munoz, and Moreno (2005) observed a positive correlation between the calcium concentration in the rooting medium of bean plants and calcium oxalate crystal formation in the leaves. In the present study, the leaves of common bean plants were found to be large accumulators of calcium (Table 3, Figure 2f). Therefore, the calcium bioavailability needs to be better assessed in the leaves of common bean plants grown with high calcium concentrations in the rooting medium.

In the grains at R9, the calcium concentration increased to 19%, from 1.41 to 1.68 g kg^{-1} DM in the same concentration range of the nutrient solution. These calcium values were higher than those obtained in the grains of 29 Andean lines with different tegument colors, cultivated in Brazil, in the United States and in Colombia (Akond, Crawford, Berthold, Talukder, & Hossain, 2011), and were considered high for common bean lines according to Ribeiro, Domingues, Zemolin, and Possobom (2013), Ribeiro et al. (2014) and Ribeiro, Jost, Maziero, Storck, and Rosa (2014). However, values for the calcium concentration in the grains varying from 0.7 to 1.4 g kg^{-1} of DM were found in 10 inbred lines of common beans grown with high calcium availability in the nutrient solution (3.85 mmol L^{-1}) by Domingues et al. (2014). Thus, in the present study, increasing the calcium concentration in the nutrient solution increased the calcium concentration in common bean grains. These data are very important because it indicates that improvement of the nutritional quality of common bean grains can be achieved by calcium management in the rooting medium.

The average calcium concentration in the leaves was 21% higher than in the grains, confirming data reported in the literature showing that the leaves of common beans and snap beans are strong calcium accumulators (Souza Junior et al., 1998; Pomper & Grusak, 2004; Favaro et al., 2007; Schmitt et al., 2013). Although no references in the literature were found regarding the use of common bean leaves as a source of calcium in human or husbandry diets, this might be a matter of interest in the future. In the present work, the calcium effect as a nutrient "per se" was evaluated because the effects of aluminum and/or other ions affecting soil chemical properties were excluded. In fact, such interactions may be the cause of the data presented in the literature showing that when calcium was applied in the soil, increases in the calcium concentration in the immature pods of snap beans have been recorded (Quintana et al., 1999).

In the present work, the ionic concentration of the nutrient solution had to be adjusted to obtain different calcium concentrations using soluble salts. As a consequence, the potassium was reduced, while the magnesium and nitrogen concentrations were increased to counterbalance the higher calcium concentrations. No disturbances in the K:Ca:Mg cationic ratio or salinity effects reducing plant growth and grain yield were recorded, and no effects on the potassium and magnesium concentrations in the leaves at R8 and the grains at R9, whether linear, quadratic or cubic, were observed (Table 3). Cabot et al. (2009) observed that an increase in calcium concentration in the nutrient solution in the presence of sodium chloride decreased the potassium and magnesium concentrations in the leaves of snap bean plants grown for 15 days. Providing potassium to snap bean plants grown in pots with soil produced a significant decrease in the calcium concentration in the leaves and stem and decreased the magnesium concentration in the pods and stem (Tüma, Skalicky, Tümová, Bláhová, & Rosülková, 2004). In the present study, all the nutrients in the nutrient solution were available for uptake by the plants, and it is unlikely that the concentrations in the plant organs were artifacts as a consequence of antagonisms and/or interactions among them.

The BRS Expedito cultivar showed higher dry mass of the roots at R6 and R9 and dry mass of the pods at R8 (Table 4). However, the Carioca cultivar had the higher number of pods per plant at R8. The identification of common bean cultivars with differential response in the growth and grain yield characteristics to different calcium concentrations in the nutrient solution is of great importance to common bean breeding because this allows the identification of a mechanism that gives the plant an ability to absorb and use calcium and thereby enables the selection of cultivars that are more efficient in calcium use.

Table 4. Mean values for the dry mass of roots at flowering (DMR R6, g), number of pods at pod filling (NP R8), dry mass of pods at pod filling (DMP R8, g) and dry mass of the root at maturation (DMR R9, g) evaluated in two common bean cultivars grown with six calcium concentrations in the nutrient solution in the second experiment.

Cultivar	DMR R6	NP R8	DMP R8	DMR R9
BRS Expedito	3.51*	17.44*	18.34*	6.24*
Carioca	2.11	21.19	15.82	4.19
Mean	2.81	19.35	17.08	5.21
C.V. (%)	23.64	15.41	15.90	16.29

*Significant at 5% by F test. ns = not significant. C.V. (%): coefficient of variation.

The results of the present work are original and can contribute to deeper knowledge of the mineral nutrition of the common bean crop. First, no data were found in the literature reporting the results of common bean plants supplied by nutrient solutions with different calcium concentrations on plant growth, grain yield and the accumulation of calcium, potassium and magnesium in parts of the plant at different phenological stages during the entire developmental cycle. Second, in this work, the effect of calcium as a nutrient was evaluated in conditions with no toxic effect of aluminum and/or chemical interactions present in the soils that negatively affect plant growth and grain yield. Third, calcium fertilization is a practice that could be previewed to increase the grain yield of this crop. This is an important conclusion of this research because current recommendations aim only to correct the pH and to neutralize the aluminum in the soil. Nevertheless, to reach this goal, more research in soil conditions must be done. The role of liming as a soil corrective had to be separate from the calcium nutrition effects. In the present research, the linear effects of calcium on plant growth, grain yield and calcium accumulation in the organs were recorded. This is not a realistic response of plant growth processes, which are mainly of the saturation type. Experiments in soil conditions should be performed to determine the maximum calcium fertilization rates to be used for this crop.

Conclusion

The dry mass of the shoot and root, the grain yield and the calcium accumulation in the leaves and grains increase linearly in common bean plants supplied with nutrient solutions with calcium concentrations between 2.20 and 4.95 mmol L^{-1}. The high calcium concentrations do not reduce the plant growth and grain yield or the accumulation of calcium, potassium and magnesium in the leaves and grains. Common bean plants grown with high calcium concentrations in the nutrient solution present higher dry mass of the shoot and root, high grain yield and high calcium concentration in the leaves and grains.

Acknowledgements

The authors would like to thank the National Council of Technological and Scientific Development (CNPq) and the Coordination for the Improvement of Higher Education Personnel (Capes) for the scholarship granted.

References

Akond, A. S. M. G. M., Crawford, H., Berthold, J., Talukder, Z. I., & Hossain, K. (2011). Minerals (Zn, Fe, Ca and Mg) and antinutrient (phytic acid) constituents in common bean. *American Journal of Food Technology, 6*(3), 235-243. doi: 10.3923/ajft.2011.235.243

Cabot, C., Sibole, J. V., Barceló, J., & Poschenrieder, C. (2009). Sodium-calcium interactions with growth, water, and photosynthetic parameters in salt-treated beans. *Journal of Plant Nutrition and Soil Science, 172*(5), 637-643. doi: 10.1002/jpln.200800124

Domingues, L. S., Ribeiro, N. D., Andriolo, J. L., Possobom, M. T. D. F., & Zemolin, A. E. M. (2014).

Selection of common bean lines for calcium use efficiency. *Ciência Agronômica, 45*(4), 767-776.

Fageria, N. K., Stone, L. F., & Moreira, A. (2008). Liming and manganese influence on common bean yield, nutrient uptake, and changes in soil chemical properties of an Oxisol under no-tillage system. *Journal of Plant Nutrition, 31*(10), 1723-1735. doi: 10.1080/01904160802324720

Favaro, S. P., Braga Neto, J. A., Takahashi, W., Miglioranza, E., & Ida, E. I. (2007). Rates of calcium, yield and quality of snap bean. *Scientia Agricola, 64*(6), 616-620. doi: 10.1590/S0103-90162007000600009

Ge, L. L., Tian, H. Q., & Russel, S. D. (2007). Calcium function and distribution during fertilization in angiosperms. *American Journal of Botany, 94*(6), 1046-1060. doi: 10.3732/ajb.94.6.1046

Gilliham, M., Dayod, M., Hocking, B. J., Xu, B., Conn, S. J., Kaiser, B. N., ... Tyermann, S. D. (2011). Calcium delivery and storage in plant leaves: exploring the link with water flow. *Journal of Experimental Botany, 62*(7), 2233-2250. doi: 10.1093/jxb/err111

Jáuregui-Zùniga, D., Ferrer, M. A., Calderón, A. A., Munoz, R., & Moreno, A. (2005). Heavy metal stress reduces the deposition of calcium oxalate crystals in leaves of *Phaseolus vulgaris*. *Journal of Plant Physiology, 162*(10), 1183-1187. doi: 10.1016/j.jplph.2005.03.002

Jost, E., Ribeiro, N. D., Maziero, S. M., Possobom, M. T. D. F., Rosa, D. P., & Domingues, L. S. (2013). Comparison among direct, indirect and index selections on agronomic traits and nutritional quality traits in common bean. *Journal of the Science of Food and Agriculture, 93*(5), 1097-1104. doi: 10.1002/jsfa.5856

Leal, R. M., & Prado, R. M. (2008). Desordens nutricionais no feijoeiro por deficiência de macronutrientes, boro e zinco. *Revista Brasileira de Ciências Agrárias, 3*(4), 301-306. doi: 10.5039/agraria.v3i4a332

Ouertatani, S., Regaya, K., Ryan, J., & Gharbi, A. (2011). Soil liming and mineral fertilization for root nodulation and growth of faba beans in an acid soil in Tunisia. *Journal of Plant Nutrition, 34*(6), 850-860. doi: 10.1080/01904167.2011.544354

Pomper, K. W., & Grusak, M. A. (2004). Calcium uptake and whole-plant water use influence pod calcium concentration in snap bean plants. *Journal of the American Society for Horticultural Science, 129*(6), 890-895.

Quintana, J. M., Harrison, H. C., Palta, J. P., Nienhuis, J., & Kmiecik, K. (1999). Calcium fertilizers fail to affect pod calcium concentration and yield of four snap bean cultivars. *HortScience, 34*(4), 646-647.

Ribeiro, N. D., Domingues, L. S., Zemolin, A. E. M., & Possobom, M. T. D. F. (2013). Selection of common bean lines with high agronomic performance and high calcium and iron concentrations. *Pesquisa Agropecuária Brasileira, 48*(10), 1368-1375. doi: 10.1590/S0100-204X2013001000008

Ribeiro, N. D., Jost, E., Maziero, S. M., Storck, L., & Rosa, D. P. (2014). Selection of common bean lines with high grain yield and high grain calcium and iron concentrations. *Revista Ceres, 61*(1), 77-83.

Ribeiro, N. D., Rodrigues, J. A., Prigol, M., Nogueira, C. W., Storck, L., & Gruhn, E. M. (2014). Evaluation of special grains bean lines for grain yield, cooking time and mineral concentrations. *Crop Breeding and Applied Biotechnology, 14*(1), 15-22. doi: 10.1590/S1984-70332014000100003

Schmitt, A., Cobert, F., Bourgeade, P., Ertlen, D., Labolle, F., Gangloff, S., ... Stille, P. (2013). Calcium isotope fractionation during plant growth under a limited nutrient supply. *Geochimica et Cosmochimica Acta, 110*(1), 70-83. doi: 10.1016/j.gca.2013.02.002

Silva, S. A., Moraes, W. B., & Souza, G. S. (2011). Doses de cálcio no crescimento do feijoeiro cultivado em solução nutritiva, na presença de alumínio. *Idesia, 29*(3), 53-58. doi: 10.4067/S0718-34292011000300006

Souza Junior, J. O., Nascimento, C. W. A., & Martinez, H. E. P. (1998). Resposta do feijoeiro cultivado em solução nutritiva a níveis de cálcio e magnésio na presença de alumínio. *Pesquisa Agropecuária Brasileira, 33*(7), 1143-1148.

Tüma, J., Skalicky, M., Tümová, L., Bláhová, P., & Rosülková, M. (2004). Potassium, magnesium and calcium content in individual parts of *Phaseolus vulgaris* L. plant as related to potassium and magnesium nutrition. *Plant Soil and Environment, 50*(1), 18-26.

14

Growth, photosynthetic activity, and potassium and sodium concentration in rice plants under salt stress

Soledad García Morales, Libia Iris Trejo-Téllez*, Fernando Carlos Gómez Merino, Camila Caldana, David Espinosa-Victoria and Braulio Edgar Herrera Cabrera

*Colegio de Postgraduados en Ciencias Agrícolas, Km 36.5, Carretera México-Texcoco, 56230, Montecillo, Texcoco, México.*Author for correspondence. E-mail: tlibia@colpos.mx*

ABSTRACT. Salt stress affects crop growth and productivity. In this study, we determined the growth, yield of photosystem II (PSII), and K^+ and Na^+ concentration in root, stem, old leaves, and young leaves of two Mexican varieties of rice, Tres Ríos and Cotaxtla. In addition, the K^+/Na^+ ratio in stem and root of both varieties was determined. The experiment was conducted in a growth chamber under controlled conditions, under a completely randomized distribution, with a 2 x 2 (Variety x Salinity) factorial arrangement and 12 replications. Plants were grown in a hydroponic solution for 15 days and then some of them were treated with 100 mM NaCl; control plants (without NaCI treatment) were grown in parallel. Salt stress caused 20 and 15% reductions in stem and root length, respectively, in the variety Tres Ríos, while in the variety Cotaxtla no significant differences were observed in these variables compared to the control. Dry matter weight decreased by 24% in the variety Tres Ríos. The quantum yield of PSII decreased by 30% the third day of treatment application, in both varieties. Na^+ concentration was significantly ($p \leq 0.05$) higher in NaCI-treated plants. In the variety Tres Ríos, the yield of PSII was completely eradicated six days after treatment implementation, while the K+ concentration in stem and older leaves also decreased and the lowest K^+/Na^+ ratio in stem was recorded, which could indicate that it is more susceptible to salinity than the variety Cotaxtla.

Keywords: *Oryza sativa*, yield of photosystem II, K^+/Na^+ ratio, salinity.

Introduction

Salinity is a major problem in irrigated and rain-fed agriculture. The irrigated farming system provides about a third of the world's food (MUNNS, 2002) and it is estimated that about 20% of the irrigated area (45 million ha) is affected by salinity (FAO, 2008). Salt stress is also a problem in rain-fed agriculture, especially in coastal areas as salt water enters them during high tide (WALIA et al., 2005). It is estimated that 2% of the rain-fed agriculture area (32 million ha) is affected by salinity (FAO, 2008). The salinity problem has been addressed through improvements in production practices and the introduction of tolerant varieties.

However, the implementation of proper irrigation management in areas affected by salinity has been economically unviable and difficult to implement on a large-scale basis (WALIA et al., 2005).

Osmotic stress and ion toxicity are effects caused by salinity. Osmotic stress is the result of the salt in the growth solution reducing the capacity of the plant to absorb water, while ion toxicity is caused by an excessive amount of salts entering the transpiration flow and damaging leaf cells. Reduced growth and photosynthesis are the main effects of salt stress (MUNNS et al., 2006). The ability of plants to grow under salinity is a feature that determines crop distribution and productivity in many areas; therefore, it is important to understand the mechanisms that confer tolerance in saline environments (PATTANAGUL; THITISAKSAKUL, 2008).

In general, rice shows variability in its sensitivity to excessive salinity at different stages of growth. It is considered a relatively saline-tolerant species in the germination stage, whereas the vegetative and early reproduction stages are the most sensitive to salinity, directly affecting yield (ZENG, 2004).

Several studies have been conducted to identify responses of rice to salinity. Walia et al. (2005) compared phenotypic responses of two *indica* rice genotypes, salt-tolerant FL478 and salt-sensitive IR29, under salinity stress during the vegetative stage of growth. Their results demonstrate that FL478 is salt tolerant and maintains a low Na^+-to-K^+ ratio in shoot tissue, and may be relatively less stressed at the cellular level compared to IR29. Moreover, Moradi and Ismail (2007) evaluated three rice varieties (IR651, IR632 and IR29) during vegetative stage under culture solution and three electrical conductivities (0, 6 and 12 dS m^{-1}). They found that tolerant rice varieties maintained a relatively higher photosynthetic function after a brief period of acclimation following exposure to salt stress. Results of three *indica* rice (Pokkali, Java and IR20) grown under 0, 50 or 100 mM NaCl in hydroponic culture, suggest that differences in the amount of Na^+ entering in shoot, and as a consequence the ability of a rice variety to withstand saline stress, are influenced by the location, amount and chemical composition of apoplastic barriers in the roots (KRISHNAMURTHY et al., 2009).

The aim of this study was to determine the effect of salt stress on the growth and photosynthetic activity of two Mexican rice plant varieties (Tres Ríos and Cotaxtla) during the vegetative stage (tillering), as well as evaluate the K^+ and Na^+ concentration in root, stem, old leaves and young leaves in plants grown under controlled conditions, and also the K^+/Na^+ ratio in shoot and root in order to identify their salinity-tolerance level.

Material and methods

We used two varieties of rice (*Oryza sativa* L. ssp. *indica*), Cotaxtla and Tres Ríos, obtained from the germplasm bank at the experimental field operated by the National Institute of Forestry, Agricultural and Livestock Research in Zacatepec, Morelos. The experiment was conducted at the Max Planck Institute of Molecular Plant Physiology in Golm, Germany, between January and February 2010.

The seeds of both varieties were surface sterilized with 70% ethanol for 5 min. and then disinfected with a solution of 3% sodium hypochlorite and a drop of Tween 20 for 30 min., then washed with sterile water five times and dried in filter paper. The dried embryos were placed in a glass bottle containing MS medium (MURASHIGE; SKOOG, 1962), supplemented with 3% (w/v) sucrose and solidified with 0.8% agar.

For germination, the glass bottles were subjected to darkness for five days at 28°C. Then the germinated seeds were transferred to a tissue culture room under controlled conditions (16h light at 140 μmol m^{-2} s^{-1}, 22°C, 8h dark and 70% relative humidity) for seven days. Twelve days after seeding, the plants were transplanted to hydroponic tanks containing 10 L of nutrient solution containing 1.43 mM NH_4NO_3, 1.00 mM $CaCl_2$ $2H_2O$, 1.64 mM $MgSO_4$ $7H_2O$, 0.13 mM K_2SO_4, 0.32 mM NaH_2PO_4 $2H_2O$, 1.00 Fe-EDTA, 7.99 μM $MnCl_2$ $4H_2O$, 0.15 μM $ZnSO_4$ $7H_2O$, 0.15 μM $CuSO_4$ $5H_2O$, 0.075 μM $(NH_4)_6Mo_7O_{24}$ $4H_2O$ and 1.39 μM H_3BO_3 (YANG et al., 1994). The pH of the nutrient solution was kept between 5.5 and 6.0. The nutrient solution was renewed every 5 d.

Two NaCI concentrations, 0 mM (control) and 100 mM, and two rice varieties (Tres Ríos and Cotaxtla) were evaluated. Salt stress was applied 15 days after transferring plants into hydroponic tanks by adding 100 mM NaCl. Non-stressed plants (control) were grown in parallel.

Photosynthetic activity was determined with the PAM-2000 portable fluorometer (Walz, Germany), without dark adaptation. The measurements were made in the second fully expanded leaf before applying the treatments, 1, 3, 5 and 6 days after initiation of salt stress. Six days after treatment application, plants were harvested and separated into shoot and root to determine their length, fresh weight and dry weight. The roots were rinsed twice to remove excess salts from the nutrient solution.

The dry weight was determined 48h after drying the samples at 70°C in an oven with forced air circulation. To determine the concentration of potassium (K^+) and sodium (Na^+), plants were divided into root, stem, old leaves and young leaves. The dry samples were subjected to wet digestion with a mixture of perchloric and nitric acids (ALCÁNTAR; SANDOVAL, 1999). The extracts obtained were read on a VARIAN™ Liberty Series II inductively coupled plasma-atomic emission spectrometer (ICP-AES), while the K^+/Na^+ ratio was obtained from the content of these ions in shoot and root, based on the concentration and dry matter weight.

The data were analyzed with SAS statistical analysis software ver. 9.0 (SAS INSTITUTE, 1990) under a completely randomized design in a 2 x 2 (variety x concentration) factorial arrangement, resulting from the combination of four treatment levels with 12 replications each. A plant was considered an experimental unit. To determine statistical differences, we performed an analysis of variance and a means comparison using Tukey's test with a significance level of 95%.

Results and discussion

Root and stem growth

Based on the variety factor, there were no significant ($p > 0.05$) differences in stem length; however, due to NaCI concentration and its interaction with the variety, there were highly significant ($p \leq 0.01$) differences. In the case of root length, the three sources of variation presented significant differences (Table 1).

Table 1. Results of analysis of variance (ANOVA) of variety, NaCl concentration (NaCl) and their interaction (Variety x NaCl) for shoot length and root length of rice plants.

Source of variation	DF	Length	
		Shoot	Root
Variety	1	ns	**
NaCl concentration	1	**	**
Variety x NaCl	1	**	*

ns = Not significant; * = Significant at 5% level; ** = Significant at 1% level.

Application of 100 mM NaCl significantly ($p \leq 0.05$) reduced shoot and root length in the variety Tres Ríos (20 and 15% respectively, compared to the control) as shown in Figure 1, while in the variety Cotaxtla there were no significant ($p > 0.05$) reductions in shoot and root length in the presence of NaCl. Similar results have been obtained in other rice varieties, namely BR11, BR23, BR29 and BR 31 (ALAM et al., 2004) treated with 0.0, 4.5, 8.5 or 12.5 dS m^{-1} (approximately equivalent to 0, 45, 85 and 125 mM NaCl, respectively). In the variety IR36, 14-days-old seedlings treated with 50 mM NaCl for 7 days showed damage symptoms caused by salinity mainly in old leaves, including rolled leaves with tips withered and necrosis(GONG et al., 2006), similar to those observed in Figure 2, which is explained by accumulation of toxic concentrations of Na^+ in older leaves (MUNNS; TESTER, 2008).

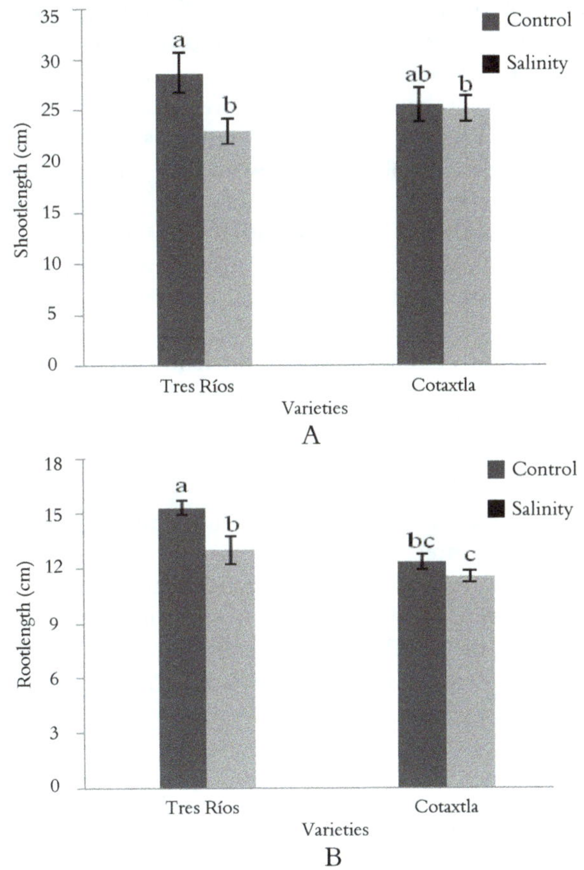

Figure 1. Effect of salinity (100 mM NaCl) on the shoot length (A) and root length (B) of rice plants of Mexican varieties (Tres Ríos and Cotaxtla), after six days of treatment.

Means with different letters, in each subfigure, are significantly different (Tukey, 0.05). Vertical bars represent standard errors (n=12).

Figure 2. Rice plants of two Mexican varieties Tres Ríos and Cotaxtla, 30 days after germination and six days after the start of salt treatment (100 mM de NaCl).

Jamil et al. (2006) found a direct relationship between reduced shoot and root length with increased salt concentration in the growth medium, which was also observed in the variety Tres Ríos in our experiments. They suggest that the above-mentioned variables are important indicators of salt stress tolerance because the root is in direct contact with the growth medium and absorbs water and other components of the media, making them available to the rest of the plant. Therefore, shoot and root length is an indicator of plant response to salinity. Furthermore, salinity problems are generally related to the presence of high concentration of Na^{+} and Cl^{-}, and excessive salt in soil or nutrient solution causes a decrease in nutrient uptake by plant such as N by bean (CHAKRABARTI; MUKHERJI, 2003) and rice (ABDELGADIR et al., 2005). Also, ionic imbalance is an effect of salt stress, which occurs in the cells due to excessive Na^{+} and Cl^{-} and reduces the absorption of other mineral nutrients, such as K^{+}, Ca^{2+} and Mn^{2+} (HASEGAWA et al., 2000; SUDHIR; MURTHY, 2004).

Biomass production

In shoot and root fresh weight, highly significant ($p \leq 0.01$) effects were found based on the source of variation, variety and NaCI concentration. Similar results were obtained in root fresh weight with the effect of the variety x NaCI interaction, contrary to that recorded in shoot where no significant difference was found (Table 2).

Table 2. Analysis of variance of variety, NaCl concentration (NaCl) and their interactions (variety x NaCl) for the shoot and root fresh weight in rice plants.

Source of variation	DF	Fresh weight	
		Shoot	Root
Variety	1	**	**
NaCl concentration	1	**	**
Variety x NaCl	1	ns	**

ns = Not significant; * = Significant at 5% level; ** = Significant at 1% level.

Salt stress (100 mM NaCl) significantly reduced shoot fresh weight in both varieties (Figure 3A). The variety Tres Ríos had 31% less shoot fresh weight and the variety Cotaxtla had 22% less, in both cases compared to the control. With respect to the variable root fresh weight (Figure 3B), only the variety Tres Ríos was affected, showing a significant ($p \leq 0.05$) decrease of 27% compared with the control.

The variety Tres Ríos had a significant reduction in shoot and root dry matter weight (Figure 4). The decrease was 24% in both variables compared with the control. This was due to the variety x concentration interaction, since no significant ($p > 0.05$) differences were found as a result of variety factor or the NaCl concentration. Similar results have been obtained in other rice varieties, BR11, BR23, BR29 and BR31 (ALAM et al., 2004), wheat (SAQIB et al., 2005) and sweet sorghum (ALMODARES et al., 2007). In the case of sweet sorghum, differences in salinity tolerance in the vegetative state are due to genetic diversity and inherited differences among varieties. It was also observed that by increasing the concentration of sodium, the fresh and dry matter weight of the plants decreased. Some species are sensitive to salinity, especially in the vegetative phase, because the mechanisms of stress tolerance have not been fully developed. However, other species may show tolerance to salinity during the vegetative phase of development (ROY et al., 2005).

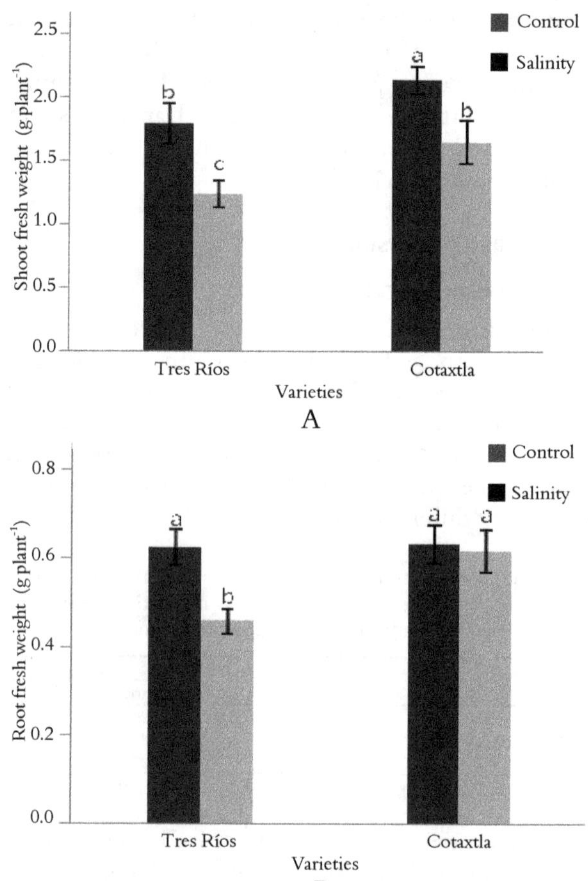

Figure 3. Shoot fresh weight (A) and root fresh weight of rice plants of two Mexican varieties (Tres Ríos and Cotaxtla), growing under salinity (100 mM NaCl) for six days.

Means with different letters, in each subfigure, are significantly different (Tukey, 0.05). Vertical bars represent standard errors (n=12).

The variety Cotaxtla showed no differences with respect to shoot and root dry matter weight, six days after the application of 0 and 100 mM NaCl. This response has been reported in other rice varieties (IR632, IR651 and IR29) in experiments in which plants subjected to salt stress for one week show no negative effects. However, after two weeks of treatment with salts, growth was significantly reduced in the three varieties (MORADI; ISMAIL, 2007).

Kumar et al. (2009) evaluated 11 varieties of rice under saline conditions and found that tolerant varieties have greater plant height and dry matter weight. In this context, the variety Cotaxtla neither had the highest shoot and root length values, nor exhibited significant ($p > 0.05$) reductions in plant length, root fresh weight, shoot dry weight and root weight, which could indicate that it is a salt-tolerant variety.

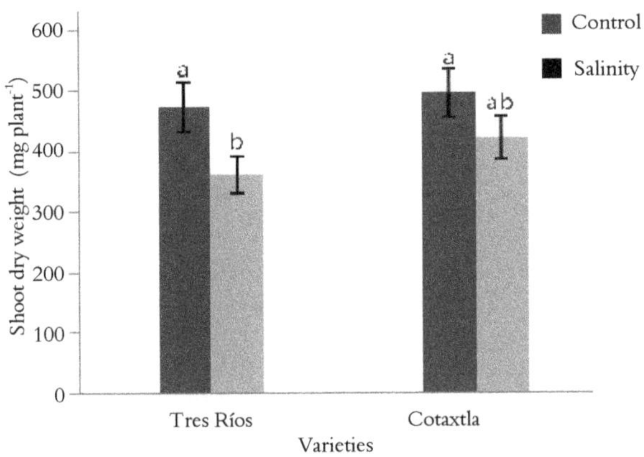

A

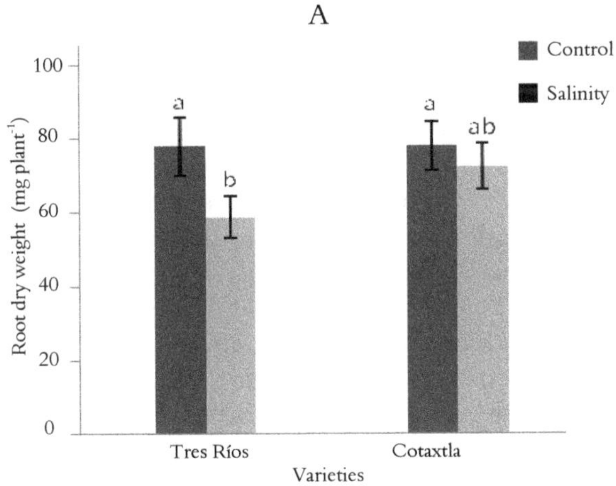

B

Figure 4. Effect of salinity on shoot dry weight (A) y root dry weight (B) of plant of the Mexican rice varieties (Tres Ríos and Cotaxtla), growing in the presence of 100 mM NaCl for six days. Means with different letters, in each subfigure, are significantly different (Tukey, 0.05). Vertical bars represent standard errors (n=12).

Photosynthetic activity

Plants subjected to salt stress have reduced photosynthetic efficiency, which seems to be associated with the complex of photosystem II (PSII) (MUNNS et al., 2006). Salinity reduces PSII activity and promotes the destruction of chlorophyll pigments by the accumulation of ions that may inhibit the quantum yield of electron transport in PSII (SUDHIR; MURTHY, 2004).

The above is seen in Figure 5, which shows the quantum yield of PSII in the second fully expanded leaf. When plants grew in the absence of salinity, both varieties had a high yield of PSII. However, when plants were subjected to salt stress (100 mM NaCl), a reduction of photosynthetic yield was observed three days after application of treatments in both varieties. Six days after application of treatments, the variety Tres Ríos showed no yield of PSII, while the variety Cotaxtla had a 90% reduction compared with the control.

These results differ from those reported in other rice varieties (IR651 and IR632) tolerant to salinity (MORADI; ISMAIL, 2007), and those obtained in sorghum (NETONDO et al., 2004). In both cases, there were no significant changes in the quantum yield of PSII under saline conditions, which may indicate that salinity does not severely affect the transport of electrons in sorghum (NETONDO et al., 2004). Rice varieties IR651 and IR632 can serve as an indirect way of determining the effect of salinity on the photosynthetic apparatus (MORADI; ISMAIL, 2007).

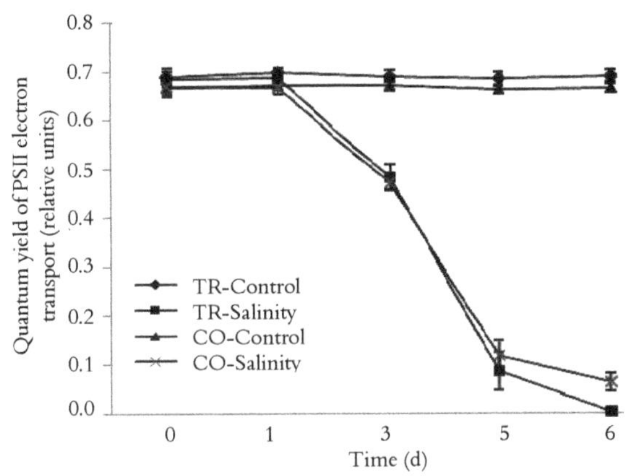

Figure 5. Changes in the quantum yield of PSII electron transport on the second fully expanded leaves of Mexican rice varieties (Tres Ríos and Cotaxtla) growing with 100 mM NaCl. Vertical bars represent standard errors (n=12).

Table 3 presents the simple effects and interaction of the factors evaluated on the quantum yield of PS II. For the variety factor, significant differences were observed on days 0, 1, 5 and 6 after the application of treatments, whereas for NaCl concentration, effects were significant at 1% on days 3, 5 and 6. Finally, the variety x NaCl concentration interaction was highly significant during the fifth and sixth days of salt stress.

Table 3. Results of analysis of variance (ANOVA) of variety, NaCl concentration (NaCl) and their interaction (Cultivar) for quantum yield of PSII in lives of rice plants.

Source of variation	DF	Time days				
		0	1	3	5	6
Variety	1	**	**	*	ns	**
NaCl concentration	1	ns	ns	**	**	**
Variety x NaCl	1	ns	ns	ns	**	**

ns = Not significant; * = Significant at 5% level; ** = Significant at 1% level.

Sodium and potassium concentration

K^+ concentration in different rice plant organs was significantly different due to the variety x NaCl concentration interaction (Table 4). In root, there were differences due to the variety factor, with Tres Ríos having a higher K^+ concentration than Cotaxtla, with and without the application of NaCl in the growth medium. In stem, the statistical differences observed are due to the NaCl concentration factor, i.e. the presence of salts significantly reduced K^+ concentration in stem. Differences were also caused by the variety x NaCl interaction, since while in the variety Tres Ríos the K^+ concentration in stem decreased significantly, it increased in the Cotaxtla variety. The same behavior was obtained in the concentration of old leaves in both varieties.

Table 4. K^+ concentration (mg kg^{-1} dry weight) on root, stem and leaves of two Mexican varieties Tres Ríos and Cotaxtla, six days after the start of salt stress (100 mM NaCl).

Varieties	NaCl (mM)	K^+ concentration (mg kg^{-1} dry weight)			
		Root	Stem	Old leaves	Young leaves
Tres Ríos	0	66.57 a	50.86 a	56.59 a	30.65 b
Tres Ríos	100	67.41 a	11.81 d	27.22 b	55.62 a
Cotaxtla	0	28.09 b	18.92 c	18.63 b	28.65 b
Cotaxtla	100	22.22 b	41.87 b	50.72 a	54.12 a
Varieties					
Tres Ríos		66.99 a	31.34 a	41.91a	43.14 a
Cotaxtla		25.16 b	30.39 a	34.67 a	41.39 a
NaCl (mM)					
0		47.33 a	34.89 a	37.61 a	29.65 b
100		44.82 a	26.84 b	38.97 a	54.87 a
$LSD_{0.05}$					
Variety (V)		4.55	22.74	23.75	8.07
NaCl		29.77	22.03	24.27	3.11
V x NaCl		7.87	6.38	22.08	6.47
CV		6.54	7.90	22.06	5.85

Means with different letters, in each column, are significantly different (Tukey, 0.05).

The reduction in the K^+ concentration obtained in the variety Tres Ríos is similar to that obtained in other rice varieties (WALIA et al., 2005) and wheat (SAQIB et al., 2005; SHARMA et al., 2005). This is due to the presence of NaCl in the growth medium, which can interfere with the absorption of cations like K^+, Ca^{2+} or Mg^{2+}, or as the result of the substitution of K^+ by Na^+ (TESTER; DAVENPORT, 2003). K^+ concentration in young leaves increased significantly by applying salinity, in both varieties, as an effect of NaCl concentration and its interaction with the variety factor.

As a result of the variety x NaCI concentration interaction (Table 5), by applying 100 mM NaCl in the hydroponic solution, Na^+ concentration was significantly higher in root, stem and leaves of both varieties, compared with the control. These results are consistent with those obtained in other rice varieties, FL478 and IR29 (WALIA et al., 2005), wheat (SAQIB et al., 2005; SHARMA et al., 2005), soybean and cucumber (DABUXILATU; IKEDA, 2005), where there was a positive relationship between salt concentration and Na^+ concentration in root and stem.

Table 5. Na^+ concentration (mg kg^{-1} dry weight) on root, stem and leaves of two Mexican rice varieties Tres Ríos and Cotaxtla, six days after the start of salt treatment.

Varieties	NaCl (mM)	Na^+ concentration (mg kg^{-1} dry weight)			
		Root	Stem	Old leaves	Young leaves
Tres Ríos	0	3.56 b	3.50 c	3.45 b	2.77 c
	100	26.37 a	13.88 b	17.15 a	9.87 b
Cotaxtla	0	2.88 b	1.41 c	1.84 b	2.17 c
	100	21.25 a	37.14 a	27.25 a	16.65 a
Varieties					
Tres Ríos		14.97 a	8.69 b	10.30 a	6.32 b
Cotaxtla		12.07 a	19.28 a	14.55 a	9.41 a
NaCl (mM)					
0		3.22 b	2.46 b	2.65 b	2.47 b
100		23.81 a	25.51 a	22.20 a	13.26 a
$LSD_{0.05}$					
Variety (V)		15.35	18.58	15.07	17.99
NaCl		5.40	11.70	6.81	3.48
V x NaCl		10.79	2.65	10.25	1.76
CV		30.54	7.25	31.56	8.55

Means with different letters, in each column, are significantly different (Tukey, 0.05).

Tester and Davenport (2003) indicate that many plants can tolerate salt stress by excluding Na^+ from the stem, or at least from the leaves, and maintaining high levels of K^+. This could be happening in the variety Cotaxtla, since in saline conditions, it had a higher concentration of Na^+ in stem than Tres Ríos, perhaps as a mechanism for excluding Na^+ from the leaves. In addition, under the same conditions, this variety had a significant increase in the K^+ concentration (Table 4) in old leaves and stem, while Tres Ríos showed reductions.

Na^+ concentration was higher in young leaves treated with salinity in both varieties. However, the variety Cotaxtla recorded a higher Na^+ concentration in young leaves compared with the variety Tres Ríos. Under some conditions, a high Na^+ concentration in stem may be beneficial in helping the plant to maintain its turgor, since the ions are sequestered in the vacuole and can thus reduce the toxicity of Na^+ (MUNNS; TESTER, 2008).

The ability of plants to discriminate between Na^+ and K^+ can be determined by an index, the K^+/Na^+ ratio (FLOWERS, 2004). Figure 6 shows the K^+/Na^+ content ratio in shoot and root of the two rice varieties studied. In general, it was found that by applying 100 mM NaCl in the hydroponic solution, the K^+/Na^+ ratio was significantly ($p \leq 0.05$)

reduced in both varieties. Similar results have been obtained in other rice varieties, FL478 and IR29 (WALIA et al., 2005) and IR36 (GONG et al., 2006), and in wheat (SAQIB et al., 2005), Sesbania rostrata and beans (JUNGKLANG et al., 2003). In shoot, the variety Tres Ríos had the lowest K^+/Na^+ ratio when the plants were subjected to stress. The information presented here is similar to that reported for other rice varieties that differ in salt tolerance (MORADI; ISMAIL, 2007). These authors found that the saline-susceptible variety IR29 had a K^+/Na^+ ratio in leaves significantly less than the IR632 and IR651 varieties (tolerant).

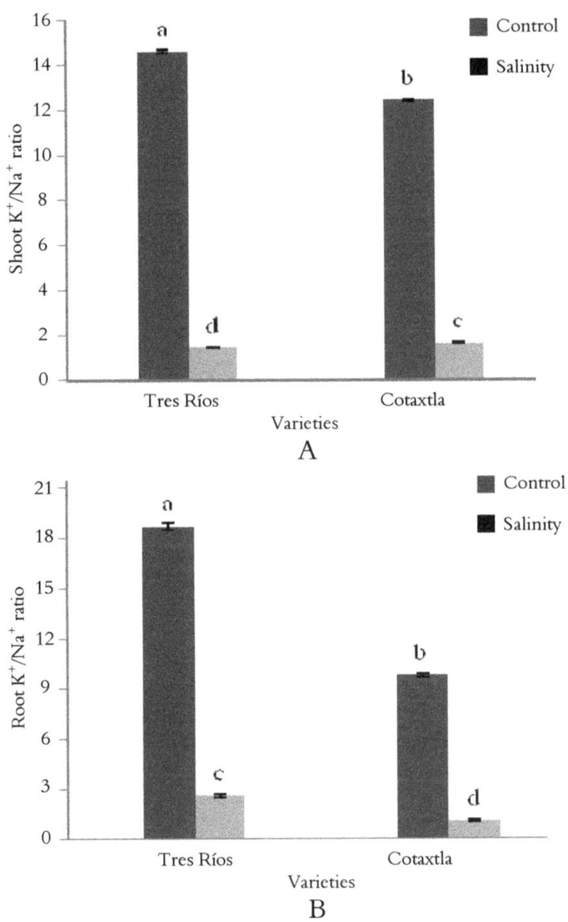

Figure 6. Effect of salinity on K^+/Na^+ ratio in shoot (A) and root (B) of Mexican rice plants Tres Ríos and Cotaxtla. After six days of exposure to 100 mM NaCl.

Means with different letters, in each subfigure, are significantly different (Tukey, 0.05). Vertical bars represent standard errors (n=4).

Conclusion

The application of 100 mM NaCl in growth medium for six days caused significant reductions in the variety Tres Ríos in shoot and root length, and fresh and dry matter weight. Moreover, it led to a significant decrease in the quantum yield of photosystem II, lower K^+ concentration in stem and leaves and, consequently, a lower K^+/Na^+ ratio in the stem, suggesting that this cultivar is more susceptible to salinity than the variety Cotaxtla, which showed greater tolerance to this stress.

Acknowledgements

We thank the National Council of Science and Technology of Mexico (CONACYT) for the award to SGM, and to LPI5 Microbial, Plant and Animal Biotechnology research line from Colegio de Postgraduados for the research facilities.

References

ABDELGADIR, E. M.; OKA, M.; FUJIYAMA, H. Nitrogen nutrition of rice plants under salinity. **Biologia Plantarum**, v. 49, n. 1, p. 99-104, 2005.

ALAM, M. Z.; STUCHBURY, T.; NAYLOR, R. E. L.; RASHID, M. A. Effect of salinity on growth of some modern rice cultivars. **Journal of Agronomy**, v. 3, n. 1, p. 1-10, 2004.

ALCÁNTAR, G. G.; SANDOVAL, V. M. **Manual de análisis químico de tejido vegetal**. Chapingo: Sociedad Mexicana de la Ciencia del Suelo A.C., 1999. (Publicación especial No. 10).

ALMODARES, A.; HADI, M. R.; DOSTI, B. Effects of salt stress on germination percentage and seedling growth in sweet sorghum cultivars. **Journal of Biological Science**, v. 7, n. 8, p. 1492-1495, 2007.

CHAKRABARTI, N.; MUKHERJI, S. Effect of phytohormone pretreatment on nitrogen metabolism in *Vigna radiata* under salt stress. **Biologia Plantarum**, v. 46, n. 1, p. 63-66, 2003.

DABUXILATU, M.; IKEDA, M. Interactive effect of salinity and supplemental calcium application on growth and ionic concentration of soybean and cucumber plants. **Soil Science and Plant Nutrition**, v. 51, n. 4, p. 549-555, 2005.

FAO-Food and Agriculture Organization. **Global network on integrated soil management for sustainable use of salt-effect soils**. Rome: FAO Land and Plant Nutrition Management Service, 2008.

FLOWERS, T. J. Improving crop salt tolerance. **Journal of Experimental Botany**, v. 55, n. 396, p. 307-319, 2004.

GONG, H. J.; RANDALL, D. P.; FLOWERS, T. J. Silicon deposition in the root reduces sodium uptake in rice (*Oryza sativa* L.) seedlings by reducing bypass flow. **Plant, Cell and Environment**, v. 29, n. 10, p. 1970-1979, 2006.

HASEGAWA, P. M.; BRESSAN, R. A.; ZHU, J.-K.; BOHNERT, H. J. Plant cellular and molecular response to high salinity. **Annual Review of Plant Physiology and Plant Molecular Biology**, v. 51, p. 463-499, 2000.

JAMIL, M.; DEOG BAE, L.; KWANG YONG, J.; ASHRAF, M.; SHEONG CHUN, L.; EUI SHIK, R. Effect of salt (NaCl) stress on germination and early seedling growth of four vegetables species. **Journal of Central European Agriculture**, v. 7, n. 2, p. 273-282, 2006.

JUNGKLANG, J.; USUI, K.; MATSUMOTO, H. Differences in physiological responses to NaCl between salt-tolerant *Sesbania rostrata* Brem. & Oberm. and non-tolerant *Phaseolus vulgaris* L. **Weed Biology and Management**, v. 3, n. 1, p. 21-27, 2003.

KRISHNAMURTHY, P.; RANATHUNGE, K.; FRANKE, R.; PRAKASH, H. S.; SCHREIBER, L.; MATHEW, M. K. The role of root apoplastic transport barriers in salt tolerance of rice (*Oryza sativa* L.). **Planta**, v. 230, n. 1, p. 119-134, 2009.

KUMAR, B. S.; TITOV, S.; MOFFAZZAL ISLAM, M.; SIDDIKA, A.; SULTANA, S.; SHAHIDUL HAQUE, M. Phenotypic and genotypic screening of rice genotypes at seedling stage for salt tolerance. **Global Journal of Biotechnology and Biochemistry**, v. 4, n. 2, p. 126-131, 2009.

MORADI, F.; ISMAIL, A. M. Responses of photosynthesis, chlorophyll fluorescence and ROS-scavenging systems to salt stress during seedling and reproductive stages in rice. **Annals of Botany**, v. 99, n. 6, p. 1161-1173, 2007.

MUNNS, R. Comparative physiology of salt and water stress. **Plant, Cell and Environment**, v. 25, n. 2, p. 239-250, 2002.

MUNNS, R.; TESTER, M. Mechanisms of salinity tolerance. **Annual Review of Plant Biology**, v. 59, n. 1, p. 651-681, 2008.

MUNNS, R.; JAMES, R. A.; LÄUCHLI, A. Approaches to increasing the salt tolerance of wheat and other cereals. **Journal of Experimental Botany**, v. 57, n. 5, p. 1025-1043, 2006.

MURASHIGE, T.; SKOOG, F. A revised medium for rapid growth and bio assays with tobacco tissue cultures. **Physiologia Plantarum**, v. 15, n. 43, p. 473-497, 1962.

NETONDO, G. W.; ONYANGO, J. C.; BECK, E. Sorghum and salinity II. Gas exchange and chlorophyll fluorescence of sorghum under salt stress. **Crop Science**, v. 44, n. 3, p. 806-811, 2004.

PATTANAGUL, W.; THITISAKSAKUL, M. Effect of salinity stress on growth and carbohydrate metabolism in three rice (*Oryza sativa* L.) cultivars differing in salinity tolerance. **Indian Journal of Experimental Biology**, v. 46, n. 10, p. 736-742, 2008.

ROY, P. K.; NIYOGI, K.; SENGUPTA, D. N.; GHOOSH, B. Spermidine treatment to rice seedlings recovers salinity stress-induced damage of plasma membrane and PM-bound H^+-ATPase in salt-tolerant and salt-sensitive rice cultivars. **Plant Science**, v. 168, n. 3, p. 583-591, 2005.

SAQIB, M.; AKHTAR, J.; QURESHI, R. H. Na^+ exclusion and salt resistance of wheat (*Triticum aestivum*) in saline-waterlogged conditions are improved by the development of adventitious nodal roots and cortical root aerenchyma. **Plant Science**, v. 169, n. 1, p. 125-130, 2005.

SAS INSTITUTE. **SAS/STAT User´s guide**. Cary: SAS Institute, 1990. (Ver. 6.0).

SHARMA, N.; GUPTA, N. K.; GUPTA, S.; HASEGAWA, H. Effect of NaCl salinity on photosynthetic rate, transpiration rate, and oxidative stress tolerance in contrasting wheat genotypes. **Photosynthetica**, v. 43, n. 4, p. 609-613, 2005.

SUDHIR, P.; MURTHY, S. D. S. Effects of salt stress on basic processes of photosynthesis. **Photosynthetica**, v. 42, n. 4, p. 481-486, 2004.

TESTER, M.; DAVENPORT, R. Na^+ transport and Na^+ tolerance in higher plants. **Annals of Botany**, v. 91, n. 5, p. 503-527, 2003.

WALIA, H.; WILSON, C.; CONDAMINE, P.; LIU, X.; ISMAIL, A. M.; ZENG, L.; WANAMAKER, S. I.; MANDAL, J.; XU, J.; CUI, X.; CLOSE, T. J. Comparative transcriptional profiling of two contrasting rice genotypes under salinity stress during the vegetative growth stage. **Plant Physiology**, v. 139, n. 2, p. 822-835, 2005.

YANG, X.; RÔMHELD, V.; MARSCHNER, H. Effect of bicarbonate on root-growth and accumulation of organic-acids in Zn-Inefficient and Zn-efficient rice cultivars (*Oryza* sativa L.). **Plant and Soil**, v. 164, n. 1, p. 1-7, 1994.

ZENG, L. Response and correlated response to salt tolerance selection in rice by yield parameters. **Cereal Research Communications**, v. 32, n. 4, p. 477-484, 2004.

15

Responses of quinoa (*Chenopodium quinoa* Willd.) seeds stored under different germination temperatures

Andressa Strenske[*], Edmar Soares de Vasconcelos, Vanessa Aline Egewarth, Neusa Francisca Michelon Herzog and Marlene de Matos Malavasi

*Centro de Ciências Agrárias, Universidade Estadual do Oeste do Paraná, Rua Pernambuco, 1777, 85960-000, Marechal Cândido Rondon, Paraná, Brazil. *Author for correspondence. E-mail: andressastrenske@hotmail.com*

ABSTRACT. In this experiment, we assessed the germination and vigor of quinoa seeds packed in paper bags and stored at room temperature for 36, 85, 119, 146, 177 and 270 days. The seeds were harvested under experimental conditions in Marechal Candido Rondon, Paraná, during the 2012/13 growing seasons. Four replicates of 100 seeds each were established for each storage time, and the seeds were evaluated, on paper, based on the BOD under the following experimental temperature conditions: alternating temperatures of 20 and 30°C and a constant temperature of 25°C. The seeds from both treatments were subject to seven-hour photoperiods and 25°C under continuous darkness. The germinated seeds were counted daily for eight days after sowing, and we evaluated the percentages of normal and abnormal seedlings and the germination index. The experimental design was completely randomized using a split-plot design. Increasing the storage time decreased the percentage of germinated seeds and seed vigor due to the increased number of abnormal seedlings. Over the 430-day study period, quinoa seed germination completely declined under the experimental conditions. The final number of germinating seeds should be evaluated 7 days after the beginning of the germination test.

Keywords: Germination test, temperature, photoperiod, seed vigor.

Introduction

Quinoa (*Chenopodium quinoa* Willd.) produces edible grains with high nutritional value for humans and animals. This grain is highly important in the human diet because it is free of gluten and used by people with allergies (Spehar, 2007), but it also contains protein, amino acids, iron, zinc and other nutrients (Gewehr, Danelli, Melo, Flôres, & Jong, 2012). The composition of the seed allows for greater storage potential than that of other oil seeds due to the greater chemical stability of the starches and lipids (Marcos Filho, 2015).

However, quinoa seeds lose viability more rapidly than cereals because of the porosity in the integument, which allows a seed to easily gain or lose moisture and may initiate germination in the panicle (Spehar, 2007).

Quinoa seed quality is influenced by the low germination rate and reduced vigor (Kappes et al., 2012), which consequently reduce longevity; i.e., lower seed quality reduces the germination percentage.

Seed quality is characterized by the sum of the genetic, physical, physiological and sanitary attributes that determine seed performance when sown or stored (Santos, Póla, Barros, & Prete, 2007). The degree of germination is influenced by intrinsic and extrinsic factors, the latter of which include temperature, relative humidity, oxygen, and the action of fungi and bacteria (Marcos Filho, 2015).

Seed germination is regulated by the interaction between physiological qualities and environmental conditions in that every plant species has specific germination requirements in terms of the availability of water, temperature, and light as well as sowing depth (Carvalho & Nakagawa, 2012).

In seed analysis, vigor tests have mainly been used to identify differences associated with the performance of batches of seed during storage and after sowing with the goal of highlighting more effective batches for stand establishment under a wide range of environmental conditions (Marcos Filho, 2015).

The germination test evaluates physiological quality, but the results do not always correspond to emergence in the field. Vigor parameters, such as germination speed indexes and emergence potential, are used to complement the information obtained in the germination test to evaluate seed performance and the development speed of normal seedlings under a wide diversity of environment conditions (Franzin, Menezes, Garcia, & Wrasse, 2004).

Considering the expansion of the area under quinoa cultivation, the growing popularity of the grain in domestic and foreign markets, and the deficit of scientific information about the physiological quality of stored seeds, the current study aimed to evaluate the influence of storage period and different temperature conditions on quinoa seed germination.

Material and methods

This experiment was carried out in the Seed Technology Laboratory located at the Marechal Candido Rondon campus of the State University of Western Paraná (UNIOESTE). Seeds were harvested during the 2012/2013 growing season on the experimental farm of Doctor Antonio Carlos dos Santos Pessoa in the western region of Paraná State, Brazil, at 24° 33' SL and 54° 31' WL and an altitude of 235 m (Instituto Agronômico do Paraná [IAPAR], 2004).

Seeds at physiological maturity were manually collected and arranged in a single layer to achieve a moisture content of approximately 14%. After processing, they were stored in Kraft paper bags under different environmental conditions for periods of 85, 119, 146, 177, and 270 days.

For the germination tests, gerbox boxes were cleaned and rinsed with distilled water, and the filter paper was autoclaved at 120°C to minimize contamination. Seeds were selected and manually graded to eliminate any that were damaged or deformed. We evaluated four replicates of 100 seeds for each storage time. The substrate in the germination boxes (11 cm x 11 cm x 3.5 cm) was moistened with water prior to the test to 2.5 times the weight of the dry paper and placed in twin BOD (Biochemical Oxygen Demand) chambers. The experimental treatments included alternating temperatures of 20 and 30°C and a constant temperature of 25°C under a photoperiod of seven hours for both thermal conditions. Furthermore, one of the BOD chambers was under a controlled temperature of 25°C under darkness.

The germinated seeds were counted from the third day after sowing, and the experiment was finished when no further radical protrusions were observed after an average of six days. In the germination test, we evaluated the percentage of normal seedlings as recommended by the Rules of Seed Germination, and we considered germinated seeds to be those that developed a primary root and a normal shoot.

The germination index was calculated as the sum of the number of seeds germinated each day divided by the number of days between sowing and germination, following the equation by Maguire (1962): $GI = (G1/N1) + (G2/N2) + (G3/N3) + \ldots + (Gn/Nn)$, where:

GI was the germination index;

G1, G2, G3, ..., Gn were the number of seedlings that germinated on the first, second, third, etc. until the nth day of counting;

N1, N2, N3, ..., Nn were the first, second, third, etc. to the nth day after sowing.

The data were submitted to a normality test, analysis of variance and a response surface using program Genes (Cruz, 2013).

Results and discussion

In the germination test with alternating temperatures between 20 and 30°C and a photoperiod of 7 hours, there was a linear decrease in the germination of quinoa seeds in relation to storage period over all of the counting days (Figure 1).

The relationship cumulative germination percentage rate with the day of counting was

0.00141 to 0.3616* counting days, according to the derived relationship between cumulative germination and the number of counting days. This variation increased when the number of counting days during the germination test increased and decreased when the seeds were stored for a longer period of time.

Seven days after sowing, the seeds stored for a 270-day period had limited cumulative germination, while the batch composed of seeds stored for shorter periods exhibited germination until the eighth day after the initiation of the germination test. This result is related to seed vigor because greater vigor increases the probability of successful germination. A similar result was observed by Nobre et al. (2013), who noted that amaranth seeds showed maximum germination and vigor to the point of maturity and then began to decline.

The storage temperature and relative humidity are the main factors affecting the reduction of physiological seed quality (Goldfarb & Queiroga, 2013). Strenske, Vasconcelos, Herzog, and Malavasi (2015) observed that the germination of stored quinoa seeds was maintained under controlled environmental conditions. The lack of controlled temperature and humidity during the storage period in this assay may explain the decline in the cumulative germination of quinoa seeds.

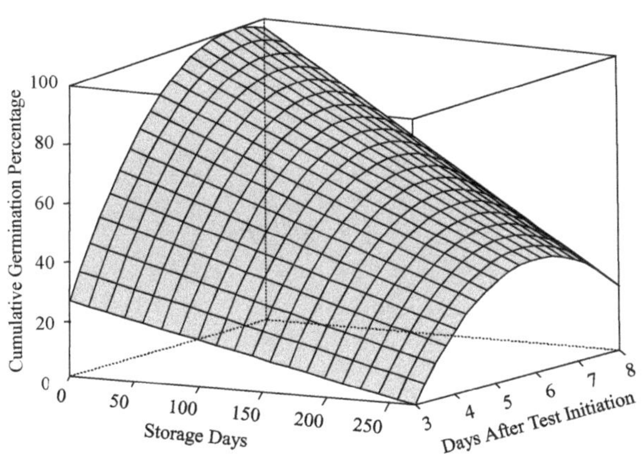

$Z = -108.51935 - 0.00141x + 60.64499y - 4.22099y^2 - 0.03616xy$ $R^2=83.4\%$

Figure 1. Cumulative germination of quinoa seeds under alternating temperature conditions of 20 and 30°C and a photoperiod of 7 hours in relation to the storage time (days after harvest) and the number of days for germination after the initiation of the germination test at Marechal Candido Rondon, Paraná State, 2013/2014.

When the quinoa seed germination test was conducted at a constant temperature of 25°C and a photoperiod of seven hours, a linear reduction in cumulative germination with the storage period was also observed, but it was less dependent on the number of counting days (Figure 2). Therefore, the rate of variation in the cumulative germination was -0.09014 -0.01901* counting days. Under these experimental conditions the value multiplied by the number of counting days is less than the length of the germination test, resulting in a lower effect of the number of counting days.

Regardless of the length of the storage period, germination occurred seven days after sowing, which is relatively fast; the protrusion of the radicle is observed within the first few days. The data obtained during the implementation of this test are similar to the observations of Oliveira and Gomes-Filho (2009) for sorghum and Bonacin, Rodrigues, Fernandes, and Rodrigues (2006) for alfalfa.

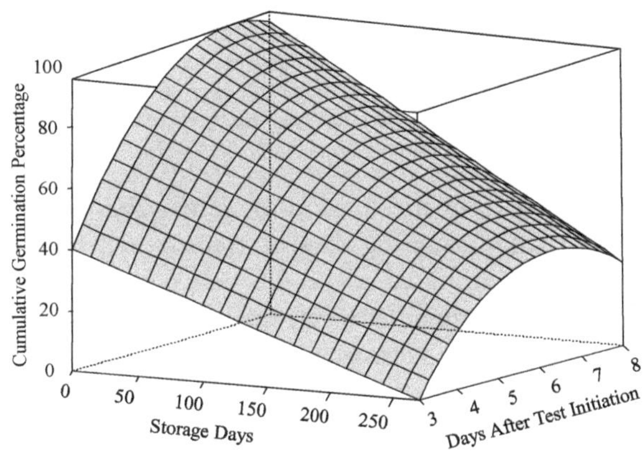

$Z = -67.04085 - 0.09014x + 47.94891y - 3.407y^2 - 0.01901xy$ $R^2 = 82\%$

Figure 2. Cumulative germination of quinoa seeds under a constant temperature of 25°C and a photoperiod of 7 hours relative to the storage time (days after harvest) and number of days for germination after the initiation of the germination test at Marechal Candido Rondon, Paraná State, 2013/2014.

When the seed germination test was conducted at a constant temperature of 25°C without a photoperiod, there was a linear reduction in the cumulative germination of quinoa seeds during the storage period that was also less dependent on the number of counting days (Figure 3) compared to the test with alternating temperatures and a seven-hour photoperiod. Thus, the rate of variation in cumulative germination percentage in relation to the number of counting days was -0.13211 - 0.01483* counting days.

Seven days after the start of the germination test, the cumulative seed germination was already limited at any storage period. In contrast, under the germination test condition of a constant temperature of 25°C without a photoperiod, the overall germination of the quinoa seeds could be assessed seven days after start of the test.

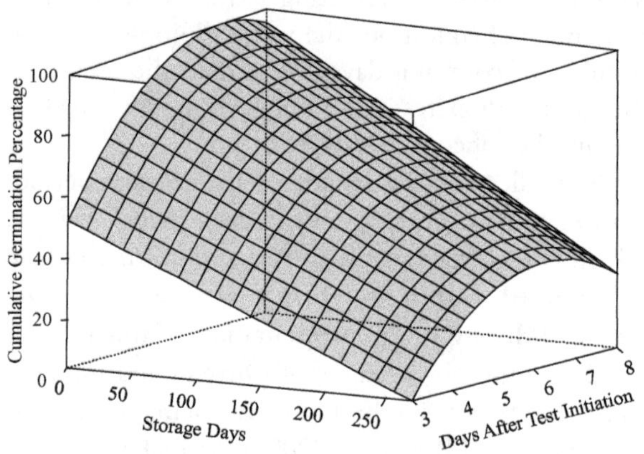

$Z = -49.22657 - 0.13211x + 43.9303y - 3.20239y^2 - 0.01483xy$
$R^2 = 84.8\%$

Figure 3. Cumulative germination of quinoa seeds under a constant temperature of 25°C without a photoperiod in relation to the storage time (days after harvest) and the number of days for germination after the initiation of the germination test at Marechal Candido Rondon, Paraná State, 2013/2014.

Starting from the maximum value obtained for the first three counting days of seeds stored for 36 days, which appeared to be the best condition for quinoa germination, 51% germination was observed in the test with a constant temperature of 25°C without a photoperiod (Figure 3) followed 42% germination of seeds placed at a constant temperature of 25°C with a seven-hour photoperiod (Figure 2). The germination test with alternating temperatures of 25 and 30°C with a seven-hour photoperiod produced the lowest germination, 34%, three days after the initiation of the test (Figure 1).

The number of abnormal plants increased by 5.8% after 100 days of storage when the quinoa seeds were subjected to alternating temperatures of 20 and 30°C and a photoperiod of seven days, and 7.6% abnormal plants were observed after 100 days of storage when the germination test consisted of a constant temperature of 25°C without a photoperiod. When the seeds were subjected to germination at a constant temperature of 25°C and a seven-hour photoperiod, the number of abnormal seedlings had a cubic relationship with the number of days of storage (Figure 4). The increase in the number of abnormal seedlings with the increase in storage period has also been observed by Nakagawa, Cavariani, and Toledo (2009) in pigeon pea seeds.

Marcos Filho (2015) found that seeds deteriorate, which is considered to be any negative change after the seed has reached its maximum quality, when exposed to long periods of storage in uncontrolled conditions. The mechanisms responsible for these changes have not yet been fully elucidated, and the sensitivity of seeds to decay in a particular environment has been attributed to genetic constitution. In addition, the temperature and relative humidity during this period are of great importance to the process of deterioration, which cannot be avoided but can be minimized through storage under appropriate conditions.

The causes of deterioration are most likely related to the loss of the integrity of the membrane system, reduced selective capacity, lipid peroxidation, solute leaching, changes in the respiratory activity of the seeds, changes in the enzymatic activity and synthesis of proteins, changes in the inability to maintain the electrochemical gradient, the loss of cellular compartmentalization and the accumulation of toxic substances (Marcos Filho, 2015). Physiological changes were also observed in this study, which included delay in germination, decreased tolerance to sub-optimal environmental conditions during germination, reduced growth and/or seedling vigor, an increased number of abnormal seedlings, increased susceptibility to pathogens attacks, uneven emergence, reduced productivity, altered seed color, decreased storage potential, complete loss of germination capacity and death.

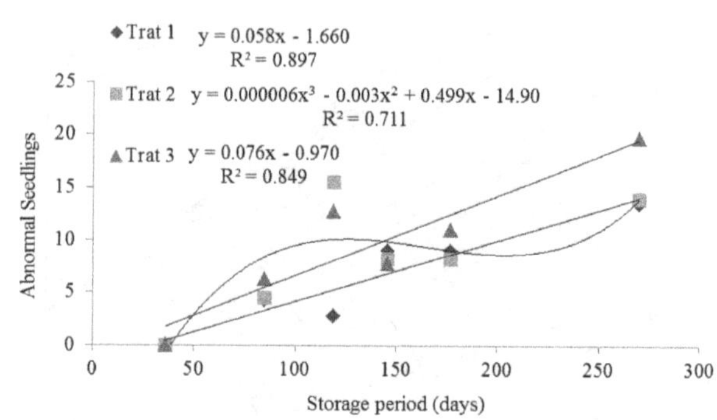

Figure 4. Relationship between the storage period and the number of seeds that produced abnormal seedlings under different germination conditions1 at Marechal Candido Rondon, Paraná State, 2013/2014. Treatment 1 – Alternating temperature between 20 and 30°C and a photoperiod of seven hours; Treatment 2 – constant temperature of 25°C and a photoperiod of seven hours; Treatment 3 – constant temperature of 25°C under darkness.

The germination values of quinoa seeds approached 90% at 36 days after harvest, indicating that the genotype does not present numbness (Figure 5), a condition presented by quinoa genotypes 2-Want and Chadmo (Ceccato, Bertero, Batlla, & Galati, 2015).

The germination of quinoa seeds decreased linearly regardless of the germination test condition, either alternating temperatures of 20 and 30°C with a photoperiod of 7 hours or a constant temperature of 25°C with or without a photoperiod of 7 hours (Figure 5). The percentage of germinated seeds decreased as the storage period increased, so it is possible to conclude that longer quinoa seed storage periods under ambient conditions result in greater deterioration. These results are supported by Strenske et al. (2015), who reported a decrease in quinoa seed germination with storage time. Under the storage conditions used in this study, it is expected that quinoa seeds completely lose their germination potential between 430 and 456 days after the seed is harvested.

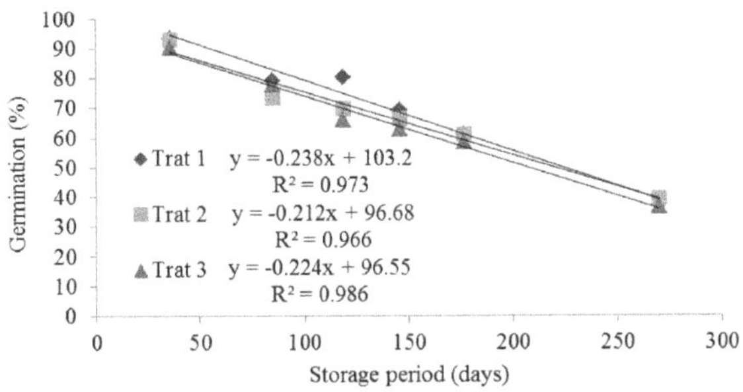

Figure 5. Relationship between the storage period and quinoa seed germination under different germination conditions at Marechal Candido Rondon, Paraná State, 2013/2014. Treatment 1 – Alternating temperature between 20 and 30°C and a photoperiod of seven hours; Treatment 2 – constant temperature of 25°C and a photoperiod of seven hours; Treatment 3 – constant temperature of 25°C under darkness.

As observed for seed germination, there was also a reduction in the germination rate (Figure 6) throughout the storage period. Under the three conditions for seed germination, alternating temperatures between 20 and 30°C with a seven-hour photoperiod or a constant temperature of 25°C with or without a seven-hour photoperiod, the loss of quinoa seed vigor was similar, so the different conditions yield similar results.

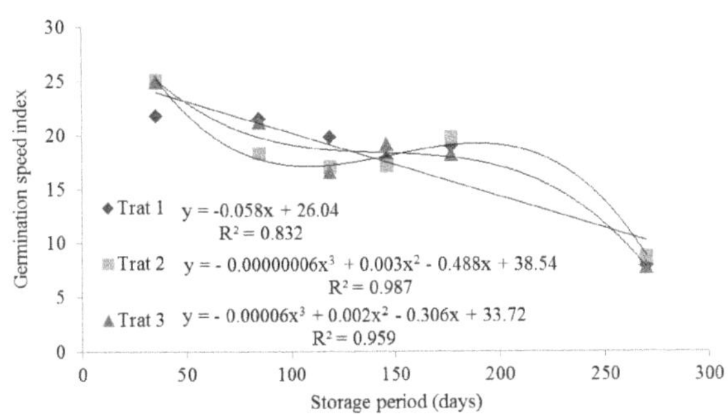

Figure 6. Germination rate of quinoa seeds with storage period under different germination conditions at Marechal Candido Rondon, 2013/2014. Treatment 1 – Alternating temperature between 20 and 30°C and a photoperiod of seven hours; Treatment 2 – temperature of 25°C and a photoperiod of seven hours; Treatment 3 – constant temperature of 25°C under darkness.

The increase in the number of abnormal seedlings and the reduction in the number of germinated seeds as well as the germination rate were expected because storage under uncontrolled environmental conditions may result in seed loss (Castellión, Mantiacevich, Buera, and Maldonado, 2010; Strenske et al., 2015), so the results for the quinoa genotype evaluated in this study were not different from previous findings. However, after 430 days, all seeds lose their germination potential, so the above values for seeds stored under uncontrolled environmental conditions were above expectations.

Conclusion

With longer storage periods under ambient conditions, the quality of quinoa seeds deteriorated as indicated by the increasing number of abnormal seedlings and the decreasing number of normal seedlings.

The final count of germinated quinoa seeds should be evaluated at 7 days after the beginning of the test.

Quinoa seeds can be stored up to 430 days under uncontrolled environmental conditions before germination completely declines.

References

Bonacin, G. A., Rodrigues, T. J. D., Fernandes, A. C., & Rodrigues, L. R. A. (2006). Germinação de sementes de alfafa submetidas a períodos de imersão em água. *Científica*, *34*(2), 150-154.

Carvalho, N. M. & Nakagawa, J. (2012). Sementes: ciência, tecnologia e produção (5ed., p. 590). Jaboticabal, SP: Funep.

Castellión, M., Matiacevich, S., Buera, P., & Maldonado, S. (2010). Protein deterioration and longevity of quinoa seeds during long-term storage. *Food Chemical*, *121*(4), 952-958.

Ceccato, D., Bertero, D., Batlla, D., & Galati, B. (2015).Structural aspects of dormancy in quinoa (*Chenopodium quinoa*): importance and possible action mechanisms of the seed coat. *Seed Science Research*, *239*(1), 1-9.

Cruz, C. D. (2013). Genes – a software package for analysis in experimental statistics and quantitative genetics. *Acta Scientiarum. Agronomy*, *35*(3), 271-276.

Franzin, S. M., Menezes, N. L., Garcia, D. C., & Wrasse, C. F (2004). Métodos para avaliação do potencial fisiológico de sementes de alface. *Revista Brasileira de Sementes*, *26*(2), 63-69.

Gewehr, M. F., Danelli, D., Melo, L. M., Flôres, S. H., & Jong, E. V. (2012) Análises químicas em flocos de quinoa: caracterização para a utilização em produtos alimentícios. *Brazilian Journal Food Technology*, *15*(4), 280-287.

Goldfarb, M., & Queiroga, V. P. (2013). Considerações sobre o armazenamento de sementes. *Revista Tecnologia & Ciência Agropecuária*, 7(3), 71-74.

Instituto Agronômico do Paraná [IAPAR]. (2004). *Cartas Climáticas do Paraná* (p. 49). Curitiba, PR: Iapar.

Kappes, C., Arf, O., Ferreira, J. P., Portugal, J. R., Alcalde, M., Arf, M. V., & Vilela, R. G. (2012). Qualidade fisiológica de sementes e crescimento de plântulas de feijoeiro, em função de aplicações de paraquat em pré-colheita. *Pesquisa Agropecuária Tropical*, *42*(1), 9-18.

Marcos Filho, J. (2015). *Fisiologia de sementes de plantas cultivadas* (2ed.). Londrina, PR: Abrates.

Nakagawa, J., Cavariani, C., & Toledo, M. Z. (2009). Germinação de sementes armazenadas de guandu. *Revista Brasileira de Sementes*, *31*(4), 43-48.

Nobre, D. A. C., David, A. M. S. S., Souza, V. N. R., Oliveira, D., Gomes, A. A. M., Aguiar, P. M., & Mota, W. F. (2013). Influência do ambiente de armazenamento na qualidade fisiológica de sementes de amaranto. *Comunicata Scientiae*, *4*(2), 216-219.

Oliveira, A. B., & Gomes-Filho, E. (2009). D E S Liens Étroits. *Revista Brasileira de Sementes*, *31*(3), 48-56.

Santos, E. L., Póla, J. N., Barros, A. S. R., & Prete, C. E. C. (2007). Qualidade fisiológica e composição química das sementes de soja com variação na cor do tegumento. *Revista Brasileira de Sementes*, *29*(1), 20-26.

Spehar, C. R. (2007). *Quinoa: alternative para diversificação agrícola e alimentar*. Planaltina, DF: Embrapa Cerrados.

Strenske, A., Vasconcelos, E. S., Herzog, N. F. M., Malavasi, M. M., (2015). Germinação de sementes de quinoa com diferentes períodos de armazenamento. *Scientia Agraria Paranaensis*, *14*(1.), 286-290.

16

Vegetative growth response of cotton plants due to growth regulator supply via seeds

João Vitor Ferrari[1*], Enes Furlani Júnior[2], Samuel Ferrari[3] and Ana Paula Portugal Gouvea Luques[2]

*[1]Programa de Pós-graduação em Agronomia, Departamento de Fitotecnia, Tecnologia de Alimentos e Sócio Economia, Faculdade de Engenharia de Ilha Solteira, Universidade Estadual Paulista "Júlio de Mesquita Filho", Av. Brasil, 56, 15385-000, Centro, Ilha Solteira, São Paulo, Brazil. [2]Departamento de Fitotecnia, Tecnologia de Alimentos e Sócio Economia, Faculdade de Engenharia de Ilha Solteira, Universidade Estadual Paulista "Júlio de Mesquita Filho", Ilha Solteira, São Paulo, Brazil. [3]Departamento de Agricultura, Universidade Estadual Paulista "Júlio de Mesquita Filho", Campus de Registro, Registro, São Paulo, Brazil. *Author for correspondence. E-mail: jao_unesp@hotmail.com*

ABSTRACT. The global cotton industry is distinguished by its numerous industrial uses of the plume as well as by high production costs. Excessive vegetative growth can interfere negatively with productivity, and thus, applying growth regulators is essential for the development of the cotton culture. The objective of this study was to evaluate the development and yield of the cotton cultivar FMT 701 with the application of mepiquat chloride to seeds and leaves. The experimental design used a randomized block design with four replications, arranged in bands. The treatments consisted of mepiquat chloride rates (MC) (0, 4, 6, 8 and 10 g a.i. kg^{-1} of seeds) applied directly to the cotton seeds and MC management by foliar spray using a 250 mL ha^{-1} rates that was administered under the following conditions: divided into four applications (35, 45, 55 and 65 days after emergence); as a single application at 70 days; and without the application of the product. The mepiquat chloride applied to cotton seeds controls the initial plant height and stem diameter, while foliar application reduces the height of the plants. After application to seed, foliar spraying MC promotes increase mass of 20 bolls, however no direct influence amount bolls per plant and yield of cotton seed. Higher cotton seed yield was obtained with a rate of 3.4 g a.i. MC kg^{-1} seeds.

Keywords: mepiquat chloride, growth, yield components.

Introduction

When cultivated under conditions with unrestricted availability of moisture, light and nutrients, especially nitrogen, the cotton plant exhibits excessive vegetative growth, which negatively interferes with productivity. Thus, applying a growth regulator is vital for the development of the cotton culture (FERRARI et al., 2008).

Additionally, under these conditions, the adequate development of the culture may not be obtained with the application of growth regulators to the leaves, primarily due to the application time or due to rainfall after the application. According to

Mateus et al. (2004), the activity of mepiquat chloride via foliar application to the cotton plants requires a minimum of 16 hours for the full effect of the product, and if there is rain during this period, an additional new application is then required.

Studies indicate that growth regulators can alternatively be applied to the seeds, with the advantages of an easier practical application and the reduction of the cotton plants' initial development after emergence (FERRARI et al., 2010; NAGASHIMA et al., 2005), without affecting root system development (ALMEIDA; ROSOLEM, 2012). Another favorable aspect is protection against adverse conditions during the first application of the growth regulator, which can prevent the product from producing the desired results (MATEUS et al., 2004).

Acquiring good productivity results is more dependent on the proper management of the culture than increasing the amount of the input. Thus, it is necessary to evaluate the physiological behaviors of the plant, especially with regard to the photosynthetic process and the partitioning of photoassimilates, which are two extremely important factors in crop yield outcome, and to compare these behaviors to those obtained using the new widely adopted techniques in the cotton culture.

Therefore, this study aimed to evaluate the growth and yield of the cotton plant cv. FMT 701 following the application of mepiquat chloride to seeds and leaves. Let there be that the management with growth regulator is extremely important for providing control of vegetative growth. Thus studies that indicate the viability of different forms to application of growth regulators it seems necessary as they may assist in search a rational cotton crop management.

Material and methods

The study was carried out in the 2008/09 and 2009/10 crop years at geographical coordinates 51°24'W, 20°20'S and an altitude of approximately 335 m in Selvíria, Mato Grosso do Sul State, Brazil. The soil in the experimental area was classified as a Latosol (Dark Red, Clayey Dystrophic Latosol) (EMBRAPA, 2006), with average annual rainfall of 1232 mm and temperature of 24.5°C.

In June 2008, the 0-0.20 m soil layer was sampled and chemically analyzed, and the results are shown in Table 1. In July 2008, 1000 kg of dolomitic limestone was applied, with 80% efficiency. After the application, the soil was prepared using a moldboard plow and grading. This area has been planted with cotton crops in conventional tillage since 2007.

The experimental design used a randomized block design with four replicates, arranged in bands. The treatments consisted of mepiquat chloride rates (0, 4, 6, 8 and 10 g a.i. kg^{-1} of seeds) applied directly to the cotton seeds and MC management through foliar applications using a 250 mL ha^{-1} rate that was administered under the following conditions: divided into four applications (35, 45, 55 and 65 days after emergence (d.a.e.)); single application at 70 d.a.e; and without application of product.

The experiment began in September 2008 and in September 2009, with millet sown as a preceding crop to obtain straw in the area; this condition was chosen because of its desirable dry matter production (GUIDELI et al., 2000) and because of its wide use as a seeding system in the large cotton producing regions of Brazil's Central-West region.

Cotton seeds were treated with carboxin+tiram (100 g a.i. per 100 kg seeds) and thiamethoxam (210 g a.i. per 100 kg seeds). After drying for 4 hours, the seeds were divided into plastic bags, and these seeds either received the application of MC (250 g a.i. L^{-1}) according to the defined rates or were not further treated. They were then agitated for 3 minutes and left to dry for 5 hours. On the same day, the dried seeds were sowed. Each experimental plot consisted of four rows that were five meters long, spaced 0.90 m apart, with two central rows in the useful area of each plot.

The cotton sowing was performed on November 17 in the first agricultural year and on the November 23 in the second year, with emergence on November 23 and 27 of the respective years. In 2008 and 2009, a tractor-drawn seeder-fertilizer applicator was used, with 200 kg ha^{-1} of NPK at a 8-28-16 proportion in the sowing furrow. The strain used was FMT 701 due to its superior performance in tests carried out in the locality. The seeding rate was 11 seeds m^{-1} with 0.90 m spacing between rows.

Table 1. Chemical analysis soil results at a depth of 0-0.20 m. Selvíria, Mato Grosso do Sul State, Brazil, 2008.

P_{resin} mg dm^{-3}	M.O. g dm^{-3}	pH ($CaCl_2$)	K	Ca	Mg	H+Al	Al	CTC	V (%)
			mmol$_c$ dm^{-3}						
12	22	4.8	4.6	19	11	22	0	58	58

Looping was performed after the emergence and set up (14 d.a.e.), with 8 m^{-1} plants remaining. The fertilizer coverage comprised 60 kg ha^{-1} of nitrogen divided into two applications: one at 35 d.a.e. using urea (45% of N) as the source and the second at 55 d.a.e. using ammonium sulfate (21% of N and 24% of S).

The application of growth regulator to the leaves was performed according to the treatments described above. The applications were performed in the morning to avoid high temperatures during the application.

Plant heights were measured using a measuring tape to measure from the ground to the top of the plant. Calipers were used to measure the stem diameter at 2 cm above the ground.

The number of reproductive branches and bolls per plants were counted at the time of harvest, in addition to 20 bolls that were randomly harvested at the middle third of the plants. The measurements were randomly obtained on five plants in the useful area of the plots. In April 2008 and April 2009, at 150 and 152 days, respectively, harvesting was manually performed in the two central rows of each plot to quantify the cotton seed production per hectare, which was weighed using a digital scale.

The data obtained in this study underwent analysis of variance using an F-test and a means comparison test (Tukey) as well as by polynomial regression at the 5% significance level using the methodology described by Banzatto and Kronka (2006). Correlation and regression analyses were performed using SISVAR software (FERREIRA, 2011). Regression equations were fitted to the significant features, and the best-fitting model was selected by the coefficient of determination.

Results and discussion

Tables 2 and 3 show the results of the plant height evaluation during the development of the cotton plants. In the assessments carried out in the early stages of development of the cotton plants (10 to 60 days in Tables 2 and 3) and in both agricultural years under study (2008/09 and 2009/10), the growth regulator rates was observed to have a significant linear effect on the reduction of plant height. These results indicate that using the product for seed treatment effectively controls initial cotton plant development, thereby facilitating the management of the cotton culture. The first months after sowing typically exhibit high rainfall, which can undermine the growth regulator applications, exceeding the time or too late for the first growth regulator application. Treating the seeds instead of the leaves also affords savings in the quantity of spraying in the production areas, resulting in reduced production costs. Similar results were found by Almeida and Rosolem (2012) in a greenhouse study evaluating the development of roots and shoots of the cotton cultivar cv. FM 993 from seeds treated with MC. They concluded that this promotes a reduction of plant height and leaf area without affecting the dry matter production of the shoots. Similarly, Nagashima et al. (2011) observed a reduction in the initial development of cotton plants after spraying MC on the seeds.

At 60 days (Table 3), the growth regulator applied to the seeds loses its residual effect and does not control the development of height of the cotton plants, therefore, the application of growth regulator via foliar spray is required. This growth regulator management was studied and was shown to be efficient because there was a reduction in plant height by using the product divided into parts, beginning at 60 days.

Similarly, a single application of the growth regulator was also efficient in reducing plant height, resulting in significant differences at 90 and 125 days in comparison with the control, with average reductions of 25.22 cm in the two years under evaluation (Table 3). The main purpose of using the growth regulator is to disrupt the biosynthesis of gibberellic acid, thus influencing dry matter production, which is related to the transport of carbohydrates because the development of the cotton plant is related to favorable weather conditions, water and also nutrient availability (ELKOCA; KANTAR, 2006; HAQUE et al., 2007; TAIZ; ZEIGER, 2013). Height control of cotton plant can be efficiently performed with the use of MC, as it inhibits cell growth (LAMAS, 2001). The use of growth regulator under Cerrado conditions is required because it can help control plant height, thereby facilitating crop management, which is especially important for mechanical harvesting.

Similarly, Ferrari et al. (2008) found that over several seasons and growth regulator rates, height differences of up to 34.56 cm between plants were observed with and without application of the product.

Table 2. Evaluation (days) of cotton plant heights (cm) in two crop years, according to the rates and application methods of the growth regulator.

p > F		10		20		30		40	
		Days after emergence							
		2008/09	2009/10	2008/09	2009/10	2008/09	2009/10	2008/09	2009/10
Application (a)		0.713	0.940	0.123	0.684	0.711	0.783	0.978	0.933
Rates (r)		0.001**	0.001**	0.001**	0.001**	0.001**	0.001**	0.001**	0.001**
a*r		0.850	0.998	0.980	0.992	0.929	0.932	0.961	0.991
	Without	6.05	7.30	10.94	11.05	30.40	32.37	54.58	54.92
Application Methods	Single	5.98	7.52	11.77	11.51	30.12	31.16	54.37	55.02
	Split	5.94	7.41	11.07	11.97	29.75	31.68	54.62	54.51
	0	9.30(1)	10.70(2)	16.00(3)	16.35(4)	37.67(5)	40.46(6)	62.75	66.23
	4	5.79	8.12	11.97	12.15	29.08	32.53	52.78	57.05
Rates	6	4.98	7.11	11.00	11.19	30.08	31.24	55.75	52.44
	8	5.24	6.16	9.39	9.51	29.81	30.32	57.53	53.62
	10	4.62	4.98	7.94	8.33	24.95	24.12	43.81	44.75
C.V.%		7.00	26.70	11.85	28.83	8.26	17.18	7.49	8.34
D.M.S.		0.32	1.52	1.03	2.55	1.91	4.19	3.14	3.51

Polynomial Equations

(1) $Y= 9.231 - 1.057x + 0.062x^2$ $R^2=0.97^{**}$ (2) $Y= 10.567 - 0.564x$ $R^2=0.98^{**}$

(3) $Y=15.693-0.7916x$ $R^2=0.98^{**}$ (4) $Y=15.960-0.795x$ $R^2=0.98^{**}$

(5) $Y= 36.631 - 1.168x$ $R^2=0.82^{**}$ (6) $Y= 39.987 - 1.473x$ $R^2=0.94^{**}$

**Significant at 1% level by Test F. Means followed by same letter vertically do not differ, by Tukey test at 5% of probability.

Table 3. Evaluation (days) of cotton plant heights (cm) in two crop years, according to the rates and application methods of the growth regulator.

p > F		60		90		125	
		Days after emergence					
		2008/09	2009/10	2008/09	2009/10	2008/09	2009/10
Application (a)		0.001**	0.001**	0.001**	0.001**	0.001**	0.001**
Rates (r)		0.001**	0.004**	0.846	0.173	0.186	0.308
a*r		0.682	0.999	0.781	0.993	0.727	0.999
	Without	83.02 a	92.68 a	111.62 a	119.68a	150.60a	146.57 a
Application Methods	Single	82.10 a	91.70 a	100.75 b	103.28 b	122.82 b	123.92 b
	Split	74.40 b	79.94 b	95.00 b	95.58 c	119.23 b	115.79 b
	0	88.08(1)	93.88(2)	101.45	110.13	132.20	133.47
	4	77.72	88.73	101.56	108.50	127.42	128.50
Rates	6	80.22	90.71	104.72	106.95	131.03	127.23
	8	83.89	87.27	102.19	103.51	134.31	130.43
	10	69.28	79.94	102.36	101.80	129.47	124.19
C.V.%		5.76	9.62	7.65	8.72	5.45	8.38
D.M.S.		3.54	6.51	6.02	7.11	5.48	8.29

Polynomial Equations

(1) $Y=87.348-1.341x$ $R^2=0.53^{**}$ (2) $Y=94.746-1.186x$ $R^2=0.77^{**}$

**Significant at 1% level by Test F. Means followed by same letter vertically do not differ, by Tukey test at 5% of probability.

The results presented in Table 4 show that the application of increased rates of growth regulator to cotton seeds reduced the stem diameter of the plants in assessments performed at 30 days. There were significant effects in the two evaluation years, with linear and quadratic fitting averages, respectively, for 2008/2009 and 2009/10. This reduction is due to the action of the product from the early stages of vegetative growth of the cotton plant, thus restricting plant development in both diameter and in height (Tables 2 and 3).

However, analyzing stem diameter averages at 80 days, it was observed that the plant growth regulator applied to seeds stopped acting to inhibit plant growth, resulting in similar values for the treatments in the two study years. However, in the 2008/09 crop year, there was a stem diameter increase for the plants that received divided MC foliar applications. One possible explanation is that product applied to the cotton leaf at different times decrease height development (Table 3, assessment at 90 days), with accumulated reserves supporting stem development instead.

In the evaluation performed at 125 days (Table 4), in the two years evaluated, no significant differences were found in the stem diameter of the cotton plant due to the application of growth regulators to the seeds and stems. Similar results were found by Cordao Sobrinho et al. (2007) that found no effect of the growth regulator in reducing the stem diameter. Thus, in contrast to plant height, stem diameter resumes its normal development after the effect of the growth regulator is complete, with no significant difference between the averages. Most likely, such dynamic stem development is necessary for the cotton plants to become resistant to growth and to withstand the weight of branches and bolls during the development progress.

An analysis of the number of reproductive branches of cotton plants indicated no changes based on the growth regulator application to the seeds or leaves (Table 4) in the two years under study. With this result, it can be observed that the growth regulator has no function in changing the number of fruiting branches per plant and that this characteristic is strongly influenced by genotype. On the other hand Bogiani and Rosolem (2009) and Rosolem et al. (2013) using rates at growth regulator in cotton plants, concluded that rates up to 22.5 and 30 g ha^{-1} (respectively) of active ingredient MC further reduction in the number of reproductive branches.

The data in Table 5 show no significant difference for the mass of 20 bolls with the treatments under study. For the growth regulator applied to the leaves and with the application methods described above, 20 bolls revealed a greater mass for the treatment in which the product was applied at different times in the assessment conducted in the 2009/10 crop year. Similarly, Teixeira et al. (2008) and Ferrari et al. (2008) observed a greater mass of 20 bolls on the cotton plants (cv. FMX 986 and Delta Opal respectively) that had received growth regulator applications at different times. For the MC rates via seeds, there was a quadratic fitting of the averages in the two years of evaluations, in which rates of 4, 6 and 8 g a.i. kg^{-1} seeds promoted the highest values. Tables 2, 3 and 4 show the interference in vegetative growth of the cotton plants, indicating a significant vegetative growth or strong initial control of the plants observed by the development of bolls.

The application of growth regulator to leaves did not interfere significantly with the average number of bolls per plant for both years under study (Table 5). Thus, the use of MC to control the growth of cotton plants did not change the number of bolls per plant. Similar results were reported by Ferrari et al. (2008), who found no significant difference for this variable.

Table 5 shows that there was a significant effect of quadratic fitting for the MC rates applied to seeds on cotton kernel yield. Where there was no treatment with phytoregulators via seeds, productivity was lower, thus verifying that as these plants had higher vegetative growth, they produced less.

Table 5 shows that the rates above 4 g a.i. kg^{-1} seeds showed a decrease in productivity, indicating excessive control of plant size. A detailed analysis showed that, in both years under study, the MC rate that provided the highest cotton yield was approximately 3.4 g a.i. kg^{-1} seeds. It was also observed that the highest rate used (10 g a.i. kg^{-1} seeds) caused a decrease in cotton kernel productivity of 17% (543 kg ha^{-1}) and 19% (594 kg ha^{-1}), respectively, for the agricultural years 2008/09 and 2009/10. Similarly, Yeats et al. (2005) observed lower yields in cotton plants that underwent soaking of the seeds with rates greater than 4 g a.i. MC kg^{-1} seeds.

Table 4. Evaluation (days) of stem diameter (cm) and number of reproductive branches of cotton plants, in two crop years, according to the rates and application methods of growth regulator.

Treatments		30		80		125		Reproductive branches (nº)	
		Days after emergence							
		2008/09	2009/10	2008/09	2009/10	2008/09	2009/10	2008/09	2009/10
Application (a)		0.042	0.956	0.008**	0.928	0.176	0.653	0.637	0.793
Rates (r)		0.001**	0.001**	0.211	0.857	0.192	0.872	0.507	0.870
a*r		0.854	0.637	0.071	0.998	0.948	0.999	0.276	0.995
	Without	7.18	8.69	17.55b	18.72	21.54	21.69	16.91	16.57
Application Methods	Single	7.15	8.81	17.71 b	18.80	22.63	22.87	17.41	17.62
	Split	6.97	8.78	19.00 a	18.30	22.86	22.90	17.18	17.51
	0	7.86(1)	9.57(2)	18.52	19.59	21.90	21.49	16.83	17.47
	4	7.08	8.79	17.65	19.07	22.14	22.54	17.33	18.29
Rates	6	7.08	9.11	18.53	18.27	23.81	22.84	17.30	17.50
	8	7.42	9.38	17.67	18.39	21.62	23.47	17.74	16.92
	10	6.06	6.96	18.07	17.71	22.24	22.08	16.62	15.98
C.V. %		8.03	14.46	6.72	23.82	10.48	20.83	9.72	30.34
D.M.S.		0.44	0.97	0.93	3.41	1.80	3.60	1.28	4.09

Polynomial Equations

(1) $Y=7.864-0.136x\ R^2=0.62^{**}$ (2) $Y= 9.372+0.162x-0.035x^2\ R^2=0.98^{**}$

** Significant at 1% level by Test F. Means followed by same letter vertically do not differ. by Tukey test at 5% of probability.

Table 5. Evaluation of 20 bolls, number of bolls per plant and cotton plant yield, cv. Delta Opal, in two crop years, according to rates and application methods of growth regulator.

Treatments		Mass 20 bolls(g)		Bolls/plant (nº)		Yield (kg ha^{-1})	
		2008/09	2009/10	2008/09	2009/10	2008/09	2009/10
Application (a)		0.186	0.044*	0.595	0.591	0.755	0.960
Rates (r)		0.048*	0.012*	0.016*	0.344	0.002**	0.010*
a*r		0.574	0.957	0.994	0.991	0.987	0.992
	Without	111.04	103.57* b	15.96	15.78	3122.29	2917.05
Application Methods	Single	110.07	108.31 ab	16.39	16.13	3112.15	2940.87
	Split	117.44	111.73 a	16.79	16.55	3192.83	2954.36
	0	112.44(1)	106.91(2)	16.53(3)	16.05	3232.08(4)	3034.56(5)
	4	116.35	111.22	16.93	17.00	3283.27	3148.47
Rates	6	113.77	113.21	16.73	16.51	3227.23	3079.29
	8	118.33	109.01	17.65	16.21	3129.43	2870.45
	10	103.36	99.00	14.05	15.03	2740.11	2554.36
C.V. %		12.01	9.26	15.63	14.54	11.77	14.40
D.M.S.		10.42	7.68	1.97	1.80	284.27	325.17

Polynomial Equations

(1) $Y=111.801+2.799x -0.339x^2\ R^2=0.63^{*}$ (2) $Y=106.404+3.317x-0.395x^2\ R^2=0.92$ **(3) $Y=16.332+0.580x -0.074x^2\ R^2=0.60^{*}$

(4) $Y=3212.515+92.651x -13.633x^2\ R^2=0.94^{**}$ (5) $Y=3031.518+86.075x-13.358x^2\ R^2=0.99^{*}$

**,*Significant at 1% and 5% levels respectively by Test F analysis of variance. Means followed by same letter vertically do not differ, by Tukey test at 5% of probability.

It was not possible to verify significant differences between the growth regulator application modes via leaf. Azevedo et al. (2004) found no difference regarding whether or not MC foliar application methods were performed.

Conclusion

The mepiquat chloride applied to cotton seeds controls the initial plant height and stem diameter, while foliar application reduces the height of the plants.

After application to seed, foliar spraying MC promotes increase mass of 20 bolls, however no direct influence amount bolls per plant and yield of cotton seed.

Higher cotton seed yields are obtained with a rate of 3.4 g a.i. MC kg^{-1} seeds.

References

ALMEIDA, A. Q.; ROSOLEM, C. A. Cotton root and shoot growth as affected by application of mepiquat chloride to cotton seeds. **Acta Scientiarum. Agronomy**, v. 34, n. 1, p. 61-65, 2012.

AZEVEDO, D. M. P.; BEZERRA, J. R. C.; SANTOS, J. W.; DIAS, J. M.; BRANDÃO, Z. N. Efeito do parcelamento do cloreto de mepiquat em algodoeiro irrigado no nordeste brasileiro. **Revista Brasileira de Oleaginosas e Fibrosas**, v. 8, n. 3, p. 823-830, 2004.

BANZATTO, D. A.; KRONKA, S. N. **Experimentação Agrícola**. Jaboticabal: FUNEP, 2006.

BOGIANI, J.; ROSOLEM, C. A. Sensibilidade de cultivares de algodoeiro ao cloreto de mepiquat. **Pesquisa Agropecuária Brasileira**, v. 44, n. 10, p. 1246-1253, 2009.

CORDAO SOBRINHO, F. P.; FERNANDES, P. D.; BELTRAO, N. E. M.; SOARES, F. A. L.; TERCEIRO NETO, C. P. C. Crescimento e rendimento do algodoeiro BRS-200 com aplicações de cloreto de mepiquat e laminas de irrigação. **Revista Brasileira de Engenharia Agrícola e Ambiental**, v. 11, p. 284-292, 2007.

ELKOCA, E.; KANTAR, F. Response of pea (*PisumsativumL.*) to mepiquat chloride under varying application rates and stages. **Journal Agronomy and Crop Science**, v. 192, n. 2, p. 102-110, 2006.

EMBRAPA-Empresa Brasileira de Pesquisa Agropecuária. **Sistema brasileiro de classificação de solos**. 2. ed. Rio de Janeiro: Embrapa, 2006.

FERRARI, S.; FURLANI JÚNIOR, E.; FERRARI, J. V.; SANTOS, M. L.; SANTOS, D. M. A. Desenvolvimento e produtividade do algodoeiro em função de espaçamentos e aplicação de regulador de crescimento. **Acta Scientiarum. Agronomy**, v. 30, n. 3, p. 365-371, 2008.

FERRARI, S.; ANDRADE JUNIOR, E. R.; BELOT, J. L.; BOLDT, A. F.; GALBIERI, F. Efeito do tratamento de sementes de algodão com cloreto de mepiquat e cloreto de clormequat sobre característica vegetativa. In. BELOT, J. L.; VILELA, P. A. (Ed.). **O sistema de cultivo adensado em Mato Grosso**: embasamento e primeiros resultados. Cuiabá: DEFANTI, 2010. p. 183-190.

FERREIRA, D. F. Sisvar: a computer statistical analysis system. **Ciência e Agrotecnologia**, v. 35, n. 6, p. 1039-1042, 2011.

GUIDELI, C.; FAVORETO, V.; MALHEIRIOS, E. B. Produção e qualidade do milheto semeado em duas épocas e adubado com nitrogênio. **Pesquisa Agropecuária Brasileira**, v. 35, n. 10, p. 2093-2098, 2000.

HAQUE, S.; FAROOGI, A. H. A.; GUPTA, M. M.; SANGWAN, R. S.; KHAN, A. Effect of ethrel, chlormequat chloride and paclobutrazol on growth and pyrethrins accumulation in *Chrysanthemum cinerariaefolium* Vis. **Plant Growth Regulation**, v. 51, n. 3, p. 263-269, 2007.

LAMAS, F. M. Estudo comparativo entre cloreto de mepiquat and cloreto de chlormequat aplicados no algodoeiro. **Pesquisa Agropecuária Brasileira**, v. 36, n. 3, p. 265-272, 2001.

MATEUS, G. P.; LIMA, E. V.; ROSOLEM, C. A. Perdas de Cloreto de Mepiquat no algodoeiro por chuva simulada. **Pesquisa Agropecuária Brasileira**, v. 39, n. 7, p. 631-636, 2004.

NAGASHIMA, G. T.; MARUR, C. J.; YAMAOKA, R. S.; MIGLIORANZA, É. Desenvolvimento de plantas de algodão provenientes de sementes embebidas com cloreto de mepiquat. **Pesquisa Agropecuária Brasileira**, v. 40, n. 9, p. 943-946, 2005.

NAGASHIMA, G. T.; SANTOS, F. T.; MIGLIORANZA, E. Respostas de cultivares de algodão ao cloreto de mepiquat aplicado via embebição de sementes. **Bragantia**, v. 70, n. 1, p. 46-49, 2011.

ROSOLEM, C. A.; OOSTERHUIS, D. M.; SOUZA, F. S. Cotton response to mepiquat chloride and temperature. **Scientia Agricola**, v. 70, n. 2, p. 82-87, 2013.

TAIZ, L.; ZEIGER, E. **Fisiologia vegetal**. Porto Alegre: Artmed, 2013.

TEIXEIRA, I. R.; KIKUTI, H.; BORÉM, A. Crescimento e produtividade de algodoeiro submetido a cloreto de mepiquat e doses de nitrogênio. **Bragantia**, v. 67, n. 4, p. 891-897, 2008.

YEATS, S. J.; CONSTABLE, G. A.; McCUMSTIE, T. Cotton growth yiel after seed treatment with mepiquat chloride in tropical winter season. **Field Crops Research**, v. 93, n. 2, p. 122-131, 2005.

17

The growth, photosynthetic pigments and essential oil composition of monocropped and intercropped lemon balm with yarrow

Thaís Cristina Silva, José Eduardo Brasil Pereira Pinto, Fernanda Ventorim Pacheco, Heitor Luiz Heiderich Roza, Diogo Mendes da Silva Lordêllo and Suzan Kelly Vilela Bertolucci*

*Universidade Federal de Lavras, Departamento de Agricultura, Cx. Postal, 3037, 37200-000, Lavras, Minas Gerais, Brazil. *Author for correspondence. E-mail: suzan@dag.ufla.br*

ABSTRACT. The purpose of this study was to assess the influence of lemon balm intercropping with yarrow on the vegetative growth, photosynthetic pigments and essential oil contents as well as the chemical composition of essential oils. Accordingly, the lemon balm and yarrow monocroppings were compared to the intercropping of both species. The plants were cultivated for 120 days under field conditions. The experimental design followed randomized blocks with seven repetitions. Growth was assessed through the accumulation of shoot dry matter and through the leaf area. The photosynthetic pigment content was set based on the fully expanded leaves from five plants in each treatment. The essential oil was extracted through the hydrodistillation of the lemon balm leaves and yarrow leaflets dehydrated in a forced ventilation oven at 40°C. The chemical analyses of the essential oils were performed using GC-FID and GC-MS. Intercropping showed a reduction of vegetative growth. However, intercropping increased the lemon balm essential oil content. On the other hand, these parameters did not change in yarrow. There were variations in the level of chemical constituents in lemon balm and yarrow essential oils, which were more evident in the major constituents.

Keywords: *Melissa officinalis* L.; *Achillea millefolium* L.; intercropping; gas chromatography.

Introduction

Melissa officinalis L. (Lamiaceae), also known as lemon balm, is an aromatic plant that has a smell similar to lemons (Reis, Pinto, Rosado, & Corrêa, 2009). It is one of the most important medicinal plants since it has antioxidant, antimicrobial, spasmolytic, astringent and specific sensorial properties (Yadegari, 2016). The essential oil produced by the species has neral (α-citral) and geranial (β-citral) as major compounds, which are attributed to the plant's medicinal properties (Luz, Silva, Habber, & Marquez, 2014; Sorensen, 2000). However, its essential oil productivity is low, and the oil content varies from 0.01% to 0.52% (Argyropoulos & Müller, 2014; Avci & Giachino, 2016; Szabó et al., 2016). The species *Achillea millefolium* L. (Asteraceae), commonly known as yarrow, has medicinal properties and is used as a painkiller, anti-inflammatory, diuretic and antispasmodic. It is also used in cosmetics and food seasonings and it is considered to be an ornamental plant (Brasil, 2009; Pinto et al., 2014).

Yarrow intercropping with lemongrass (*Cymbopogon citratus* DC.) or rosemary (*Rosmarinus officinalis* L.) has increased the essential oil yield, and this was previously observed by Santos et al. (2009) and Arashiro et al. (2011).

Intercropping can be defined as a cultivation system based on the simultaneous growth of two or more plant species in the same area (Gao et al., 2009). This method is more acceptable for plant cultivation according to the agro-ecological models because it presents countless environmental, productive and economic advantages due to the increased productivity and decreased number of weeds, plagues and diseases (Grangeiro et al., 2008; Wang et al., 2014).

The intercropping of medicinal plants aimed at increasing essential oil production in low productivity species may be an alternative for family farmers. However, the awareness regarding the influence of intercropping of medicinal plants on the production of active ingredients is limited, and previous studies focused on greenery species (Chamoli, Varshney, Srivastava, Pandey, & Dayal, 2013; Singh, Singh, Ram, Yadav, & Chanotiya, 2013). In addition, these studies that focused on the use of medicinal plants did not demonstrate the influence of this cultivation technique on the quality of the essential oil.

Therefore, by considering the value given to the use of plants for medicinal purposes and the increased demand for raw materials, studies that are able to generate knowledge regarding intercropping of medicinal plants are needed because these studies will demonstrate the advantages of this technique for vegetal production. Accordingly, the purpose of the current study is to assess the influence of yarrow-lemon balm intercropping on the growth, photosynthetic pigment production and content, and qualitative and quantitative chemical composition of essential oils.

Material and method

The experiment was conducted in the experimental site at the Agricultural Department (DAG) of Federal University of Lavras (UFLA) from November 2013 to March 2014. The experimental site is located at 21° 14' S and 45° 00' W at an altitude of 918 m. The Cwa type is the prevailing climate for the area, which consists of dry winters and warm summers (Sá Júnior, Carvalho, Silva, & Alves, 2012).

The soil type in the experimental site is classified as Dystrophic Red Latosol (oxisol) (Santos, 2013). The soil was analyzed in the Soil Chemical Analyses Laboratory of the UFLA Soil Science Department in order to establish the soil's fertility attributes. The physical-chemical properties of the soil (0-20 cm) are as follows: pH: 5.7, K: 142.0 mg dm^{-3}, P: 2.84 mg dm^{-3}, Ca: 2.60 $_c$mol dm^{-3}, Mg: 0.90 cmol$_c$ dm^{-3}, Al: 0.10 cmol$_c$ dm^{-3}, H+Al: 2.90 cmol$_c$ dm^{-3}, SB: 3.86 cmol$_c$ dm^{-3}, t: 3.96 cmol$_c$ dm^{-3}, T: 6.76 cmol$_c$ dm^{-3}, V: 57.16%, M: 2.535, organic matter: 2.87 dag kg^{-1}, and P-Rem: 24.34 mg L^{-1}. The climatological data for the assessed period was provided by the climatology station at the Engineering Department of UFLA, which had a mean record for a minimum temperature of 18.1°C, maximum temperature of 29.5°C, rainfall of 139.5 mm, and relative humidity of 69.5%.

The seedlings were produced from yarrow and lemon balm plant matrices found at the Medicinal Garden of UFLA. The lemon balm vegetative propagation was conducted by using apical cuttings, and the yarrow vegetative propagation was completed through the use of stolons. The seedlings were transferred to the field and then planted 30 days after cultivation with a 0.40 x 0.40 m spacing in November 2013. One kilogram of bovine manure was added to each pit (15 x 15 x 20 cm) at planting. The culture handlings necessary for plant maintenance were performed throughout the experiment.

The experimental design consisted of randomized blocks with four treatments placed in seven blocks with nine useful plants from each species per experimental unit including the border. The treatments consisted of lemon balm monocropping, yarrow monocropping and lemon balm/yarrow intercropping.

Harvest occurred 120 days after cultivation. Lemon balm stems and leaves and yarrow petioles and leaflets were individually dehydrated in a forced ventilation oven at 40°C until they reached a constant weight. Growth was assessed in nine lemon balm and yarrow plants per treatment by using shoot dry matter. The leaf area of the lemon balm was also determined by 50 leaves plant^{-1}, which was standardized from the third pair of leaves using Li-Cor model Li-3100 leaf area meter (Li-Cor, Inc. Lincoln, Nebraska, USA).

The analyzed photosynthetic pigments were carotenoids and a, b and total chlorophylls. The extraction was performed according to the methodology described by Lichtenthaler and Buschmann (2001). The fully expanded leaves located in the third node of the five plants were collected per treatment. To extract the pigments, 200 mg of fresh leaves were homogenized with 20 mL of 80% acetone (v/v). The mixture was then filtered through glass wool, and the volume was increased to 30 mL with the addition of 80% acetone. Immediately following this procedure the absorbance was measured at 663 nm, 646 nm and 470 nm. All procedures were done in the dark to preserve the integrity of the pigments. The results were expressed as milligrams of pigment per gram of foliar tissue fresh matter (FM).

The data were subjected to an F test analysis of variance and the means were compared by using the Scott-Knott test as well as through orthogonal contrasts via the Scheffé test at 5% probability using the SISVAR® software version 5.3 (Ferreira, 2010). The performed orthogonal contrasts were Y1, lemon balm monocropping versus lemon balm intercropping, and Y2, yarrow monocropping versus yarrow intercropping.

To proceed with the essential oil extraction, 45 g of lemon balm dry leaves and 45 g of yarrow leaflets (without the petiole) were subjected to hydrodistillation in a modified Clevenger for 90 minutes. The essential oil was collected and purified through a liquid-liquid partition with dichloromethane (3 x 5 mL). The collected organic portion was treated with anhydrous magnesium sulfate. The salt was removed by simple filtration and dry organic fraction at room temperature in a gas exhaustion chapel. The essential oil content (mg 100 g^{-1} leaf dry matter) and yield (mg $plant^{-1}$) were determined. Next, the samples were stored under refrigeration at 4°C until the chemical analyses were conducted.

The chemical composition analyses were performed by using samples composed of aliquots of equal volumes of essential oil from the repetitions in each treatment. The chromatographic analyses were conducted in the Agilent 5890A chromatographer equipped with a flame ionization detector (GC-FID) and with an Agilent 5975C mass selective detector (GC-MS), using an electron-impact ionization of 70 eV in scan model at a 1.0 scan s^{-1} speed with a mass-acquisition interval of 40 - 400 m/z.

The essential oil analyses were performed in a HP-5MS column (30 cm long x 250 μm internal diameter x 0.25 μm thick). Helium was used as the carrier gas with a flow of 1.0 mL $min.^{-1}$. The temperatures in the injector and in the detector were 220 and 240°C, respectively. For the lemon balm oil, the initial temperature in the oven was 60°C with a temperature ramp of 3°C $min.^{-1}$ up to 150°C. This ramp was followed by the temperature ramp of 10°C min^{-1} up to 250°C. For the yarrow oil, the initial temperature in the oven was 60°C with temperature ramp of 3°C $min.^{-1}$ up to 200°C. This ramp was followed by a temperature ramp of 10°C $min.^{-1}$ up to 270°C, and the yarrow oil was maintained under isothermal conditions for 1 minute. The essential oil samples from both species were diluted with ethyl acetate (1%, v/v). One microliter of each sample was automatically injected using the split mode at a ratio of 1:50.

The chemical components were identified using the comparison of their retention indices concerning the co-injection of an n-alkanes standard solution (C_8-C_{20}, Sigma-Aldrich®, St. Louis, USA) and through the comparison of mass spectra in the database of the NIST/EPA/NHI (Nist, 2008) library. The retention indices were calculated using the Van den Dool and Kratz (1963) equation. The attributions of the chromatographic peaks were set based on the retention indices described in the literature (Adams, 2007).

Result and discussion

The means of the analyzed growth variables of the lemon balm were significantly influenced by the intercropping system when the variables were compared against the monocropping system (Table 1). The intercropping system presented the lowest leaf area (486.76 cm^2), stem dry matter (24.43 g) and leaf dry matter (27.99 g) values. On the other hand, there were no significant differences in the yarrow growth variables.

Table 1. Mean leaf area★ and shoot dry matter values, and the orthogonal contrast estimates of the lemon balm and yarrow monocropping and intercropping in the field.

Plant	Cultivation	Leaf area (cm^2)	Dry matter (g $plant^{-1}$)	
			Stem	Leaf
Lemon Balm	Monocropping	537.88	35.13	36.12
Lemon Balm	Intercropping	486.76	24.43	27.99
Contrast	MS *vs* MC	51.121**	10.694	8.121**
Yarrow	Monocropping	-	41.11	21.43
Yarrow	Intercropping	-	42.60	22.28
Contrast	MFS *vs* MFC	-	-1.497	-0.856
CV (%)		10.85	24.65	17.99

** Statistically significant in the Scheffé test at 5% probability. MS (lemon balm monocropping), MC (lemon balm intercropping), MFS (yarrow monocropping), MFC (yarrow intercropping).

Brandão, Silva, Jaruche, Santos, & Martins (2014) have analyzed yarrow intercropping in pots planted with lemon balm and lettuce. They found satisfactory results in the dry matter production, and their results demonstrated that the field conditions may lead to the competition for resources. A study conducted using yarrow/rosemary intercropping demonstrated that rosemary reduced growth (Arashiro et al., 2011). Other reports using yarrow intercropping have also found reduced growth of other aromatic species (Corrêa Júnior, Ming, &

Scheffer, 1998, Santos et al., 2009; Scheffer & Corrêa Júnior; 2006). However, Arashiro et al. (2011) discovered an increase in rosemary growth when it was intercropped with yarrow in fertilized treatments because it increased the competition for nutrients. Thus, the negative effect of intercropping with other aromatic species, such as rosemary and lemon balm, on the yarrow growth variables may be related to an interspecific competition associated with nutrient availability.

The photosynthetic pigment production was also influenced by the intercropping of both studied species. The intercropping system has promoted an increase in chlorophyll *a* (0.38 to 0.66), chlorophyll *b* (0.57 to 0.76), total chlorophyll (0.95 to 1.43) and carotenoid (0.43 to 0.50) amounts in lemon balm, and there was a reduction in the ratio of b/a (Table 2). Such a reduction results from more production of chlorophyll *a*. The yarrow pigments have also presented a significant increase in the intercropping system, except for the amount of chlorophyll *a*, which remained the same.

Table 2. Mean quantification values of photosynthetic pigments and the orthogonal contrast estimates in the lemon balm and yarrow monocropping and intercropping in the field.

Plant	Cultivation	Chlorophyll (mg/g MF^{-1})				Carotenoid
		a	*b*	*b/a*	Total	(mg g^{-1} MF)
Lemon balm	Monocropping	0.38	0.57	1.47	0.95	0.43
Lemon balm	Intercropping	0.66	0.76	1.17	1.43	0.50
Contrast	MS *vs* MC	0.286**	0.196**	-0.301**	0.483**	0.138**
Yarrow	Monocropping	0.63	0.61	1.00	1.23	0.56
Yarrow	Intercropping	0.65	1.13	1.81	1.78	0.58
Contrast	MFS *vs* MFC	0.21	0.526**	0.809**	0.555**	-0.789**
CV (%)		13.96	15.91	22.39	10.7	11.34

** Statistically significant in the Scheffé test at 5% probability. MS (lemon balm monocropping), MC (lemon balm intercropping), MFS (yarrow monocropping), MFC (yarrow intercropping).

Plants often cultivated in shady environments receive longer wavelengths in the far-red region. Therefore, in order to increase light absorption and maintain the energetic balance between photosystems, the plants invest in more chlorophyll *b*, total chlorophyll and the ratio of *b/a* (Cao, 2000). Accordingly, the intercropping of the studied species may have promoted a competition for light, and it may have promoted the shady plant features in yarrow. Although the chlorophyll *b* contents were higher in lemon balm, the plant showed features of plants cultivated under sunlight with a low ratio of *b/a*, more carotenoid production and a smaller leaf area. Thus, it seems that intercropping may have promoted competition for light, nutrients and water, which has stressed the lemon balm.

Intercropping increased the content and yield of lemon balm essential oils (Table 3). On the other hand, yarrow did not show significant differences between the monocropping and intercropping scenarios. The increased essential oil content in plants intercropped with yarrow was also observed in lemongrass and in rosemary (Arashiro et al., 2011; Santos et al., 2009). In addition, Corrêa Júnior et al. (1998) have stated that yarrow presents allelopathy over aromatic plants and that it favors the production of essential oils in these plants.

Table 3. Mean values of the content and yield of the essential oils and the orthogonal contrast estimates of lemon balm and yarrow monocropping and intercropping in the field.

Plant	Cultivation	Content (%)	Yield	
			(g plant^{-1})	(g ha^{-1})
Lemon balm	Monocropping	0.111	0.041	2561
Lemon balm	Intercropping	0.221	0.061	1898
Contrast	MS *vs* MC	0.110**	-0.019**	663.714
Yarrow	Monocropping	0.109	0.045	2816
Yarrow	Intercropping	0.121	0.052	1608
Contrast	MFS *vs* MFC	0.012	-0.006	1208,14**
CV (%)		21.75	25.25	30.73

** Statistically significant in the Scheffé test at 5% probability. MS (Lemon balm monocropping), MC (lemon balm intercropping), MFS (yarrow monocropping), MFC (yarrow intercropping).

The increased essential oil production may be associated with the deviation of carbon skeletons from photosynthesis to produce the secondary metabolites in lemon balm. Since the production of chlorophylls a and b improved, there is a higher proportion of PSI and PSII, respectively (Nakazono, Dacosta, Futatsugi, & Dias, 2001). Thus, the intercropping system has apparently promoted a balanced ratio between the photosystems in the lemon balm because there was increase in both chlorophylls *a* and *b*. This response in the intercropping conditions must have helped increase the photosynthetic rates and promote higher growth; this is a fact that was not observed in the present study. Accordingly, the photosynthetic products may have been used in secondary metabolism at the expense of growth and may be the response of lemon balm to survive in competitive environments.

There were no qualitative differences in the chemical composition of lemon balm essential oils between the monocropping and intercropping systems (Tables 4 and 5, respectively). Regardless of the cultivation system, the chemical nature of the lemon balm constituents was mostly monoterpene, and the lemon balm presented neral (29.68 - 27.29%) and geraniol (41.04 - 38.03%) as major constituents. However, the monocropping presented higher total monoterpene and citral (neral + geraniol) contents, with differences regarding the intercropping of 3.03 and 5.4%, respectively. The prevalence of monoterpene, geraniol and neral compounds as major constituents was recognized by other authors for the species (Brant et al., 2009; Martins & Pastori, 2004).

Table 4. Chemical composition of lemon balm leaf essential oil in monocropping and intercropping.

Constituents	Area (%) ± SD	
	Monocropping	Intercropping
Geranial	41.04 ± 1.95	38.03 ± 4.36
Neral	29.68 ± 0.77	27.29 ± 2.16
Citronellal	6.54 ± 2.47	3.89 ± 0.61
Trans-6-hydroxy-α-terpineol	2.23 ± 0.74	5.29 ± 2.28
Geranic acid	1.96 ± 0.24	2.18 ± 0.23
Methyl geranate	1.24 ± 0.44	1.74 ± 0.20
Isogeranial	1.06 ± 0.19	0.70 ± 0.15
Methyl citronellate	0.92 ± 0.19	0.61 ± 0.06
Valerate citronellate	0.90 ± 0.00	3.89 ± 1.83
Geranyl acetate	0.82 ± 0.25	0.64 ± 0.17
Iso-pinocamphone	0.73 ± 0.14	0.49 ± 0.10
Citronelol	0.51 ± 0.07	0.40 ± 0.05
Geranyl vinyl ether	0.49 ± 0.14	0.35 ± 0.09
Photocitral A	0.40 ± 0.15	0.32 ± 0.08
Prenyl angelate	0.31 ± 0.01	0.35 ± 0.05
1,8-Cineole	0.30 ± 0.11	0.26 ± 0.07
Trans-geraniol	0.29 ± 0.08	0.19 ± 0.00
Cis-pinocamphone	0.28 ± 0.09	0.29 ± 0.04
Linanool	0.23 ± 0.06	0,22 ± 0,06
α-Cyclogeraniol	0.23 ± 0.06	nd
Total Monoterpenes	90.16	87.13
Caryophyllene oxide	7.02 ± 1.52	7.32 ± 1.56
4(12),8(13)-Carophyl dien-5α-ol	0.30 ± 0.11	0.33 ± 0.10
Humulene epoxide II	0.22 ± 0.00	0.22 ± 0.01
Total Sesquiterpenes	7.54	7.87
Octen-3-ol	0.29 ± 0.12	0.22 ± 0.05
Benzaldehyde	0.16 ± 0.03	nd
Non-identified	nd	4.10 ± 0.83
Others	0.45	4.32
Total Area	98.15	99.91

SD: Standard deviation (n = 3); nd =non-detected.

Table 5. Chemical composition of yarrow leaflet essential oil in monocropping and intercropping.

Constituents	Area (%) ± SD	
	Monocropping	Intercropping
Borneol	9.74 ± 1.49	9.29 ± 1.37
Terpinen-4-ol	6.96 ± 1.40	6.76 ± 1.15
1,8-Cineol	3.18 ± 0.89	5.29 ± 1.85
α-Terpineol	2.66 ± 0.41	2.59 ± 0.35
Trans-sabinene hydrate	1.64 ± 0.22	1.58 ± 0.15
Sabinene	1.53 ± 1.02	1.70 ± 0.24
Cis-sabinene hydrate	1.62 ± 0.20	1.60 ± 0.19
α-Pinene	1.28 ± 0.48	2.23 ± 0.00
D-Limonene	0.75 ± 0.25	1.22 ± 0.48
Bornyl acetate	0.67 ± 0.09	0.64 ± 0.10
trans-2-Mentenol	0.51 ± 0.09	0.51 ± 0.08
α-Terpinene	0.34 ± 0.09	0.51 ± 0.14
Eugenol	0.29 ± 0.04	0.25 ± 0.04
Culminaldehyde	0.21 ± 0.01	0.24 ± 0.03
Total Monoterpenes	31.38	34.41
Chamazulene	33.81 ± 2.93	26.25 ± 4.75
β-Caryophyllene	12.35 ± 1.77	11.56 ± 1.82
Caryophyllene oxide	6.89 ± 0.73	6.04 ± 0.66
Spathulenol	2.47 ± 0.19	2.18 ± 0.30
α-Caryophyllene	1.55 ± 0.21	1.45 ± 0.23
D-Germacrene	0.57 ± 0.18	0.61 ± 0.17
Total Sesquiterpenes	57.64	48.09
Non-identified	1.77 ± 0.06	1.51 ± 0.25
1-Octen-3-ol	1.32 ± 0.06	0.94± 0.44
1-Nonen-3-ol	0.50 ± 0.07	nd
Non-identified	nd	6.91 ± 3.67
Others	3.59	9.36
Total Area	92.61	91.86

SD: Standard deviation (n = 3); nd = non-identified.

Regarding the quantitative analyses of yarrow leaf chemical composition, the most evident variations between the assessed cultivation systems were a 9.55% difference in the total sesquiterpenes and a 7.56% difference in the chamazulene content. The main constituents of this oil were the following: chamazulene (33.81 - 26.25%), β-caryophyllene (12.35 - 11.56%), borneol (9.74 - 9.29%), 4-terpineol (6.96 - 6.76%) and caryophyllene oxide (6.89 - 6.04%). The presence of these compounds as main constituents in yarrow was already described in other studies (Nadim, Mali, Ahmad, & Bakshi, 2011; Rahimmalek et al., 2009).

Intercropping led to a reduced amount of monoterpene compounds identified as major compounds in the lemon balm, as well as a reduction of the major constituents of sesquiterpene in yarrow. The medicinal properties attributed to both species are associated with the presence of these major compounds (Reis et al., 2009). Although intercropping has promoted an increase in essential oil production, the technique that promoted essential oil quality decreased in both species under field conditions. Thus, the intercropping between yarrow and lemon balm is not indicated under field conditions. In addition, the results found in this study have created the need for carefully assessing the quality of essential oils in future studies regarding yarrow intercropping with different aromatic species.

Conclusion

Intercropping of lemon balm and yarrow decreases growth in lemon balm, but it has no significant effect on yarrow.

The photosynthetic pigment contents change by the crop systems. Intercropped lemon balm exhibits an increase content of chlorophylls *a*, *b*, and total, and in carotenoids. While, intercropped yarrow has a significant increase in chlorophylls *b* and total, remaining the same content of chlorophyll *a* and carotenoids in comparison with monocropping system.

The content and yield of lemon balm essential oil increases in the intercropping system, whereas yarrow does not change the essential oil content. The amount of major constituents in both species drop in the intercropping system.

Acknowledgements

The authors thank Fundação de Amparo à Pesquisa do Estado de Minas Gerais (FAPEMIG), Coordenação de Aperfeiçoamento de Pessoal de Nível Superior (CAPES) and Conselho Nacional de

Desenvolvimento Científico e Tecnológico (CNPq) for the financial support and scholarships.

References

Adams, R. P. (2007). *Identification of essential oil components by gas chromatography/mass spectrometry*. Illinois, US: Allured.

Arashiro, M. P., Ziroldo, D. F., Yamaguchi, M. U., Sartor, C. F. P., Patroni, S. M. S., D'Oliveira, P. S., & Cortez, L. E. R. (2011). Cultivo consorciado entre *Achillea millefolium* L. e *Rosmarinus officinalis* L. e seu efeito no rendimento do óleo essencial, biomassa e atividade antimicrobiana. *Revista Saúde e Pesquisa*, *4*(3), 373-380.

Argyropoulos, D., & Müller, J. (2014). Changes of essential oil content and composition during convective drying of lemon balm (*Melissa officinalis* L.). *Industrial Crops and Products*, *52*, 118-124. doi: 10.1016/j.indcrop.2013.10.020

Avci, A. B., & Giachino, R. R. A. (2016). Harvest stage effects on some yield and quality characteristics of lemon balm (*Melissa officinalis* L.). *Industrial Crops and Products*, *88*, 23-27. doi: 10.1016/j.indcrop.2016.01.002

Brandão, D. S., Silva, P. H. L., Jaruche, Y. G., Santos, R. R., & Martins, E. R. (2014). Produção de biomassa e do rendimento do óleo essencial de melissa em cultivo solteiro e consorciado com mil-folhas e alface. *Ciência Rural*, *44*(9), 1513-1518.

Brant, R. S., Pinto, J. E. B. P., Rosa, L. F., Albuquerque, C. J. B., Ferri, P. H., & Corrêa, R. C. (2009). Crescimento, teor e composição do óleo essencial de melissa cultivada sob malhas fotoconversoras. *Ciência Rural*, *39*(5), 1401-1407.

Brasil. Ministério da Saúde. (2009). *Publica a Relação Nacional de Plantas Medicinais de Interesse ao SUS. RENISUS, 19 de fevereiro de 2009*. Retrieved on May 9, 2012 from http://189.28.128.100/portal/saude/profissional/vi

Cao, K. F. (2000). Leaf anatomy and chlorophyll content of 12 woody species in contrasting light conditions in a Bornean heath forest. *Canadian Journal of Botany*, *78*(10), 1245-1253.

Chamoli, M., Varshney, V. K., Srivastava, P. K., Pandey, R., & Dayal, R. (2013). Assessment of biomass yield, essential oil and β-asarone content of *Acorus calamus* L. intercropped with *Morus alba* L. *Journal of Essential Oil-Bearing Plants*, *16*(6), 763-770.

Corrêa Júnior, C., Ming, L. C., & Scheffer, M. C. (1998). *Cultivo de plantas medicinais condimentares e aromáticas*. Curitiba, PR: Emater.

Ferreira, D. F. (2010). *SISVAR: sistema de análise de variância: versão 5.3* [Software]. Lavras, MG: UFLA.

Gao, Y., Duan, A., Sun, J., Li, F., Liu, Z., Liu, H., & Liu, Z. (2009). Crop coefficient and water-use efficiency of winter wheat/spring maize strip intercropping. *Field Crops Research*, *111*(1-2), 65-73.

Grangeiro, L. C., Negreiros, M. Z., Santos, A. P., Costa, L. M., Silva, A. R. C., & Lucena, R. R. M. (2008). Crescimento e produtividade de coentro e rabanete em função da época de estabelecimento do consórcio. *Ciência e Agrotecnologia*, *32*(1), 55-60.

Lichtenthaler, H. K., & Buschmann, C. (2001). Chorophylls and carotenoids: measurement and characterization by UV VIS spectroscopy. In R. E. Wrolstad (Ed.), *Current protocols in food analytical chemistry* (p. 431-438). Madison, US: John Wiley & Sons.

Luz, J. M. Q., Silva, S. M., Habber, L. L., & Marquez, M. O. M. (2014). Produção de óleo essencial de *Melissa officinalis* L. em diferentes épocas, sistemas de cultivo e adubações. *Revista Brasileira de Plantas Medicinais*, *16*(3), 552-560.

Martins, M. B. G., & Pastori, A. P. (2004). Anatomia foliar com ênfase nos tricomas secretores e análise cromatográfica do óleo essencial de *Melissa officinalis* L. (Lamiaceae). *Revista Brasileira de Plantas Medicinais*, *6*(2), 77-82.

Nakazono, E. M., Dacosta, M., Futatsugi, K., & Paulilo, M. T. S. (2001). Early growth of *Euterpe edulis* Mart., in different light environments. *Revista Brasileira de Botânica*, *24*(2), 173-179.

Nadim, M. M., Mali, K. A. A., Ahmad, J., & Bakshi, S. K. (2011). The essential oil composition of *Achillea millefolium* L. cultivated under tropical condition in India. *World Journal of Agricultural Sciences*, 7(5), 561-565.

National Institute of Standards and Technology [NIST]. (2008). *PC version 2.0 of the NIST/EPA/NIH mass spectral library* [Gaithersburg Software]. Gaithersburg, MD: U.S. Department of Commerce.

Pinto, J. E. B. P., Ferraz, E. O., Bertolucci, S. K. V., Silveira, H. R. O., Santos, A. R., & Silva, G. M. (2014). Produção de biomassa e óleo essencial em mil folhas cultivadas sob telas coloridas. *Horticultura Brasileira*, *32*(3), 321-326.

Rahimmalek, M., Tabatabaei, B. E. S., Etemadi, N., Goli, S. A. H., Arzani, A., & Zeinali, H. (2009). Essential oil variation among and within six *Achillea* species transferred from different ecological regions in Iran to the field conditions. *Industrial Crops and Products*, *29*(2-3), 348-355.

Reis, E. S., Pinto, J. E. B. P., Rosado, L. D. S., & Corrêa, R. M. (2009). Teor e composição química do óleo essencial de *Melissa officinalis* L. *in vitro* sob a influência do meio de cultura. *Acta Scientiarium. Agronomy*, *31*(2), 331-335.

Sá Júnior, A., Carvalho, L. G., Silva, F. F., & Alves, M. C. (2012). Application of the Koppen classification for climatic zoning in the state of Minas Gerais, Brazil. *Theoretical* and Applied *Climatology*, *108*(1-2), 1-7.

Santos, H. G. (2013). *Sistema brasileiro de classificação de solos* (3a ed.). Brasília, DF: Embrapa.

Santos, A., Paduan, R. H., Gazin, Z. C., Jacomassi, E., D'Oliveira, P. S., Cortez, D. A. G., & Cortez, L. E. R. (2009). Determinação do rendimento e atividade antimicrobiana do óleo essencial de *Cymbopogon citratus* (DC.) Stapf em função de sazonalidade e consorciamento. *Revista Brasileira de Farmacognosia, 19*(2A), 436-441.

Scheffer, M. C., & Corrêa Júnior, C. (2006). *Boas Práticas Agrícolas (BPA) de plantas medicinais, aromáticas e condimentares*. Brasília, DF: MAPA/SDC.

Singh, M., Singh, U. B., Ram, M., Yadav, A., & Chanotiya, C. S. (2013). Biomass yield, essential oil yield and quality of geranium (*Pelargonium graveolens* L. Her.) as influenced by intercropping with garlic (*Allium sativum* L.) under subtropical and temperate climate of India. *Industrial Crops and Products, 46*, 234-237. doi: 10.1016/j.indcrop.2013.01.032

Sorensen, J. M. (2000). Melissa officinalis L. *The Internacional Journal of Aromatherapy, 10*(1/2), 7-15.

Szabó, K., Malekzadeh, M., Radácsi, P., Ladányi, M., Rajhárt, P., Inotai, K., ... Németh, E. (2016). Could the variety influence the quantitative and qualitative outcomeof lemon balm production? *Industrial Crops and Products, 83*, 710–716. doi: 10.1016/j.indcrop.2015.12.027

Yadegari, M. (2016). Effect of micronutrients foliar application and biofertilizeres on essential oils of lemon balm. *Journal of Soil Science and Plant Nutrition, 16*(3), 702-715.

Van den Dool, H., & Kratz, P. D. (1963). A generalization of the retention index system including linear temperature programmed gas liquid partition chromatography. *Journal of Chromatography, 11*(1), 463-471.

Wang, M., Wu, C., Cheng, Z., Meng, H., Zhang, M., & Zhang, H. (2014). Soil Chemical Property Changes in Eggplant/Garlic Relay Intercropping Systems under Continuous Cropping. *Plos one, 9*(10), 1-13. doi: 10.1371/journal.pone.0111040

18

In vitro growth of sweet potato fed with potassium phosphite

Edison Hitoshi Hirosse, José Eduardo Creste, Ceci Castilho Custódio and Nelson Barbosa Machado-Neto*

*Laboratório de Cultura de Tecidos, Faculdade de Ciências Agrárias, Universidade do Oeste Paulista, Rod. Raposo Tavares, Km 572, 19067-175, Presidente Prudente, São Paulo, Brazil. *Author for correspondence. E-mail: nbmneto@unoeste.br*

ABSTRACT. The main objective of this work was to verify the effect of the substitution of phosphate by its related compound phosphite on the *in vitro* growth of sweet potato nodal segments. Segments from a previous culture were incubated in MS media with the following combinations of phosphate:phosphite: Control (100 / 0%); T1 (87.5 / 12.5%); T2 (75 / 25%); T3 (50 / 50%) and T4 (0 / 100%) in a 5 x 5 factorial arrangement, with five treatments (Control, T1, T2, T3 and T4) and five evaluation periods (7, 14, 21, 28 and 35 days). The design was completely random, with three replicates of two explants per period. The evaluations were made weekly by counting the number of shoots and the length and dry weight of shoots and roots. Increasing the phosphite concentration decreased all attributes measured, including growth velocity, length and dry weight. These results indicate that phosphite cannot be used as substituted of phosphate in sweet potato tissue cultures.

Keywords: *Ipomoea batatas*, nutrition, plantlet, shoot, root.

Introduction

Phosphorus (P) is one of the main nutrients for plants. It is a component of the nucleotide triphosphates (NTP) generated during respiration and photosynthesis and is a component of DNA. It is important in flowering, fruiting and root development. When plants lack phosphorus, their leaves acquire a bluish-green colour and then purple tones, followed by yellowing (ABEL et al., 2002; MALAVOLTA et al., 2002). The primary difference between phosphite (designated Phi in this paper) and phosphate (designated Pi in this paper) is that the Phi has a hydrogen atom in place of one oxygen atom (McDONALD et al., 2001). To be metabolised, Pi reacts with the enzyme phosphatase. This enzyme recognises three of the four oxygen atoms and binds to them. The last atom of oxygen becomes available to react with other catalysts.

Phosphite has only three atoms of oxygen. One hydrogen atom takes the place of the other oxygen. For this reason, continuation of the metabolism of Phi is not possible. Thus, Phi cannot participate in the same biochemical reactions as Pi. Phi is discarded by most of the enzymes involved in transfer reactions of phosphorus (PLAXTON, 1998). However, Orbovi et al. (2008) have suggested that Phi could replace Pi for plant nutrition.

Phosphites are liquid products originating from the neutralisation of phosphorous acid (H_3PO_3) with a base. The base used may be sodium, potassium, ammonium hydroxide or other compounds. The base most commonly used for this purpose is potassium hydroxide. This base generates the compound potassium phosphite. Potassium phosphite has excellent plant health facilitation qualities and fungicidal activity. It either acts directly on the fungi or activates the defence mechanisms of plants to induce the production of phytoalexins (RATJEN; GERENDÁS, 2009).

Phi is quickly absorbed by the roots, leaves and trunk bark and therefore entails a lower energy requirement for absorption by the plant. As an effective chelating agent, it favours the absorption of K, Ca, B, Zn, Mo, Mn and other nutrients. It can be mixed with other products. Some formulations of phosphites can reduce the pH of a solution, thereby improving the efficiency of some herbicides (LOVATT; MIKKELSEN, 2006) and improving plant defences (ORBOVI et al., 2008).

One of the reasons for the ineffectiveness of potassium phosphite for elevating productivity is that products based on phosphorous acid are not considered to be good sources of phosphorus. Generally, these products fail to supply the demand for inorganic phosphate for use in energy transport (STEHMANN; GRANT, 2000).

Phosphites probably do not have sufficient nutritional effects to increase crop productivity. Although Phi compounds are marketed as fertilisers containing phosphorus, an increase in productivity is not expected from application of this nutrient. McDonald et al. (2001), Thao and Yamakawa (2008, 2009), and Thao et al. (2008, 2009) have all argued that Phi cannot be used as a phosphorus source, either by crops or by growing cells. According to Singh et al. (2003), cell cultures grown using Phi as a nutrient have exhibited increased cell death, protein deterioration and DNA damage, perhaps owing to the erroneous incorporation of Phi in place of Pi. In contrast, Orbovi et al. (2008) assert that Phi could be used by plants as a nutrient.

Other effects of Phi application mentioned in the literature include the nutritional balance of plants, better ripening and fruit quality, and superior postharvest quality (BRACKMANN et al., 2004; LOBATO et al., 2008; MOOR et al., 2009; NOJOSA et al., 2005).

The sweet potato (*Ipomoea batatas* L. (Lam)) is native to Central and South America, from the Yucatan Peninsula, Mexico, to Colombia. Its use dates back more than ten thousand years, according to analysis of dried potatoes found in caves located in Chilca Canyon, Peru and evidence found in archaeological writings in the Mayan region of Central America (PEARSALL, 2008).

The sweet potato is grown as an annual crop. It stores food reserves in its roots. These roots have considerable alimentary, industrial and high-calorie (carbohydrate) potential. They contain iron, calcium and phosphorus, and they are rich in vitamins A, B and C (SILVA et al., 2002). The sweet potato responds to Pi provision by increasing the accumulation of starch in its reserve tissues (OLIVEIRA et al., 2005).

Plant tissue culturing involves a set of techniques having broad applications in agriculture. These techniques first isolate small fragments of living tissue called explants. Explants may be a fragment of leaf, root, or stem that can respond to the conditions of the inducing medium. Explants with in vitro regeneration capability are isolated from a plant, disinfected and grown aseptically for indefinite periods in an appropriate culture medium (TORRES et al., 1998). The goal is to obtain a new plant identical to the original, i.e., to make a clone. Cloning is defined as the asexual propagation of cells or organisms to obtain a new individual while maintaining the donor's genotype (TORRES et al., 1998).

The objective of this work was to verify the effects of phosphite on the growth of the sweet potato when phosphite was substituted partially or totally for a phosphate source.

Material and methods

The work was conducted in the Tissue Culture Laboratory of the Faculdade de Ciências Agrárias da Universidade do Oeste Paulista (UNOESTE, Presidente Prudente, São Paulo State).

The study used a cultivar of sweet potato (*Ipomoea batatas* (L.) Lam) clone 1208 donated by the Germplasm Bank of the Agência Paulista de Tecnologia dos Agronegócios (APTA), Pólo Regional da Alta Sorocabana, Presidente Prudente-SP. The potatoes were collected, washed under running tap water with commercial soap, and treated with a disinfecting solution containing 1% Benomyl. The potatoes were placed in plastic trays containing washed sand to allow budding and the formation of nodal segments.

The stems were washed and disinfected with a commercial solution of 20% sodium hypochlorite (NaClO) and 100 μL of Triton X100 for 20 minutes, rinsed with autoclaved distilled water, and then cut into segments having two nodes. The culture medium used was MS (MURASHIGE; SKOOG, 1962), jellified with 2 g L^{-1} Phytagel and 30 g L^{-1} sucrose, pH adjusted to 5.9, aliquoted in 50 mL flasks and autoclaved at 121°C for 20 min. The plants obtained from this first culture provided the segments for the experiment.

Treatments were conducted using the changes in molar concentrations of phosphorus produced by the exchange of hydrogenous dipotassium phosphate (K_2HPO_4) in equimolar amounts by potassium phosphite (K_3PO_3). For this purpose, stock solution I of MS medium, which contains KNO_3 and KH_2PO_4, was divided into MSIa, containing 9.5 g of KNO_3 100 mL^{-1} water, and MSIb, containing only phosphate (850 mg KH_2PO_4 100 mL $water^{-1}$). Thus, during the process of exchange, the phosphite was balanced with nitrogen, phosphorus and potassium in the solution with nitric acid, MSIa, MSIb and potassium phosphite, as shown in Table 1.

Table 1. Chemical composition of the culture media used, with corrections to the final concentration of potassium, nitrogen and phosphorus. The initial solution consisted of 0:30:20 (N:P:K) potassium phosphite.

Treatments	MS solutions used		Complements
Control	0% HPO_3^{-2} 100% PO_4^{-3}	Complete MS full strength (I to VII)	None
T1	12.5% HPO_3^{-2} 87.5% PO_4^{-3}	MS (II to VII)	0.25 g Phosphite 19.3 mL MSIa 60.5 μL HNO_3 17.5 mL MSIb
T2	25% HPO_3^{-2} 75% PO_4^{-3}	MS (II to VII)	0.5 g Phosphite 18.6 mL MSIa 121 μL HNO_3 15 mL MSIb
T3	50% HPO_3^{-2} 50% PO_4^{-3}	MS (II to VII)	1.0 g Phosphite 15.9 mL MSIa 242.5 μL HNO_3 10 mL MSIb
T4	100% HPO_3^{-2} 0% PO_4^{-3}	MS (II to VII)	2.0 g Phosphite 485 μL HNO_3 10.4 mL MSIa

The plantlets obtained by micropropagation were removed from the culture bottles under laminar flow and sectioned into nodal segments containing two nodes. These segments were transferred to bottles, each containing two explants and the culture media with different concentrations of Phi. The bottles were sealed with plastic wrap and incubated in a growth room at 25 ± 3°C with a photoperiod of 16 hours for 35 days. After this period, the bottles were opened and the plantlets were removed and washed in distilled water to remove excess medium. The plantlets were then stretched and measured using a ruler. The measurements taken included the greatest lengths of shoots and roots separately. The number of shoots and roots per explant were counted every seven days.

To assess the dry weight, plant material was placed in paper bags and dried in an oven at 65°C for 72 hours. The material was kept in a desiccator to cool and then weighed on an analytical balance accurate to 0.00001 g (NAKAGAWA, 1999).

A 5 x 5 factorial scheme with 5 treatments (T, T1, T2, T3 and T4) and 5 evaluation periods (7, 14, 21, 28 and 35 days) was used. The experiment used a completely randomised design with three replicates per trial period. Each plot consisted of two explants. Regression equation curves were also calculated, and the statistical significance of the regressions was assessed. Data analysis was performed with the aid of SISVAR software (FERREIRA, 2008).

Results and discussion

One of the main difficulties facing successful establishment of the in vitro culture was disinfection. The branches collected from the mother plant, especially those collected from field material (CHAVES et al., 2005; ERIG; SCHUCH, 2003; PINTO et al., 2002), exhibited high levels of contamination. Such contamination does not occur in new sprouts that grow from potatoes placed in humid sand.

Growth occurred in all treatments with differing concentrations of Phi. The control grew more at all times. T4, with a higher concentration of Phi, had the lowest growth found, either for shoots or for roots (Figure 1A and B).

Different concentrations of Phi altered development. Reduced plantlet growth was observed along with rapid decay. The control and the treatment with the lowest dose of Phi (T1) maintained plant growth until 29 days (maximum point). The other three treatments shortened the time of growth to 22, 23 days, and no growth, respectively, for the higher Phi doses (Figure 1A). Shoot length also decreased from 7.13 cm in the control treatment to 2.7 cm in the treatment with 50% Phi (T3).

The control exhibited maximum root growth until the 30th day. The roots of plantlets treated with the lowest dose of Phi kept growing until the 32nd day. The treatments with higher Phi concentrations altered root growth by shortening the time of growth to 24 days, 20 days and no growth, respectively (Figure 1B). Root length also decreased from 24.08 cm in the control to 9.57 cm in the T3 treatment (50% Phi).

Shoots developed in all treatments (Figure 2), including those with the various Phi concentrations. The control, T1 and T2 exhibited the highest shoot numbers, and they differed very little (1.13 to 1.1). The T3 and T4 treatments, which had the highest Phi concentrations, produced the lowest numbers of shoots (0.78 and 0.53, respectively).

The Phi concentrations affected the shoot numbers of the plantlets. These treatments reduced the shoot numbers and did so rapidly. The control and the T1 treatment (lower Phi dose) maintained the same numbers of shoots on the plants until the 26th and 27th days, respectively. Treatments T2 and T4 shortened this time to 24 days. T3 maintained growth until the 25th day. T1 exhibited faster initial development, but this was not subsequently maintained. This result may have been obtained because the Phi and the Pi have similar initial uses by the plants, whereas during the subsequent growth the availability of Pi was not sufficient to allow the plantlets to maintain their initial growth rate (Figure 2).

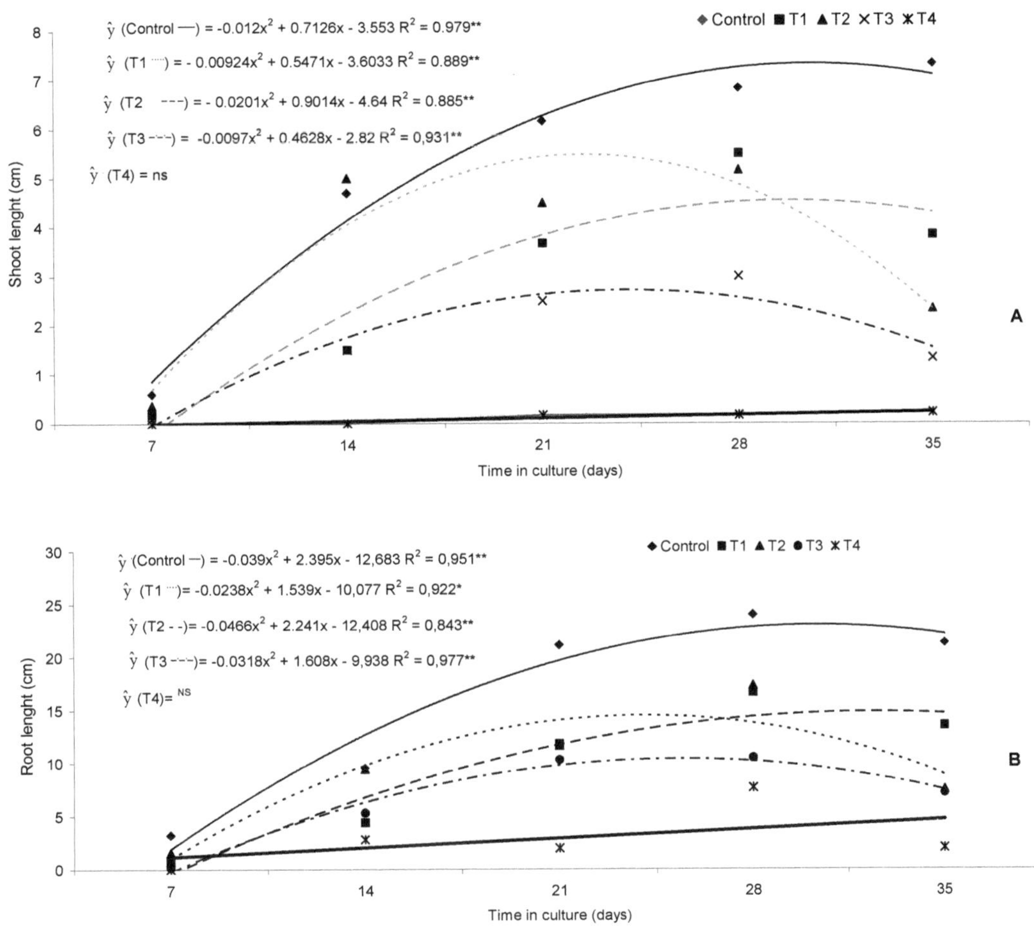

Figure 1. Sweet potato plantlet shoot (A) and root (B) length in media containing different doses of phosphorus (phosphite and phosphate sources) during 35 days of growth.

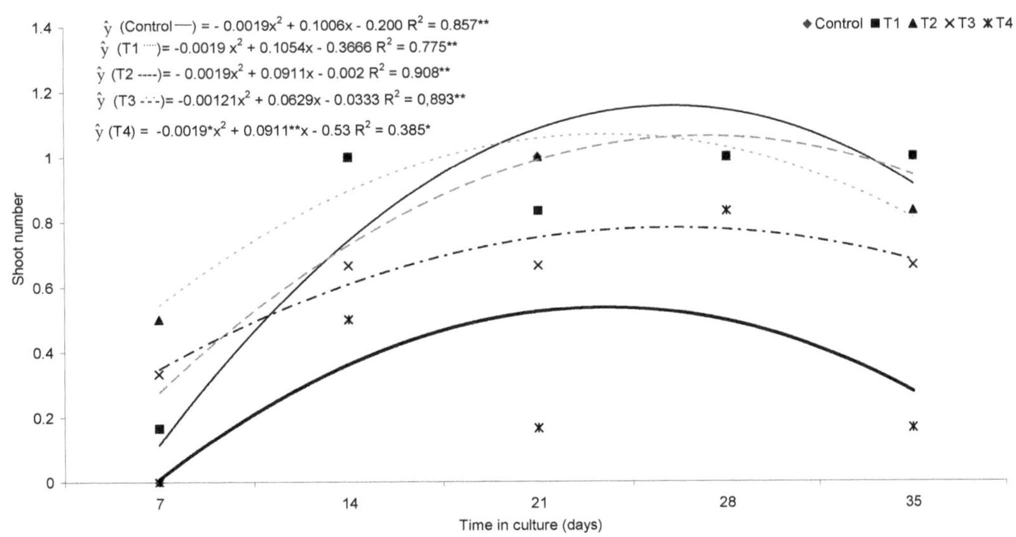

Figure 2. Shoot numbers in sweet potato plantlets in media containing different doses of phosphorus (phosphite and phosphate sources).

Dry mass accumulation (Figure 3A and B) was high for the control treatment, but it decreased as the Phi dose increased. Maximum growth was reached at 28 and 30 days for shoots and roots (Figure 3A and B), respectively, in the control. T1, the lowest Phi dose treatment, showed a maximum shoot dry mass of 0.013 g at 33 days after inoculation. T2 exhibited a short duration of dry mass accumulation (23 days) and 0.015 g of maximum shoot dry mass. T3 exhibited a longer period of dry mass increase (26 days), but the accumulation of dry mass exhibited in this treatment was the lowest observed (0.0057 g) (Figure 3A).

The root dry mass (Figure 3B) showed an increase of 0.00046 g day^{-1} for treatment T1. The root dry mass reached its growth maximum at 21 days after inoculation in treatment T2, 0.007 g. T3 exhibited the lowest dry matter accumulation, 0.00021 g day^{-1} (Figure 3B).

The decreases observed in all measurements show that Phi cannot be used as a phosphorus source. This effect is linked to the high demand for Pi in carbohydrate metabolism. Substitution of Pi by Phi interferes with the capacity for synthesis of substances that function in storage or structure (starch, sucrose and cellulose). The cell maintains the functioning of its basal metabolism as a primary priority, and accordingly, synthesis is limited under these conditions. The mechanism giving rise to this effect is that Phi, unlike Pi, cannot be used as an essential nutrient. Phosphate has important cellular functions (MALAVOLTA et al., 1997). Phi can be used only after conversion to microbial phosphate.

In addition, Phi is a potent inhibitor of the main route of energy production in fungi: the respiration pathway (STEHMANN; GRANT, 2000). In this case, the tissues are very tender, and uptake of Phi is not limited. The plant can absorb Phi but cannot use it for the most basic functions, namely the transport of energy through nucleotide triphosphates (NTP), or the genetic processes (replication, transcription and translation) that require a supply of appropriate NTP.

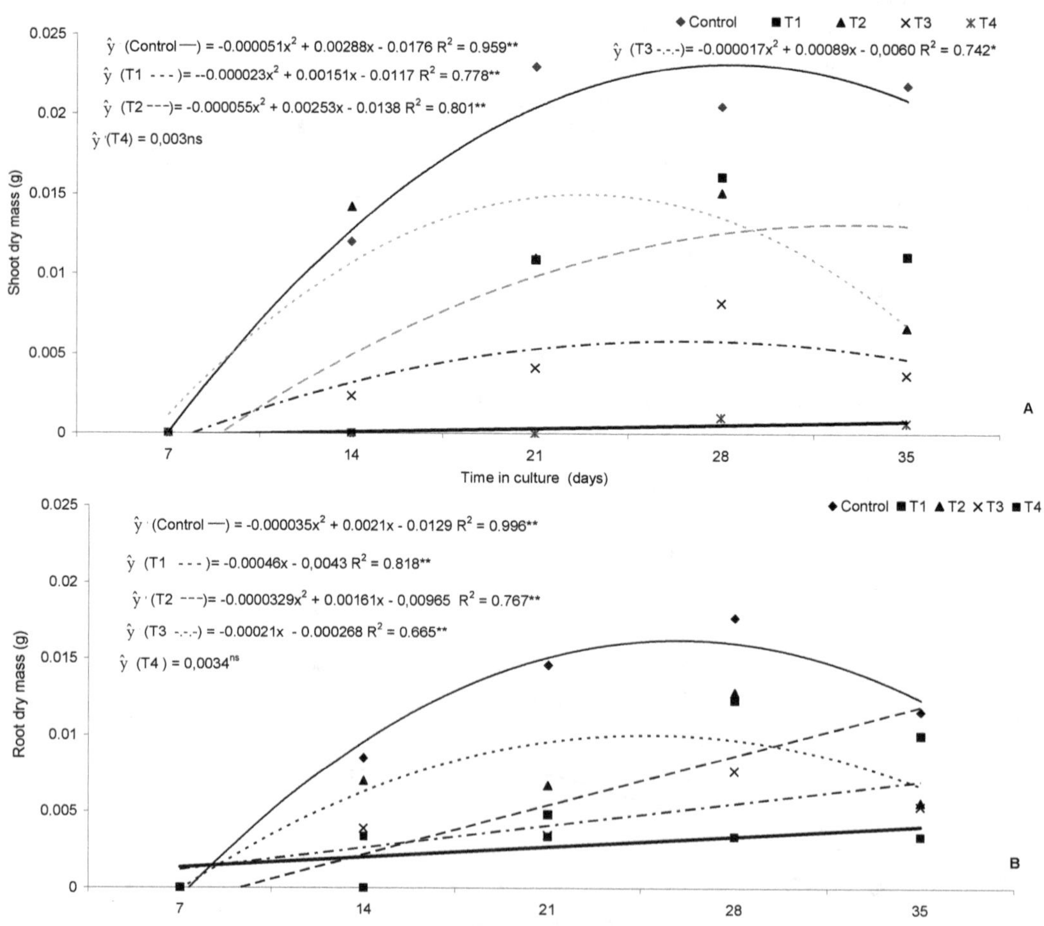

Figure 3. Shoot (A) and root (B) dry mass in sweet potato plantlets in media containing different doses of phosphorus (phosphite and phosphate sources) during 35 days of growth.

Thao et al. (2008, 2009) have investigated different cultures and have shown that Phi is not an appropriate phosphorus source for plant nutrition. Their data dramatically reveal the negative effects of this salt on plant nutrition.

In older tissues, the effects are less drastic than in growing tissues, in which Phi represents the sole source of phosphorus. Less drastic effects are exhibited because the mature tissues can translocate the existing phosphorus and thereby reduce the effects of Phi on developing tissues. In cell cultures of *Brassica napus* subjected to phosphorus deficiency, the older cells accelerate their decay and death, thereby providing phosphate for growing cells (SINGH et al., 2003).

Conclusion

The replacement of phosphate by phosphite in all the doses studied here inhibited the growth of sweet potato explants.

Phosphite cannot be used as a phosphorus source for *in vitro* cultures of the sweet potato.

References

ABEL, S.; TICCONI, C. A.; DELATORRE, C. A. Phosphate sensing in higher plants. **Physiologia Plantarum**, v. 115, n. 1, p. 1-8, 2002.

BRACKMANN, A.; GIEHL, R. F. H.; SESTARI, I.; STEFFENS, C. A. Fosfitos para o controle de podridões pós-colheita em maçãs Fuji durante o armazenamento refrigerado. **Ciência Rural**, v. 34, n. 4, p. 1039-1042, 2004.

CHAVES, A. C.; SCHUCH, M. W.; ERIG, A. C. Estabelecimento e multiplicação in vitro de *Physalis peruviana* L. **Ciência e Agrotecnologia**, v. 29, n. 6, p. 1281-1287, 2005.

ERIG, A. C.; SCHUCH, M. W. Tipo de explante e controle da contaminação e oxidação no estabelecimento *in vitro* de plantas de Macieira (*Malus domestica* Borkh.) Cvs. Galaxy, Maxigala e Mastergala. **Revista Brasileira de Agrociência**, v. 9, n. 3, p. 221-227, 2003.

FERREIRA, D. F. SISVAR: um programa para análises e ensino de estatística. **Revista Symposium**, v. 6, n. 2, p. 36-41, 2008.

LOBATO, M. C.; OLIVIERI, F. P.; GONZÁLEZ ALTAMIRANDA, E. A.; WOLSKI, E. A.; DALEO, G. R.; CALDIZ, D. O.; ANDREU, A. B. Phosphite compounds reduce disease severity in potato seed tubers and foliage. **European Journal of Plant Pathology**, v. 122, n. 3, p. 349-358, 2008.

LOVATT, C. J.; MIKKELSEN, R. L. Phosphite fertilizers: What are they? Can you use them? What can they do? **Better Crops**, v. 90, n. 4, p. 11-13, 2006.

MALAVOLTA, E.; VITTI, G. C.; OLIVEIRA, S. A. **Avaliação do estado nutricional das plantas**: princípios e aplicações, 2. ed. Piracicaba: Potafós, 1997.

MALAVOLTA, E.; FAVARIN, J. L.; CABRAL, C. P.; HEINRICHS, R.; SILVEIRA, J. S. M. Repartição de nutrientes nos ramos, folhas e flores do cafeeiro. **Pesquisa Agropecuária Brasileira**, v. 37, n. 7, p. 1017-1022, 2002.

McDONALD, A. E.; GRANT, B.; PLAXTON, W. C. Phosphite (phosphorous acid): Its relevance in the environment and agriculture and influence on plant phosphate starvation response. **Journal of Plant Nutrition**, v. 24, n. 10, p. 1505-1519, 2001.

MOOR, U.; PÕLDMAA, P.; TÕNUTAREA, T.; KARPA, K.; STARASTA, M.; VOOL, E. Effect of phosphite fertilization on growth, yield and fruit composition of strawberries. **Scientia Horticulturae**, v. 119, n. 3, p. 264-269, 2009.

MURASHIGE, T.; SKOOG, F. A revised medium for rapid growth and bio-arsays with tabacco tissue cultures. **Physiologia Plantarum**, v. 15, n. 3, p. 473-497, 1962.

NAKAGAWA, J. Testes de vigor baseados no desempenho das plântulas In: KRZYZANOWSKI, F. C.; VIEIRA, R. D.; FRANÇA NETO, J. B. (Ed.). **Vigor de sementes**: conceitos e testes. Londrina: Abrates, 1999. p. 1-24.

NOJOSA, A. B. A.; RESENDE, M. L. V.; RESENDE, A. V. **Indução de resistência de plantas a patógenos e insetos**. Usos de fosfitos e silicatos na indução de resistência. Piracicaba: Fealq, 2005.

OLIVEIRA, A. P. J.; SILVA, E. L.; PEREIRA, W. E.; BARBOSA, L. J. N. Produção de batata-doce e teor de amido nas raízes em função de doses de P_2O_5. **Acta Scientiarum. Agronomy**, v. 27, n. 4, p. 747-751, 2005.

ORBOVI, V.; SYVERTSEN, J. P.; BRIGHT, D.; VAN CLIEF, D. L.; GRAHAM. J. H. Citrus seedling growth and susceptibility to root rot as affected by phosphite and phosphate. **Journal of Plant Nutrition**, v. 31, n. 4, p. 774-787, 2008.

PEARSALL, D. M. Plant domestication and the shift to agriculture in the Andes. In: SILVERMAN, H.; ISBELL, W. H. (Ed.) **The handbook of South American archaelogy**. New York: Springer Science, 2008. p. 105-120.

PINTO, G.; VALENTIM, H.; COSTA, A.; CASTRO, S.; SANTOS, C. Somatic embryogenesis in leaf callus from a mature *Quercus suber* L. tree. ***In Vitro* Cellular and Development Biology Plant**, v. 38, n. 6, p. 569-572, 2002.

PLAXTON, W. C. Metabolic aspects of phosphate starvation in plants, phosphorus. In: LYNCH, J. P.; DEIKMAN, J. (Ed.). **Plant biology**: regulatory roles in molecular, cellular, organismic, and ecosystem processes. Rockville: American Society of Plant Physiologists, 1998. p. 229-241.

RATJEN, A. M.; GERENDÁS, J. A critical assessment of the suitability of phosphite as a source of phosphorus **Journal of Plant Nutrition and Soil Science**, v. 172, n. 6, p. 821-828, 2009.

SILVA, J. B. C.; LOPES, C. A.; MAGALHÃES, J. S. Cultura da batata-doce. In: CEREDA, M. P. (Ed.). **Agricultura**: tuberosas amiláceas Latino-Americanas. São Paulo: Cargill, 2002. v. 2, p. 449-503.

SINGH, V. K.; WOOD, S. M.; KNOWLES, V. L.; PLAXTON, W. C. Phospite accelerates programmed cell death in phosphate starved oilseed rape (*Brassica napus*) suspension cell cultures. **Planta**, v. 218, n. 2, p. 233-239, 2003.

STEHMANN, C.; GRANT, B. R. Inhibition of enzymes of the glycolytic pathway and hexose monophosphate bypass by phosphonate **Pesticide Biochemistry and Physiology**, v. 67, n. 1, p. 13-24, 2000.

THAO, H. T. B.; YAMAKAWA, T. Growth of celery (*Apium graveolens* var. *dulce*) as influenced by phosphate. **Journal of the Faculty of Agriculture**, v. 53, n. 2, p. 375-378, 2008.

THAO, H. T. B.; YAMAKAWA, T. Phosphite (phosphorous acid): Fungicide, fertilizer or bio-stimulator? **Soil Science and Plant Nutrition**, v. 55, n. 2, p. 228-234, 2009.

THAO, H. T. B.; YAMAKAWA, T.; SHIBATA, K. Effect of phosphite-phosphate interaction on growth and quality of hydroponic lettuce (*Lactuca sativa*). **Journal of Plant Nutrition and Soil Science**, v. 172, n. 3, p. 378-384, 2009.

THAO, H. T. B.; YAMAKAWA, T.; MYINT, A. K.; SARR, P. S. Effects of phosphite, a reduced form of phosphate, on the growth and phosphorus nutrition of spinach (*Spinacia oleracea* L.) **Soil Science and Plant Nutrition**, v. 54, n. 5, p. 761-768, 2008.

TORRES, A. C.; CALDAS, L. S.; BUSO, J. A. **Cultura de tecidos e transformação genética em plantas**. Brasília: Embrapa/SPI/CNPH, 1998. v. 1.

19

Application of lime, phosphogypsum and fertilization rates affect soil fertility and common bean development in no-tillage system in a Cerrado Oxisol

Maria da Conceição Santana Carvalho and Adriano Stephan Nascente*

*Empresa Brasileira de Pesquisa Agropecuária, Embrapa Arroz e Feijão, Km12, Zona Rural GO-462, 75375-000, Santo Antônio de Goiás, Goiás, Brazil. *Author for correspondence. E-mail adriano.nascente@embrapa.br*

ABSTRACT. There is a lack of information about the effects of interactions among lime, phosphogypsum (PG) and fertilization on soil fertility and their effects on common bean yields in the tropics. This study aimed to determine the effects of annual application of PG and limestone and rates of fertilization on the chemical attributes of soil, fertilization efficiency, yield components and the grain yield of the common bean. The study was performed for three growing seasons in an acidic Oxisol. The experimental design was randomized blocks in a 4x4 factorial scheme. The treatments consisted of lime, lime + PG, PG, and control (without corrective application) with four fertilization rates. In an acidic and low-fertility Cerrado soil, limestone or limestone + PG applied annually at 1/3 of the total rate in a no tillage system increased the common bean yield and the efficiency of the fertilization, although the improvement of the soil acidity indicators did not reach the desired levels. The application of soluble fertilizers to provide adequate and balanced amounts of nutrients provided greater yield gains and improved soil fertility compared to surface liming without fertilization, but the combination of the two practices resulted in the greatest benefits to both plant and soil fertility.

Keywords: *Phaseolus vulgaris*; soil acidity; calcium carbonate; liming; phosphogypsum.

Introduction

The common bean (*Phaseolus vulgaris* L.) has great social and economic importance in several countries as an important source of protein. Despite its importance, technology is not frequently used by dry bean farmers, resulting in a global average grain yield of only 866 kg ha^{-1} (FAO, 2014). However, there are reports of yield ranging from 3,023 to 3,500 kg ha^{-1} (Nascente, Kluthcouski, Crusciol, Cobucci, & Oliveira, 2012). The common bean is a crop considered highly demanding of soil fertility due to the short life cycle and the superficial, underdeveloped root system. Thus, correction of soil acidity and a balanced supply of nutrients are key factors in achieving significant increases in grain yield (Fageria, Baligar, & Jones, 2011).

In a no-tillage system (NTS), soil acidity has been corrected by lime application on the surface without incorporation (Conyers, Heenan, McGhie, & Poile, 2003; Caires, Garbuio, Churka, Barth, & Corrêa, 2008; Churka Blum, Caires, & Alleoni, 2013). However, the limestone reaction is usually limited to the site of its application, since it has low soil mobility, which leads to questions about the viability of this practice to correct soil acidity in deeper layers (Caires et al., 1998; Soratto & Crusciol, 2008a). Previous studies conducted in Cerrado soil showed that the common bean crop responds strongly to improvement in soil acidity indexes (Fageria, 2008). Optimum values for maximum common bean grain yield in the 0 - 0.20 m soil layer were pH (in water) 6.5 and base saturation of 67% (Fageria, 2008). However, in the Cerrado region, it is common to find cropping areas with 30 - 35% base saturation in the 0 - 0.20 m layer, since other crops rotated with common beans, such as soybean and corn, are less sensitive to soil acidity indexes. In those cases, the required amounts of limestone to raise base saturation to 65 - 70% may be greater than 3 t ha^{-1}, and a single application without incorporation may cause nutritional imbalances due to a possible super-liming effect on the topsoil (Padua, Silva, & Melo, 2006). An alternative is to split the calculated dose into two or three applications, but there is a lack of information about the efficiency of this practice for the common bean crop in the short and medium term.

It is also known that liming is one of the most important practices employed to increase the efficiency of fertilizer use, since the availability of most nutrients in the soil is optimal with pH (in water) between 6.0 and 7.0, except for cationic micronutrients, whose availability decreases with increasing pH (Fageria & Nascente, 2014). In addition, the neutralization of exchangeable Al and the supply of Ca favor the growth and distribution of roots in the soil profile, facilitating the uptake of nutrients applied through fertilizers. However, the application of high doses of fertilizers, mainly nitrogen fertilizers, causes soil acidification (Barbosa Filho, Fageria, & Silva, 2004). In recent years, it has been observed that the rates of fertilizer application to crops cultivated in Cerrado soils have been increasing, while liming has often been neglected. This behavior in tropical regions can be explained in part by the insufficiency of research results showing the interaction between these factors on soil fertility, plant nutrition and crop productivity in the different production systems.

For soils from the Cerrado region, the application of phosphogypsum (PG) as a subsoil improver is recommended when, in the 0.20 - 0.40 m and 0.40 - 0.60 m soil layers, the aluminum saturation is higher than 20% or the Ca content is lower than 5.0 $mmol_c$ dm^{-3} (Sousa & Lobato, 2004). However, there is a lack of information on the response of the common bean to the application of PG alone or in combination with dolomitic limestone and their interactions with the application of increasing rates of fertilization in Cerrado soils in common bean crops. It is important to understand these interactions in order to establish management strategies that improve nutrient-use efficiency in long-term crop production, focusing on sustainable intensification.

This study was conducted to answer the following questions: First, in clay Cerrado soil cultivated in NTS, is the annual application of 1/3 of the total amount of limestone and PG on the soil surface without incorporation an effective practice to improve the soil acidity indicators to the levels required by the common bean crop and to increase the efficiency of fertilizer use? Second, how does the common bean grown with irrigation in a no-tillage system respond to different levels of fertilization in a Cerrado soil, with or without the annual application of limestone and PG. Third, what are the effects on soil fertility up to a depth of 0.40 m? Therefore, the study aimed to determine the effect of annual application of PG and limestone on the soil surface without incorporation and of rates of fertilization on the chemical attributes of the top layer of the soil, as well as on fertilization efficiency, yield components and the grain yield of irrigated common bean crops grown in a no-tillage system in the tropical Cerrado region.

Material and method

Site description

The experiment was conducted over three growing seasons (2011, 2012, and 2013) at the Capivara Farm, located in the city of Santo Antonio de Goiás, Goiás State, Brazil. The geographical coordinates of the site are 16° 28' 00" S, 49° 17' 00" W. The average daily temperature and precipitation during the experiment were monitored (Figure 1).

The experimental area was cultivated for five years under a NTS, with corn (*Zea mays* L.) grown in summer and common bean (*Phaseolus vulgaris* L.) grown with irrigation in winter. The soil was classified as a clay-textured Typic Acric Red-Yellow Latossol (Oliveira & Rodrigues, 2012). Before the experiment began (May 2011), the chemical characteristics of the soil were determined (at depths of 0 - 0.20 m and 0.20 - 0.40 m) to calculate the requirements for liming and the addition of phosphogypsum (Table 1). The soil analyses were performed following the methods described by Donagema, Campos, Calderano, and Teixeira (2011).

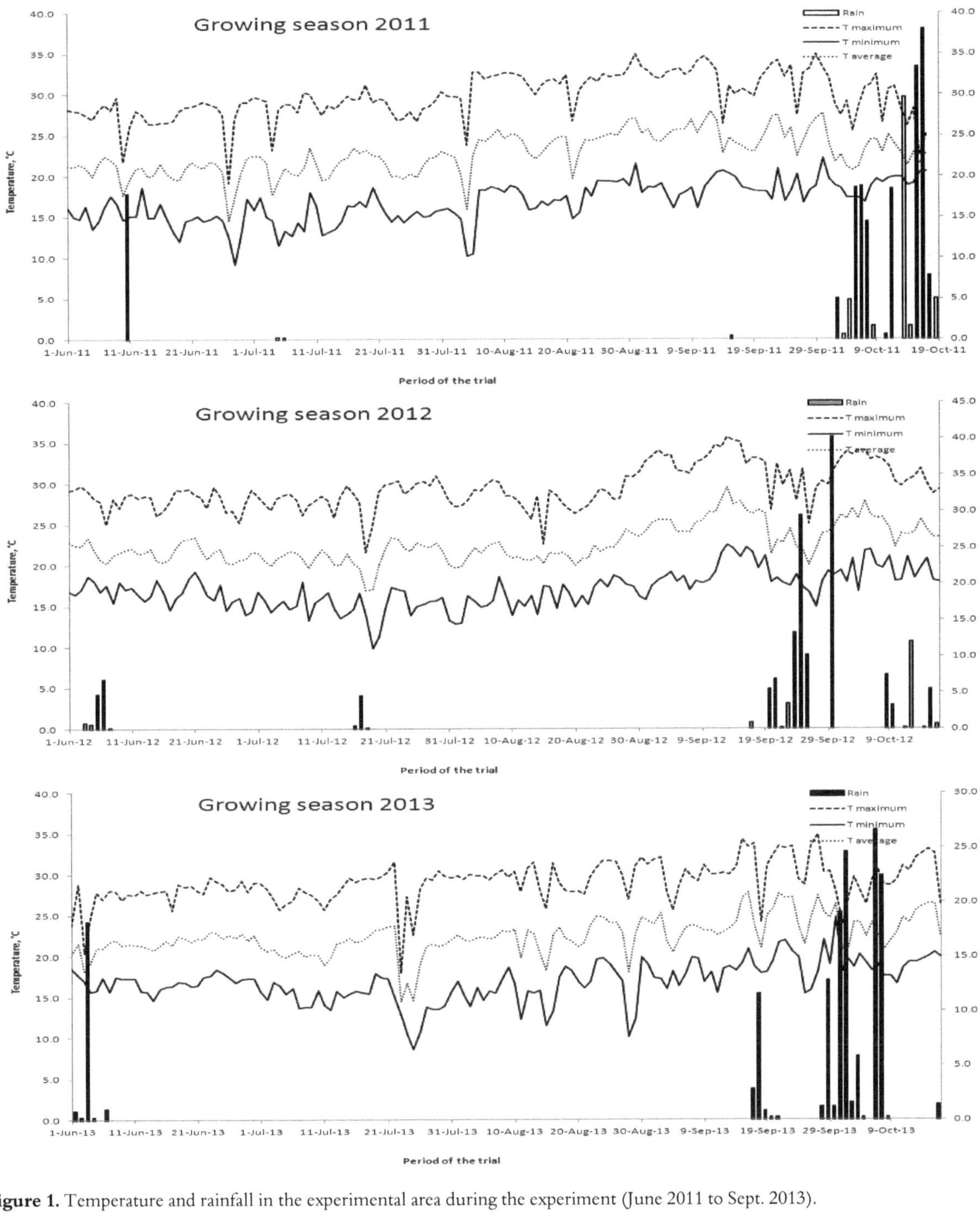

Figure 1. Temperature and rainfall in the experimental area during the experiment (June 2011 to Sept. 2013).

Table 1. Chemical characteristics of the soil before beginning the experiment, May 2011.

Depth	Ca	Mg	Al	H+Al	K	CEC[1]	pH[4]
(m)	----------------------			cmol$_c$ dm^{-3}	----------		
0 - 0.20	0.72	0.37	0.38	6.5	0.16	7.78	5.0
0.20 - 0.40	0.53	0.17	0.42	7.1	0.14	7.94	4.8
Depth	V[2]	SOM[3]	P	Zn	Cu	Fe	Mn
(m)	%	g kg^{-1}	----------------------			mg dm^{-3}	----------
0 - 0.20	16.0	18.3	6.0	3.1	2.1	58.7	11.7
0.20 - 0.40	10.6	17.0	2.3	3.1	2.1	58.7	11.7

[1]Cation exchange capacity. [2]Percent base saturation. [3]Soil organic matter. [4]pH in water. [5]P Mehlich-1.

Experimental design and treatments

The experimental design was a randomized complete block layout arranged in a 4x4 factorial design with four replications. The treatments consisted of four types of soil amendment [limestone, limestone + phosphogypsum, phosphogypsum, and control (no amendment application)] with four fertilization rates [0, 50, 100, and 150% of the fertilization recommended for common beans with high expected grain yield (Sousa & Lobato, 2004)]. Thus, the following quantities were applied: 0, 60, 120, and 180 kg ha^{-1} N (urea); 0, 60, 120, and 180 kg ha^{-1} P_2O_5 (55% simple superphosphate – 16% P_2O_5, 12% S, 20% Ca and 45% triple superphosphate – 46% P_2O_5, 14% Ca); and 0, 30, 60, and 90 kg ha^{-1} K_2O (potassium chloride – 60% K_2O), respectively, for the four rates of fertilizer. Phosphorus and K were applied on the day of sowing, and the application of N was divided among the day of sowing, five days after emergence and the V4 growth stage (third trifoliate leaf).

The lime application rate (5.0 Mg ha^{-1}) was calculated to increase the base saturation to 70% at a depth of 0 - 0.20 m. The amount of phosphogypsum (PG) used (2.5 Mg ha^{-1}) was determined by the clay content of the soil (500 g kg^{-1}) at a depth of 0.20 - 0.40 m as recommended by Sousa and Lobato (2004). These amendments were broadcasted on the surface of the soil without incorporation. The applications of lime and PG were divided into three applications as follows: on November 11th, 2010 (2 Mg ha^{-1} of lime and 1.0 Mg ha^{-1} PG), on November 11th, 2011 (2 Mg ha^{-1} of lime and 1.0 Mg ha^{-1} of PG) and on October 19th, 2012 (1 Mg ha^{-1} of lime and 0.5 Mg ha^{-1} of PG). The limestone used contained 26% Ca and 7.2% Mg. Its effective neutralizing value was 86.6%. The PG contained 21.8% Ca, 17.4% S, 1% P_2O_5 and traces of Si, Al, Mg, Na, Fe, Ce, Ti, La, K, Sr, Zr, and Pr.

Common bean cultivation (winter 2011, 2012 and 2013)

The sowing of the common bean cultivar 'Perola' was performed mechanically on June 6th, 2011, June 21st, 2012, and June 6th, 2013 using no-till graining with a row spacing of 0.45 m and a density of 12 seeds m^{-1}. Each plot consisted of 10 rows that were each 5 m long. The usable area of each plot consisted of the 8 central rows, with 0.5 m on each side disregarded. Crop management was performed according to the crop recommendations to keep the area free of weeds, diseases and insects. An irrigation sprinkler system was used for water management.

Summer crops

In November 2010, November 2011, and November 2012, soybeans were cultivated in the same area and fertilized with 0, 40, 80, and 120 kg ha^{-1} P_2O_5 (simple superphosphate) and 0, 40, 80, and 120 kg ha^{-1} of K_2O (potassium chloride) in the plots corresponding to common bean fertilization rates of 0, 50, 100, and 150%, respectively (Carvalho & Nascente, 2014).

Soil characterization

A galvanized steel auger 4.5 cm in diameter was used to sample soil at depths of 0 to 0.10 m, 0.10 to 0.20 m, and 0.20 to 0.40 m in November 2013, 36 months after the first liming. For each soil layer, 15 random subsamples were collected from each plot and combined to form a composite sample. These samples were air dried, sieved (2 mm mesh) and analyzed to determine the pH ($CaCl_2$ 0.01 mol L^{-1}), organic matter content, potential acidity (H + Al), Al, Ca, Mg, exchangeable K, and P (Mehlich-1). Percent base saturation (V) was calculated. Soil analyses were performed following the methods described by Donagema et al. (2011).

Yield components, grain yield, and efficiency of fertilizer use

The common bean crop was harvested on September 14th, 2011, October 2nd, 2012, and September 12th, 2013 from the usable area, first by hand and then using the thresher model BC 80 III. The common bean grain was weighed, and the yields were corrected to a moisture content of 130 g kg^{-1} and then converted to kg ha^{-1}. Agronomic characteristics, the number of pods per plant and the number of grains per pod, were evaluated for 10 randomly chosen plants per plot, along with the 100-grain weight (calculated from 20 random samples per plot, adjusted to a moisture content of 130 g kg^{-1}). To calculate the efficiency of fertilizer use, we summed the average yields in each treatment (control, lime, lime + PG, and PG) from the three growing seasons (2011, 2012, and 2013) and divided this sum by the amount of fertilizer used on the common bean crop during the three growing seasons.

Statistical analyses

An analysis of variance was performed for all variables. The soil corrective and level of fertilizer were considered fixed effects. Blocks, years and all interactions were considered random effects. A comparison of means was performed with a Tukey test ($p \leq 0.05$). To analyze the efficiency of fertilizer use, we used an LSD test at $p \leq 0.05$. A regression analysis was used for quantitative data (fertilizer levels). These analyses were performed using SAS statistical software.

Result

Alterations in soil chemical attributes

Thirty-six months after the first soil amendment application, the correctives significantly affected the soil chemical attributes up to 0.40 m depth (Table 2). The largest changes, as expected, occurred in the superficial soil layers (0-0.10 m and 0.10-0.20 m) in which the application of limestone, with or without phosphogypsum (PG), increased the pH, increased the Ca and Mg content, and reduced the Al and H + Al content, thus resulting in an increase in soil base saturation.

Table 2. Effects of the application of lime (L) and/or phosphogypsum (PG), fertilization rates and soil depth layers on the values of pH in water, Ca, Mg, Al, H+Al, K, cation exchange capacity (CEC) and base saturation (V) in the soil 36 months (November 2013) after the first application of the soil correction. Crop managed under the no-tillage system.

Soil corrective	Depth 0-0.10 m								
	pH	Ca	Mg	Al	H+Al	K	CEC	P	V
			----------cmol$_c$ dm^{-3}----------					mg kg^{-1}	%
Control	5.33 b	1.24 c	0.41 c	0.12 a	5.1 a	0.18 a	7.0 b	11.2 a	26.2 b
Limestone	6.14 a	2.38 a	1.08a	0.00 b	3.3 b	0.19 a	6.9 b	12.1 a	52.9 a
L + PG	6.07 a	2.51 a	0.88 b	0.00 b	3.2 b	0.18 a	6.8 b	11.4 a	53.2 a
Phosphogypsum	5.37 b	1.60 b	0.38 c	0.10 a	5.3 a	0.19 a	7.5 a	9.8 a	29.2 b
Source of variation	F probability								
Soil corrective (SC)	<0.001	<0.001	<0.001	<0.001	<0.001	0.0323	0.0013	0.0433	<0.001
Fertilization rate (F)	<0.001	0.0744	0.0143	0.1829	<0.001	<0.001	0.0002	<0.001	0.0041
SC x F	0.7860	0.7820	0.5170	0.7660	0.7604	0.6906	0.5380	0.6763	0.8278
Soil corrective	Depth 0.10-0.20 m								
Control	5.02 b	0.76 c	0.23 b	0.34 a	5.3 ab	0.14 a	6.5 b	29.4 a	17.6 b
Limestone	5.29 a	1.19 ab	0.56 a	0.13 b	4.7 bc	0.13 a	6.6 ab	29.1 a	28.4 a
L + PG	5.27 a	1.41 a	0.47 a	0.13 b	4.4 c	0.14 a	6.4 b	31.6 a	31.5 a
Phosphogypsum	4.95 b	0.98 bc	0.22 b	0.34 a	6.0 a	0.14 a	7.3 a	29.5 a	18.6 b
Source of variation	F probability								
Soil corrective (SC)	<0.001	<0.001	<0.001	<0.001	<0.001	0.0431	0.0095	0.0341	<0.001
Fertilization rate (F)	0.0142	0.3186	0.4606	0.2002	0.1330	<0.001	0.0347	<0.001	0.7874
SC x F	0.9622	0.8927	0.8197	0.6583	0.4339	0.8532	0.6052	0.6712	0.6071
Soil corrective	Depth 0.20-0.40 m								
Control	5.02 b	0.66 b	0.18 b	0.35 a	5.2 a	0.11 a	6.2 a	4.0 a	15.2 b
Limestone	5.03 ab	0.79 ab	0.32 a	0.29 ab	5.1 a	0.09 b	6.3 a	4.0 a	19.2 ab
L + PG	5.08 a	0.91 a	0.30 a	0.23 b	4.9 a	0.09 b	6.2 a	3.4 a	20.8 a
Phosphogypsum	5.04 ab	0.86 ab	0.21 b	0.30 ab	5.4 a	0.10 ab	6.5 a	3.5 a	17.9 ab
Source of variation	F probability								
Soil corrective (SC)	0.0451	0.0177	<0.001	0.0306	0.2368	0.0120	0.0343	0.6067	0.0057
Fertilization rate (F)	0.0213	0.1298	0.2771	0.4925	0.8743	0.0949	0.9958	<0.001	0.2640
SC x F	0.7598	0.6557	0.4041	0.8844	0.8056	0.7886	0.8328	0.7286	0.5581

[1]Means followed by the same letter in a column within each depth do not differ by the Tukey test at $p < 0.05$.

In the 0.20 - 0.40 m layer, the treatment with limestone + PG was the most effective at reducing soil acidity, resulting in the highest values of pH, Ca content and base saturation as well as the lowest value of exchangeable Al content. These results differed from the control, but were similar to limestone alone and PG alone (Table 2). The application of PG caused an increase in Ca content in the 0 - 0.10 m layer that differed from the control, increases of CEC in the 0 - 0.10 m soil layer that differed from all other treatments, and increases of CEC in the 0.10 - 0.20 m soil layer that differed from both the lime + PG treatment and the control. The available K content in the soil surface layers was not altered by the application of the correctives, but decreased significantly in the 0.20 - 0.40 m layer in limestone treatments (with or without PG) compared to the control. Fertilization rates affected pH, H + Al, base saturation, CEC, and Mg, K, and P contents available in the 0 - 0.10 m layer (Table 2). The pH, CEC, and Mg were adjusted to the linear regression model, and the other attributes were adjusted to the quadratic polynomial regression model (Figure 2). In the 0.10 - 0.20 m layer, significant effects were observed for pH, CEC, P, and K (Table 2), in which pH and CEC were adjusted to the linear regression model and P and K were fitted to a quadratic regression model (Figure 3). In the 0.20 - 0.40 m layer, fertilization rates affected only the pH and available P content; pH was adjusted to the linear regression model, and the available P was adjusted to a quadratic regression model (Figure 4).

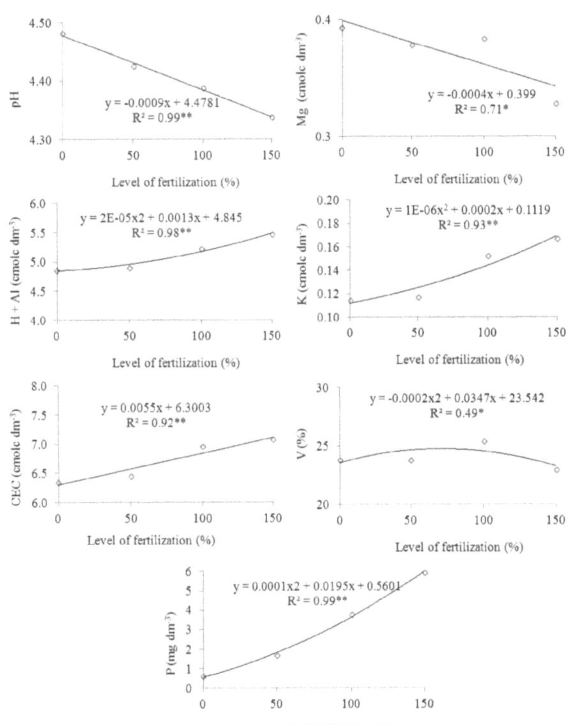

Figure 2. Chemical attributes of the soil (pH in water, Ca, Mg, Al, H+Al, CEC, V, K, and P) in the 0 - 0.10 m soil layer was a function of fertilization rates. Soil was measured 36 months after the first application of lime and phosphogypsum under the no-tillage system.

Yield components and grain yield

There was no interaction between soil amendments and fertilization rates for yield components or grain yield, but there were isolated effects of soil amendments and fertilization rates in the three growing seasons (Table 3). In the first growing season (2011), among the soils amendments, treatments with limestone provided the highest number of pods per plant, differing from the control, and grain yield of common bean, differing from both the control and PG alone.

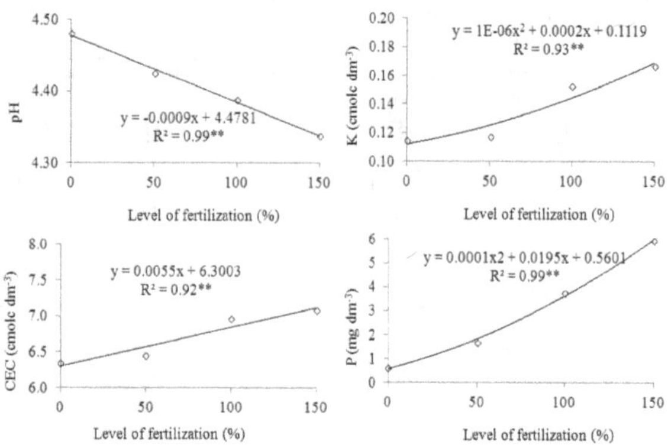

Figure 3. Chemical attributes of the soil (pH in water, CEC, K, and P) in the 0.10 - 0.20 m soil layer as a function of fertilization rates. Soil was measured 36 months after first application of lime and phosphogypsum under the no-tillage system.

The application of PG alone did not affect any of the variables (Table 3). Fertilization rates produced significant effects on the plant population and the number of pods per plant with linear adjustment and on number of grains per pods and grain yield with quadratic adjustment (Figure 6). In 2012, the application of limestone + PG provided the highest grain yield; while the grain yield produced with the application of limestone + PG was not significantly different from the treatment with limestone alone, it did differ from the control and the application of PG alone. Fertilization rates produced significant effects on the plant population with linear adjustment and on the number of pods per plant, the number of grains per plant and the grain yield with quadratic adjustment (Figure 6). In 2013, there was a significant effect of fertilization rates on all variables except plant population (Table 3), with quadratic adjustment for all four variables (Figure 7). The number of grains per pod was higher in the limestone treatment, differing from the control and the application of limestone + PG. Grain yield was higher with the application of limestone and limestone + PG, differing from the control treatment.

Table 3. Effects of application of lime (L) and/or phosphogypsum (PG), fertilization rates and growing season on the plant population (POP), number of pods plant^{-1} (NPP), number of grains pod^{-1} (NGP), 100-grain weight (W100) and grain yield (YIELD) of the common bean crop. Crop cultivated under the no-tillage system.

	Growing season 2011				
Soil corrective	POP	NPP	NGP	W100	YIELD
	n° x 1000	n°	n°	g	kg ha^{-1}
Control	284.4 a	6.67 b	4.10 a	23.5 a	1801 b
Limestone	278.8 a	9.04 a	4.05 a	23.6 a	2176 a
L + PG	272.0 a	8.57 ab	3.65 a	24.3 a	2106 ab
Phosphogypsum	278.8 a	8.58 ab	3.85 a	24.9 a	1826 b
Source of variation	F probability				
Soil corrective (SC)	0.4300	0.0224	0.1781	0.3414	0.0076
Fertilization (F)	<0.001	<0.001	<0.001	0.2508	<0.001
SC x F	0.1405	0.7899	0.6554	0.7405	0.7424
Correctives	Growing season 2012				
Control	251.5 a	11.07 a	4.27 a	27.3 a	2772 b
Limestone	245.6 a	12.03 a	4.30 a	28.1 a	2942 ab
L + PG	249.2 a	11.83 a	4.07 a	28.3 a	3015 a
Phosphogypsum	246.8 a	10.62 a	4.20 a	27.1 a	2783 b
Source of variation	F probability				
Soil corrective (SC)	0.7530	0.1763	0.2313	0.0242	0.0054
Fertilization (F)	<0.001	<0.001	<0.001	0.1376	<0.001
SC x F	0.3528	0.5394	0.5722	0.7534	0.0558
Correctives	Growing season 2013				
Control	230.1 a	11.52 a	2.78 b	27.47 a	2079 b
Limestone	225.0 a	12.39 a	3.09 a	27.93 a	2236 a
L + PG	236.1 a	11.93 a	2.78 b	27.26 a	2246 a
Phosphogypsum	236.1 a	11.25 a	2.90 ab	28.47 a	2148 ab
Source of variation	F probability				
Soil corrective (SC)	0.6608	0.6220	0.0880	0.3261	0.0456
Fertilization (F)	0.3064	<0.001	0.0075	<0.001	<0.001
SC x F	0.0592	0.5454	0.8949	0.3832	0.0923

[1]Means followed by the same letter in a column do not differ by the Tukey test at $p < 0.05$.

The increases in grain yield resulting from the application of fertilizer were much higher than the increases due to the application of limestone alone or limestone + PG (Figure 8). In the absence of fertilization, the application of limestone and limestone + PG provided an increase of 72 and 49%, respectively, in the accumulated grain yield, while in the absence of correctives, the application of 50 and 100% of the recommended fertilization provided an increase of 232 and 375%, respectively (Figure 8A). The greatest increases in accumulated grain yield occurred with the combination of limestone application with or without PG and fertilizer, clearly indicating an additive effect among these factors (Figure 8A). Considering the average of all levels of fertilization, the increase in accumulated grain yield due to the application of limestone or lime + PG was 11%, while increases due to application of fertilizer, regardless of the correctives, were 165, 249, and 250% for the 50, 100, and 150% of recommended fertilizer levels, respectively (Figure 8B). Figure 9 shows a clear relationship between soil P availability and common bean grain yield, in which a variation of the soil P levels from 5 to 30 mg dm^{-3} was associated with a variation from 13 to 100% in the relative yield of common bean.

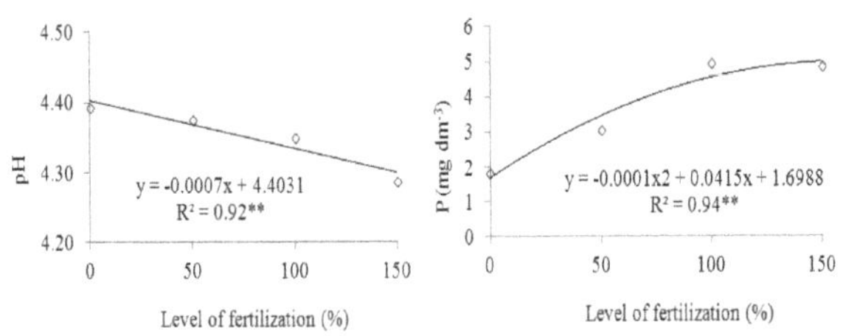

Figure 4. Chemical attributes of the soil (pH in water and P) in soil depths of 0.20 - 0.40 m as a function of fertilization rates. Soil measured 36 months after the first application of lime and phosphogypsum under the no-tillage system.

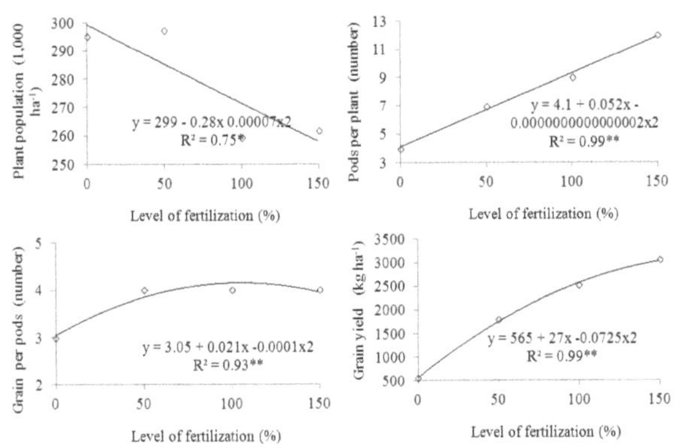

Figure 5. Plant population, grain yield, pods per plant and grains per pods of the common bean crop as a function of the level of fertilization. Growing season 2011.

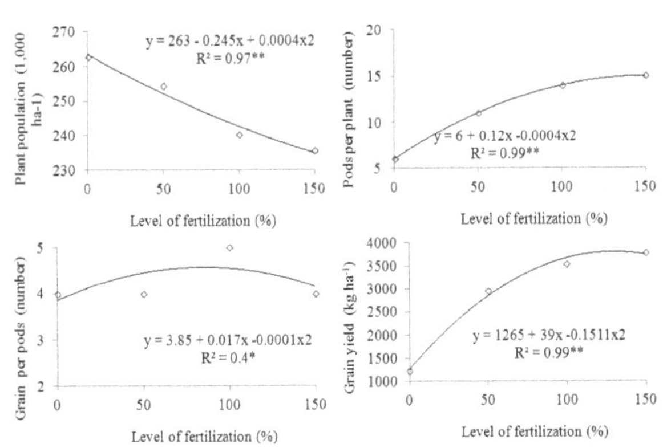

Figure 6. Plant population, grain yield and yield components of the common bean crop as a function of the rates of fertilization. Growing season 2012.

Efficiency of fertilizer use

The treatments with lime, regardless of the addition of PG, increased the efficiency of fertilizer use (Table 4). For every kilogram of fertilizer applied, 11.53 kg of common bean grains were obtained after lime application and 11.56 kg after application of lime with PG. These values differed from the control (10.56 kg of common bean grains) and from the application of PG alone (10.87 kg of common bean grains).

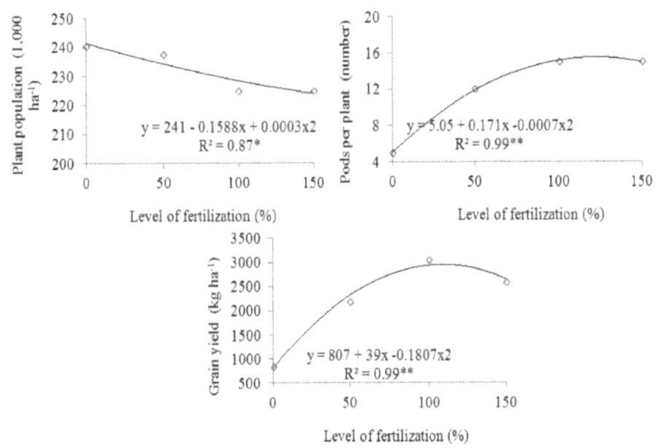

Figure 7. Plant population, grain yield and yield components of the common bean crop as a function of the rates of fertilization. Growing season 2013.

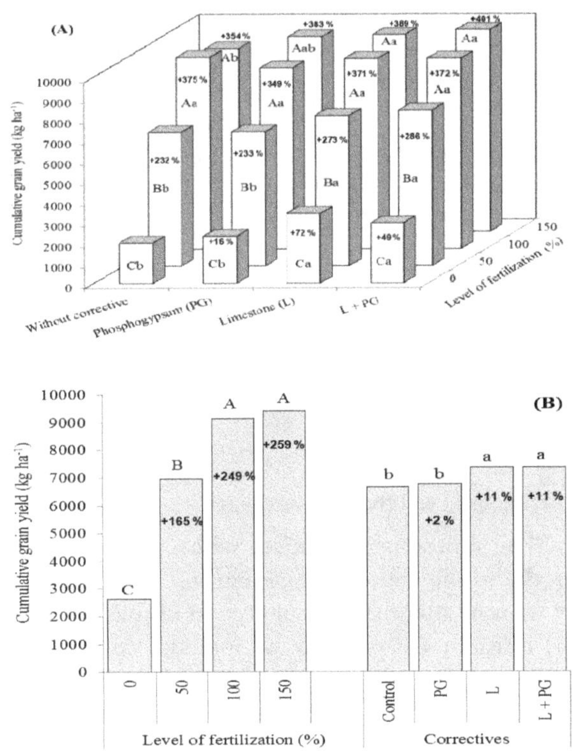

Figure 8. Cumulative common bean grain yield in 2011, 2012, and 2013 growing seasons as a function of the combination between correctives and rates of fertilization (A) and of each isolated factor (correctives or fertilization) (B). The values inside or above the bars correspond to the percent of increase compared to the control (without corrective and/or fertilization). Means followed by the same capital letter among levels of fertilization or the same small letter among correctives do not differ by the Tukey test at $p < 0.05$.

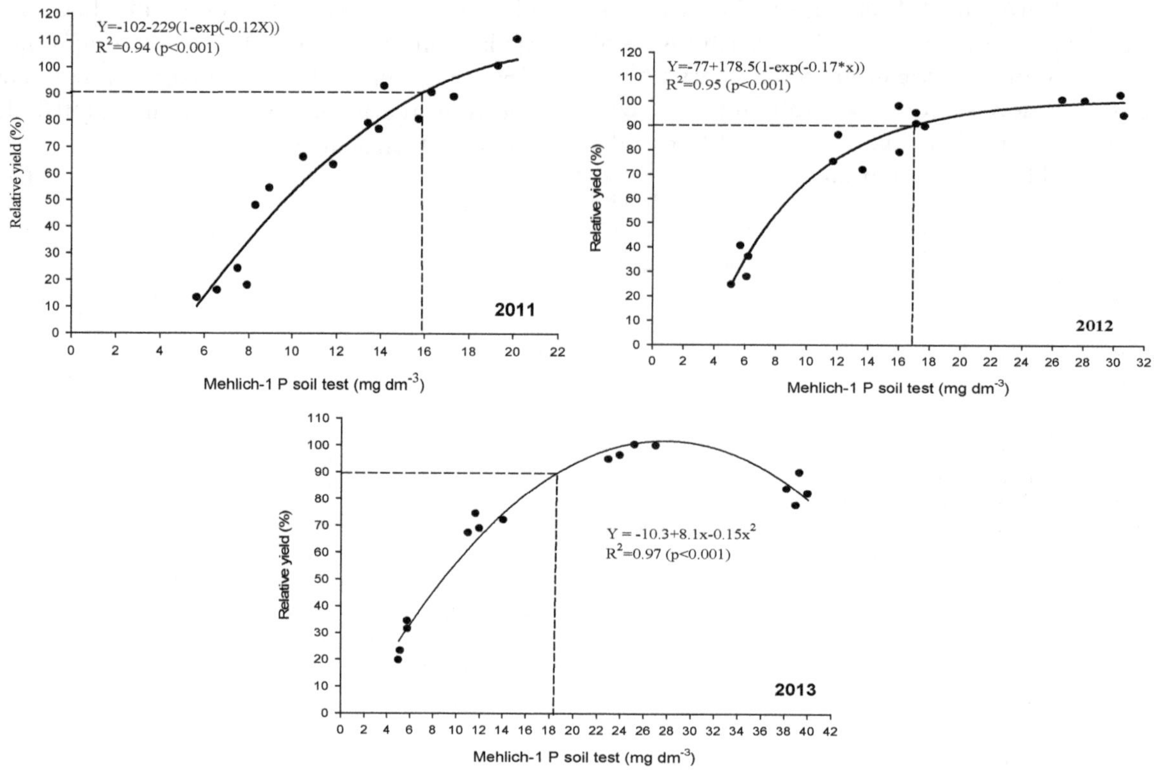

Figure 9. Relationship between the relative yield (RY) and soil Mehlich-extractable P at 0 - 0.20 m soil depth in 2011, 2012, and 2013 growing season. The data points represent the means of each rate of fertilization in each corrective treatment. Dashed lines represent the equivalent soil P level for 90% of the RY.

Table 4. Efficiency of NPK fertilizer use in the common bean crop cultivated under the no-tillage system as a function of lime (L) and phosphogypsum (PG) application.

	No corrective	Limestone	L + PG	PG
SSASY – A[1]	6653	7354	7367	6756
N - K_2O-P_2O_5 applied by fertilizer - B[2]	630	630	630	630
Fertilizer efficiency use - A/B	10.56 B[3]	11.67 A	11.69 A	10.72 B

[1]SSASY - Sum of average of the common bean grain yield from application of rates 0, 50, 100, and 150% of the recommended fertilization in three growing seasons (2011, 2012, and 2013). [2]Average of the amount of N, P_2O_5, and K_2O applied via fertilizer at doses 0, 50, 100 and 150% of the recommended fertilization in three growing seasons (2011, 2012 and 2013). Thus, it was applied 0, 60, 120 and 180 kg ha^{-1} for N, totaling 360 kg ha^{-1} of N; 0, 40, 80, and 120 kg ha^{-1}, for P_2O_5 and K_2O totaling 240 kg ha^{-1} of each nutrient, i.e., applied 840 kg N + K_2O + P_2O_5. The average was 210 kg ha^{-1} (840/ 4 fertilizer rates) for each year time 3 (growing season) = 630. [3]Means followed by the same letter in the row do not differ by LSD test at $p < 0.05$..

Discussion

Alterations in soil chemical attributes

The limestone treatment, with or without PG, produced the greatest changes in soil acidity indexes 36 months after the first application (Table 2). When the research was planned, we intended to reach the base saturation of 70% in the 0 - 0.20 m layer, but the highest value we achieved was 53% in the 0 - 0.10 m layer. Considering the average between 0 - 0.10 m and 0.10 - 0.20 m, the highest base saturation achieved was only 41 - 42%. Thus, splitting the rate of liming to apply 1/3 of the total annually was not a fully effective practice for correcting the soil acidity, considering the conditions of this experiment. However, our data do not allow us to determine whether the application of the same 5 Mg ha^{-1} of limestone all in the first year would be enough to reach the aimed base saturation. This could be a question investigated in future research.

Results from the field in subtropical regions have shown that the movement of the limestone when applied on the soil surface is very low (Conyers et al., 2003; Caires et al., 2008; Soratto & Crusciol, 2008a; 2008b; Churka Blum et al., 2013). However, after some time, the limestone can reach deep layers. According to Caires, Alleoni, Cambri, and Barth (2005), the effects of surface liming on all three acidity-related variables (pH, Al, and basic cations) were significant at depths of 0 - 0.05 m and 0.05 - 0.10 m from 1 yr after the first application onward and at a depth of 0.10 - 0.20 m from 2.5 yr after the first application onward. Our results showed that in a tropical region after 36 months it is likely that movement of the lime to deep layers (0.10 – 0.20 m and 0.20 – 0.40 m) occurred that produced changes in the chemical attributes, although not at the magnitude that we expected. In the evaluation done 24 months after lime application, there were only effects in the 0 - 0.10 m layer (Carvalho & Nascente, 2014).

The application of PG, which is not a soil acidity corrective, did not result in significant alterations of the pH of the soil or on other chemical attributes (Table 2). If applied in large quantities, this soil conditioner can cause leaching of bases (Ca, Mg, and K), especially Mg, to deeper layers (Ramos, Korndorfer, Pereira, & Camargo, 2006; Caires et al., 2008). In our experiment, we found higher amounts of Ca in the 0.10 - 0.40 m layer only when PG was applied with limestone, probably because Ca from PG moved to layers below 0.40 m. This could be an important strategy in tropical soil to enrich deeper layers with Ca and to stimulate root development. The greater root development could contribute to the plants reaching water in deep layers and could reduce problems caused by a period with no rain (Crusciol et al., 2016).

The pH level in all layers decreased with increases in the amount of fertilizer applied (Figure 2). An explanation of this pH decrease could be because of the use of nitrogen fertilizers that contain or produce $N\text{-}NH_4^+$, such as ammonium sulfate and urea. In the soil, this ion rapidly oxidizes to nitrate, releasing H^+ and producing significant decreases in pH (Fageria et al., 2011; Nascente et al., 2012).

The Mg content decreased with increasing levels of fertilization. This finding can be explained by competition with K for soil adsorption sites. Soil K content increased with increasing levels of fertilization (Figures 2 and 3), probably because the quantities of K-fertilizer applied were higher than the quantities exported in the grains, which is approximately 14 kg ha^{-1} of K_2O in each 1,000 kg of grains. According to Fageria et al. (2011), the application of K to soils directly affects Mg, as it competes for the same adsorption sites in the soil. Thus, it is probable that greater amounts of K caused the release of Mg from soil colloids and, possibly, the leaching of this nutrient to deeper layers. In addition, the reduction of the soil Mg level can be explained by exportation into grains, once the grain yield was increased as the rate of fertilization increased, since the fertilizers applied do not contain Mg.

In the soil layers evaluated, phosphorus also increased significantly with increasing fertilization rates (Figures 2, 3, and 4). This result was expected because at higher rates of fertilization, the amounts of phosphorus applied were higher than the quantities exported in grains. The quantity of P exported is approximately 9 kg ha^{-1} of P_2O_5 in each 1,000 kg of grains produced. This can be used as a strategy for corrective fertilization in areas with low soil P tests, considering the low mobility of P into the soil, especially in tropical soils such as Oxisols, which results in accumulation of this nutrient in the soil.

Yield components, grain yield, and efficiency of fertilizer use

Liming increased the common bean grain yield in all growing seasons (Table 4). This result is a consequence of the changes in soil properties caused by the application of limestone to the soil, such as increased pH, availability of Ca and Mg, base saturation and CEC (Table 2).

Over all growing seasons, application of PG alone did not increase the common bean grain yield. Other researchers have also reported similar results in areas with annual crops when the soil was under a NTS (Caires, Chueiri, Madruga, & Figueiredo, 1998, Caires, Blum, Barth, Garbuio, & Kusman, 2003; Carvalho & Nascente, 2014). The use of PG is more recommended when you have, in soil depths of 0.20 to 0.40 m, aluminum saturation higher than 20% or if the calcium contents are below 5 $mmol_c$ dm^{-3} (Sousa & Lobato, 2004). Under these conditions, the use of PG would stimulate root development in deeper soil layers, facilitating plants reaching water in case of a period of no rain, and allowing plants to obtain more nutrients and develop better (Sousa & Lobato, 2004; Caires, Barth, & Garbuio, 2006). However, in an NTS, the greatest quantity of nutrients remains in the top soil, as observed in our trial (Table 2). Additionally, our trial was irrigated, so it is likely that in our trial, PG did not provide increases in the soil profile that allowed better development of the common bean plants. Moreover, the low Mg content in the soil (Table 1) may have influenced the absence of a response to the application of PG alone, since, as already demonstrated in other works (Oliveira & Pavan, 1996; Michalovicz et al., 2014; Carvalho & Nascente, 2014), the application of PG may result in preferential leaching of Mg and a reduction of the nutrient content in the leaves. Increased fertilization rates produced increases in the yield components and grain yield in all growing seasons. Therefore, this confirms that the common bean is highly responsive to fertilization (Fageria et al., 2011; Nascente et al., 2012). From the equations displayed in Figures 5, 6, and 7, the maximum common bean grain yield was achieved when the rates of fertilization were 206, 129, and 109% in 2011, 2012, and 2013, respectively. This reduction in the rate of fertilizers required by the common bean is a result of the accumulation of nutrients in the soil (Figures 2, 3, 4, and 9) due to the residual effect of previous fertilization. Thus, these results reinforce the importance of monitoring the soil fertility status as a tool to support the planning of fertilization in cropping systems in order to adjust possible nutritional imbalances.

The equations displayed in Figure 9 show that the equivalent soil P (Mehlich-1) level for 90% of the relative P production was 16, 17 and 18 mg dm^{-3} in growing seasons 2011, 2012, and 2013, respectively. These values are higher than the 12 mg dm^{-3} reported by Sousa and Lobato (2004) as the critical level for irrigated cropping systems in clay Oxisols in the Cerrado region at a 0 to 0.20 m depth, indicating that the common bean is a lower P-acquisition efficiency crop compared to soybean or maize. Föhse, Claassen, and Jungk (1988) demonstrated that the common bean is a low P-uptake efficiency species, as determined by both a low root-shoot ratio and a low influx rate. In 2013, the mathematical model that best expressed this relationship was the quadratic model (Figure 9), suggesting a negative interaction effect could be occurring with other nutrients, such as zinc, when the soil P content was higher than 30 mg dm^{-3}.

The effects of fertilization rates were of greater magnitude than the effects of liming. This could be because the changes in soil attributes up to a depth of 0.20 m that were caused by the application of limestone to the soil surface were of low magnitude for a highly demanding crop such as the common bean (Table 2). For example, the optimal pH in the 0 - 0.20 m layer for developing common bean crops ranges from 5.8 to 6.3 (Fageria et al., 2011). Fageria (2008) verified that the maximum common bean yield was achieved with a base saturation of 67% and a pH of 6.5. In our work, these values were not reached. Even so, it was observed that the response to fertilization was increased by surface-applied lime, improving the efficiency of fertilizer use. This indicates that the better strategy for farmers to increase fertilizer use efficiency and, consequently, to increase common bean grain yield in NTS is to avoid the decrease of base saturation to rates lower than 50%, since the effect of surface application of limestone without incorporation is slow.

Conclusion

The annual application of 1/3 of the total rate of limestone or limestone + PG improved the soil acidity indexes up to a depth of 0.20 m 36 months after the first application. The application of PG alone did not result in significant alterations in soil chemical attributes and common bean grain yield.

Increasing levels of NPK and with fertilizers also containing Ca and S provided significant increases in grain yield and alterations in soil fertility. The common bean productivity increases in response to fertilization were even greater when combined with the application of limestone alone or limestone + PG.

Acknowledgements

To the National Council for Scientific and Technological Development (CNPq) for an award for excellence in research of the second author.

References

Barbosa Filho, M. P., Fageria, N. K., & Silva, O. F. (2004). Fontes e métodos de aplicação de nitrogênio em feijoeiro irrigado submetido a três níveis de acidez do solo. *Ciência e Agrotecnologia, 28(4), 785-792.* doi: 10.1590/S1413-70542004000400009

Caires, E. F., Alleoni, L. R. F., Cambri, M. A., & Barth, G. (2005). Surface application of lime for crop grain production under a no-till system. *Agronomy Journal, 97*(3), 791-798. doi: 10.2134/agronj2004.0207

Caires, E. F., Barth, G., & Garbuio, F. J. (2006). Lime application in the establishment of a no-till system for grain crop production in Southern Brazil. *Soil & Tillage Research, 89*(1), 3-12. doi: 10.1016/j.still.2005.06.006

Caires, E. F., Blum, J., Barth, G., Garbuio, F. J., & Kusman, M.T. (2003). Alterações químicas do solo e resposta da soja ao calcário e gesso aplicados na implantação do sistema de plantio direto. *Revista Brasileira de Ciência do Solo, 27*(2), 275-286. doi: 10.1590/S0100-06832003000200008

Caires, E. F., Chueiri, W. A., Madruga, E. F., & Figueiredo, A. (1998). Alterações de características químicas do solo e resposta da soja ao calcário e gesso aplicados na superfície em sistema de cultivo sem preparo de solo. *Revista Brasileira de Ciência do Solo, 22*(1), 27-34. doi: 10.1590/S0100-06831998000100004

Caires, E. F., Garbuio, F. J., Churka, S., Barth, G., & Corrêa, J. C. L. (2008). Effects of soil acidity amelioration by surface liming on no-till corn, soybean, and wheat root growth and yield. *European Journal of Agronomy, 28*(1), 57-64. doi: 10.1016/j.eja.2007.05.002

Carvalho, M. C. S., & Nascente, A. S. (2014). Limestone and phosphogypsum effects on soil fertility, soybean leaf nutrition and yield. *African Journal of Agricultural Research, 9*(17), 1366-1383. doi: 10.5897/AJAR2014.8626

Churka Blum, S., Caires, E. F., & Alleoni, L. R. F. (2013). Lime and phosphogypsum application and sulfate retention in subtropical soils under no-till system. *Journal of Soil Science and Plant Nutrition, 13*(2), 279-300. doi: 10.4067/S0718-95162013005000024

Conyers, M. K., Heenan, D. P., McGhie, W. J., & Poile, G. P. (2003). Amelioration of acidity with time by limestone under contrasting tillage. *Soil & Tillage Research*, *72*(1), 85-94. doi: 10.1016/S0167-1987(03)00064-3

Crusciol, C. A. C., Artigiani, A. C. C. A., Arf, O., Carmeis Filho, A. C. A., Soratto, R. P., Nascente, A. S., & Alvarez, R. C. F. (2016). Soil fertility, plant nutrition, and grain yield of upland rice affected by surface application of lime, silicate, and phosphogypsum in a tropical no-till system. *Catena*, *137*(2), 87-99, 2016. doi: 10.1016/j.catena.2015.09.009

Donagema, G. K., Campos, D. V. B., Calderano, S. B., & Teixeira, W. G. (2011). *Manual de métodos de análise de solo* (2a ed.). Rio de Janeiro, RJ: Embrapa Solos.

Fageria, N. K. (2008). Optimum soil acidity indices for dry bean production on an Oxisol in no tillage system. *Communications in Soil Science and Plant Analysis, 39*(5-6), 845-857. doi: 10.1080/00103620701880909

Fageria, N. K., & Nascente, A. S. (2014). Management of soil acidity of South American soils for sustainable crop production. *Advances in Agronomy*, *128*(1), 221-275. doi: 10.1016/B978-0-12-802139-2.00006-8

Fageria, N. K., Baligar, V. C., & Jones, C. A. (2011). *Growth and mineral nutrition of field crops*. Boca Raton, FL: CRC Press.

Föhse, D., Claassen, N., & Jungk, A. (1988). Phosphorus efficiency of plants. In: External and internal P requirement and P uptake efficiency of different plant species. *Plant and Soil*, *110*(1), 101-109.

Food and Agriculture Organization of the United Nations [FAO]. (2014). *FAOSTAT – production: crops*. Retrieved on Oct. 29, 2016 from http://faostat.fao.org

Michalovicz, L., Müller M. M. L., Foloni, J. S. S., Kawakami, J., Nascimento, R., & Kramer, L.F.M (2014). Soil fertility, nutrition and yield of maize and barley with gypsum application on soil surface in no-till. *Revista Brasileira de Ciência do Solo, 38*(5), 1496-1505. doi: 10.1590/S0100-06832014000500015

Nascente, A. S., Kluthcouski, J., Crusciol, C. A. C., Cobucci, T., & Oliveira, P. (2012). Fertilization of common bean cultivars in tropical lowlands. *Pesquisa Agropecuária Tropical*, *42*(4), 407-415. doi: 10.1590/S1983-40632012000400003

Oliveira, E. L., & Pavan, M. A. (1996). Control of soil acidity in no-tillage system for soybean production. *Soil &Tillage Research*, *38*(1-2), 47-57. doi: 10.1016/0167-1987(96)01021-5

Oliveira, V. A., & Rodrigues, C. (2012). *Levantamento detalhado dos solos da Fazenda Capivara, município de Santo Antônio de Goiás-GO*. Goiânia: Viasat Geotecnologia e Meio Ambiente.

Pádua, T. R. P., Silva, C. A., & Melo, L. C. A. (2006). Calagem em latossolo sob influência de coberturas vegetais: neutralização da acidez. *Revista Brasileira de Ciência do Solo*, *30*(5), 869-878. doi: 10.1590/S0100-06832006000500013

Ramos, L. A., Korndorfer, G. H., Pereira, H. S., & Camargo, M. S. (2006). Reatividade de corretivos da acidez e condicionadores de solo em colunas de lixiviação. *Revista Brasileira de Ciência do Solo, 30*(5), 849-857. doi: 10.1590/S0100-06832006000500011

Soratto, R. P., & Crusciol, C. A. C. (2008a). Atributos químicos do solo decorrentes da aplicação em superfície de calcário e gesso em sistema plantio direto recém-implantado. *Revista Brasileira de Ciência do Solo*, *32*(2), 675-688. doi: 10.1590/S0100-06832008000200022

Soratto, R. P., & Crusciol, C. A. C. (2008b). Nutrição e produtividade de grãos da aveia-preta em função da aplicação de calcário e gesso em superfície na implantação do sistema plantio direto. *Revista Brasileira de Ciência do Solo, 32*(2), 715-725. doi: 10.1590/S0100-06832008000200026

Sousa, D. M. G., & Lobato, E. (2004). *Cerrado: correção do solo e adubação* (2a ed.). Brasília, DF: Embrapa Cerrados.

20

Germination and biochemical components of *Salvia hispanica* L. seeds at different salinity levels and temperatures

Emanoela Pereira de Paiva, Salvador Barros Torres*, Tatianne Raianne Costa Alves, Francisco Vanies da Silva Sá, Moadir de Sousa Leite and Jeferson Luiz Dallabona Dombroski

*Centro de Ciências Agrárias, Departamento de Ciências Agronômicas e Florestais, Universidade Federal Rural do Semi-Árido, Av. Francisco Mota, 572, Costa e Silva, 59625-900, Mossoró, Rio Grande do Norte, Brazil. *Author for correspondence. E-mail: sbtorres@ufersa.edu.br*

ABSTRACT. Most plant species are susceptible to the effects of salinity, such as increases in osmotic potentials and deleterious ionic effects, which in turn affect water absorption in plants and, consequently, compromise germination and seedling growth. Hence, this study aimed to evaluate the effects of salinity on the germination, initial growth, and physiological and biochemical components of *S. hispanica* seedlings at different temperatures. The experimental design was completely randomized, with treatments distributed in a 5 x 4 factorial scheme with five saline concentrations (0.0 (control), 4.5, 9.0, 13.5, and 18.0 dS m^{-1}) four temperature regimes (20, 25, 30, and 20-30°C), and four replicates of 50 seeds per treatment. The experiment evaluated the germination, growth, biochemical components (chlorophyll *a*, chlorophyll *b*, total carotenoids, amino acids, proline and sugars) and phytomass accumulation of *S. hispanica* seedlings. Saline levels higher than 4.5 dS m^{-1} together with treatment temperatures of 30 or 20-30°C negatively affected the germination, vigor, growth and biochemical components of the seedlings. The 25°C treatment temperature promoted the best conditions for the development of *S. hispanica* seedlings up to the saline level of 9.0 dS m^{-1}.

Keywords: Lamiaceae; chia; salt stress; thermal stress.

Introduction

Salvia hispanica L. (Lamiaceae), known as chia, is an herbaceous plant that contains essential oils in its leaves, stems and seeds. The species is cultivated mainly in Mexico, Bolivia, Ecuador, and Guatemala (Capitani et al., 2015; Imram et al., 2016). In Brazil, *S. hispanica* cultivation has aroused the interest of producers because of an increasing demand for functional food products. Its seeds are recommended to reduce cholesterol levels, combat free radicals, control diabetes and help enable weight loss due to their high fiber content (approximately 30%) and low glycemic index; in addition, they are a source of omega-3 and omega-6 fatty acids (Ali et al., 2012).

From an agronomic point of view, studies on *S. hispanica* are very scarce, especially with respect to seed technology. Thus, it is necessary to know the factors that limit germination and seedling development to allow the formulation of management strategies for *S. hispanica* cultivation. Among these factors, salinity stands out because *S. hispanica* seeds are particularly vulnerable to its effects. Under high salinity conditions, there is an initial decrease in water absorption and, consequently, a reduction in germination (Falk & Munné-Bosch, 2010; Taiz & Zeiger, 2013). Therefore, the reduction in germination power in salt-stressed plants compared with control plants serves as an indication of the tolerance index of the species to salinity, elucidating the possible responses and tolerance of the plants to salt stress in different development stages (Flowers & Colmer, 2008; Munns & Tester, 2008; Zhang, Zhang, Lü, Zhou, & Han, 2014).

In studies conducted on *Salvia aegyptiaca*, Gorai, Gasmi, and Neffati (2011) observed that salt stress reduced the species germination speed and percentage. These same authors concluded that *S. aegyptiaca* has the capacity to tolerate moderately saline conditions, especially when temperature conditions are adequate. In addition, Dal'Maso et al. (2013), while analyzing the effect of salinity on the germination and initial development of *S. hispanica* seeds, observed that an increase in the concentration of potassium chloride reduced germination, the germination speed index and seedling growth, thus finding that *S. hispanica* seedlings are subject to limitations on growth, development and survival during salinity stress conditions.

Among environmental factors, temperature is the main factor responsible for determining seed germination rate, and an optimal temperature leads to the maximum percentage of fast seed germination, while maximum and minimum temperatures result in low germination percentages or embryo death (Meiado, Albuquerque, Rocha, Rojas-Aréchiga, & Leal, 2010). In addition to influencing germination, temperature affects the initial development of seedlings and directly affects their phenology because they have neither the ability of the seeds to tolerate adverse environmental conditions nor the physical robustness acquired with age (Miranda, Correia, Almeida-Cortez, & Pompelli, 2014; Sanchez et al., 2014). Thus, the germination period involves a series of transformations that depend on favorable environmental conditions, leading to high mortality rates when conditions are unfavorable in this development stage (Silva, Duarte, Lopes, Moraes, & Pereira, 2008; Oliveira, Souza, Carvalho, & Souza, 2015).

Given the above observations, this study aimed to evaluate the effects of salinity on the germination and initial growth of *S. hispanica* seedlings at different temperatures.

Material and method

The experiment was carried out at the Laboratory of Seed Analysis of the Plant Science Department of the Federal Rural University of the Semi-Arid Region (UFERSA), Mossoró, Rio Grande do Norte State, Brazil, using *S. hispanica* seeds from a commercial production field located in the municipality of Santana do Livramento, Rio Grande do Sul State (30° 53' 27" S, 55° 31' 58" W at 208 m altitude). The seeds were manually processed, placed in transparent plastic bags (0.15-mm thick) and stored in a cold and dry chamber (10 ± 2°C and 50% of the relative humidity of the environment) during the experimental period.

Salt stress was simulated using sodium chloride (NaCl) as a solute diluted in distilled water at the following concentrations: 0.0 (control), 4.5, 9.0, 13.5, and 18.0 dS m^{-1}. The electrical conductivity of the solutions was measured using a conductivity meter. For the control, distilled water was used to moisten the substrate.

The experimental design was completely randomized, with treatments distributed in a factorial scheme of 5 x 4, with five saline concentrations, four temperatures, and four replicates of 50 seeds.

The germination test was conducted in Biochemical Oxygen Demand (B.O.D.) germination chambers regulated at temperatures of 25, 30, 35, and 20-30°C with a photoperiod of eight hours. The seeds were sown on paper towels (Germitest®) previously moistened for each saline concentration at the proportion of 2.5 times its dry weight.

After sowing, the germination test was monitored for eight days, which is considered the normal period for seedlings to produce their primary root and shoots (Brasil, 2009); germination results are expressed as percentages. Concurrent to the germination test, the germination speed index was determined, the seedlings were evaluated daily from the beginning of germination until the eighth day after sowing, and the germination index was calculated according to the equation proposed by Maguire (1962).

At the end of the germination test, the normal seedlings of *S. hispanica* were evaluated for growth by determining the shoot length (from the base to the apex using a ruler graduated in mm) and primary root length (from the base to the root tip). After the measurements, the seedlings were divided into roots and shoots, dried in a forced-air oven (65°C) until they reached a constant weight and weighed on a precision analytical scale (0.001 g). Total dry matter was determined by the sum of the values of root and shoot dry matter.

The contents of chlorophyll *a* and *b* and carotenoids were determined through the extraction of chlorophyll in acetone (80%) and quantification with spectrophotometry. The absorbances of the samples were recorded with a spectrophotometer at 470, 646.8, and 663.2 nm, and the contents of chlorophylls and carotenoids (g pigment kg^{-1} DM) were obtained according to Lichthenthaler (1987) using the following equations: 1) chlorophyll *a* = 12.25 ABS663.2 – 2.79 ABS646.8; 2) chlorophyll *b* = 21.50 ABS646.8 – 5.10 ABS663.2; 3) total carotenoids = (1000 ABS470 – 1.82 chlorophyll *a* – 85.02 chlorophyll *b*)/198.

On the eighth day of the experiment, samples of shoot and root dry matter from plants grown at the different levels of saline and thermal stress were collected to determine total soluble sugars, free amino acids and proline. The content of total soluble sugars was determined by the anthrone method (Yemm & Willis, 1954), and the results are expressed as μmol GLU g^{-1} of fresh matter. To quantify the contents of amino acids, absorbance at 570 nm was measured by the acid ninhydrin method (Yemm & Cocking, 1955) using glycine as the standard substance, and the results are expressed as μmol TFAA g^{-1} of matter. Proline was quantified according to the methodology described by Bates, Waldren, and Teare (1973), and the results are expressed as PRO g^{-1} of fresh matter.

The data were subjected to analysis of variance by F test at the 0.05 probability level. Depending on the significance, the data were subjected to polynomial regression analysis ($p \leq 0.05$) using the statistical program SISVAR® (Ferreira, 2011).

Result and discussion

The interactions between the saline levels and the imposed temperatures significantly influenced the germination, growth, physiology and biochemical component variables of the *S. hispanica* seedlings. Seed germination decreased linearly with the increase in salinity at 20, 20-30, and 25°C, on the order of 0.53, 0.54, and 0.44% per dS m^{-1}, respectively. For 30°C, there was a slight reduction in germination up to the salinity level of 13.5 dS m^{-1}. However, increases in salinity above this level drastically reduced the germination of *S. hispanica* seeds, reaching 0% at the saline level of 18 dS m^{-1} (Figure 1A).

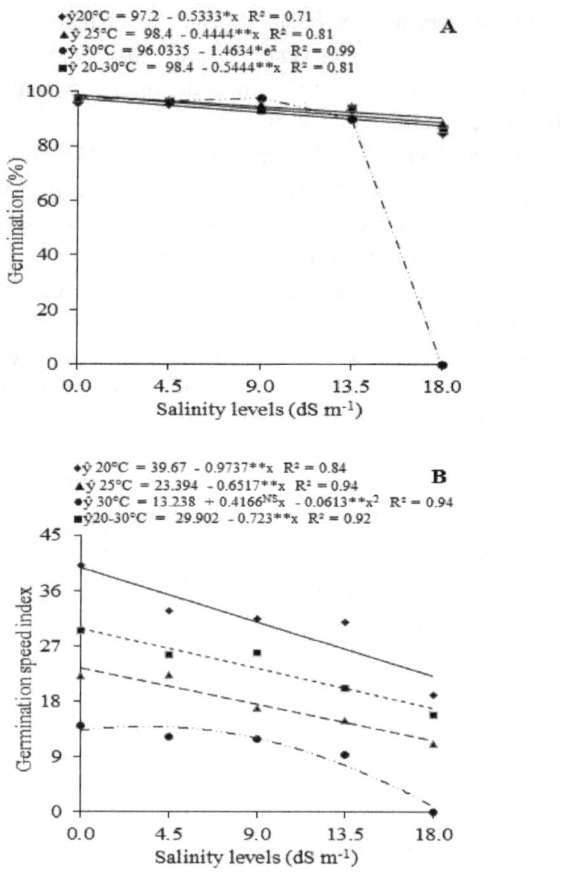

Figure 1. Germination (A) and germination speed index (B) of *Salvia hispanica* L. seeds at different levels of salinity and temperature. NS,★★, and ★ = non significant, significant at $p < 0.01$ and $p < 0.05$ propabaility level, respectively.

For the germination speed index, the results were similar to those of the germination percentage, with linear decreases in the vigor of *S. hispanica* seeds at temperatures of 20, 20-30, and 25°C corresponding to reductions of 0.97, 0.72, and 0.65 in the germination speed index per dS m^{-1}, respectively. At 30°C, there was a quadratic response, with a maximum germination speed index at the estimated salinity of 3.4 dS m^{-1}; however, above this saline level, the index decreased and reached zero at the salinity level of 18 dS m^{-1} (Figure 1B). Seed germination at 30°C was low and consequently resulted in lower vigor at the highest saline levels. Under high temperature conditions, there is an increase in seed metabolism, leading to greater soaking during the germination process. Thus, due to the high concentrations of salts present in the water and in the substrate, the absorption of the saline solution caused the seedlings to have inadequate supplies of nutrients because of the ionic imbalance resulting from excess Na^+ and Cl^- ions (Munns & Tester, 2008; Taiz & Zaiger, 2013; Lopes, Nascimento, Barbosa, & Costa, 2014).

Evaluating the germination of *S. hispanica* seeds in a saline medium, Dal'Maso et al. (2013) found reductions in germination from at osmotic potential of -0.9 MPa and higher, with germination percentages of 39% at -0.9 MPa and 10% at -1.2 MPa. Stefanello, Neves, Abbad, and Viana (2015), also researching *S. hispanica* seeds, observed that germination and vigor were compromised by an increase in the concentration of salts. Similar results were found in broccoli seeds by Lopes et al. (2014); in tomato seeds by Silva Júnior et al. (2014); in beet seeds by Bernardes, Mengarda, Lopes, Nogueira, and Rodrigues (2015); and in cabbage by Oliveira et al. (2015).

The shoot and root lengths of *S. hispanica* seedlings were drastically reduced by the increase in salinity in the substrate (Figure 2A and B). For seedling shoots, there were linear reductions of 0.11, 0.15, 0.16, and 0.14 cm per dS m^{-1} at temperatures of 20, 20-30, 25, and 30°C, respectively (Figure 2A).

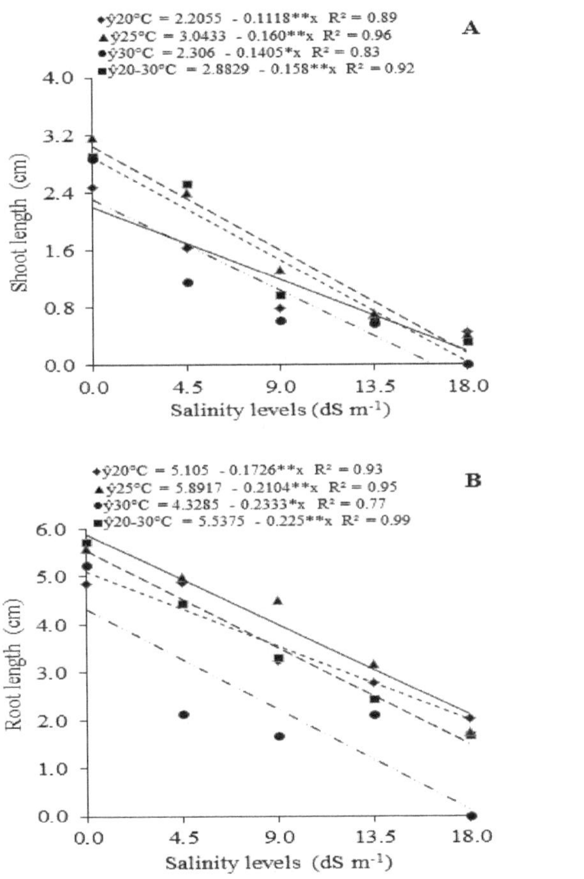

Figure 2. Shoot length (A) and root length (B) of *S. hispanica* seedlings at different levels of salinity and temperature. ^{NS},$\star\star$, and $\star$ = non significant, significant at $p < 0.01$ and $p < 0.05$ propabaility level, respectively.

For root length, there were reductions of 0.17, 0.22, 0.21, and 0.23 cm dS^{-1} m^{-1} at temperatures of 20, 20-30, 25, and 30°C, respectively (Figure 2B). The greatest reductions in seedling growth occurred at 30 and 20-30°C. In addition, at the highest temperature (30°C) associated with the highest salinity stress (18 dS m^{-1}), the growth of *S. hispanica* seedlings decreased, indicating that the association of salt stress with thermal stress causes deleterious effects on the seeds. Germination and seedling growth are processes that require energy from the reserves of the seeds (Zhang et al., 2014).

Under normal conditions, a large portion of the energy is extracted from the reserves and consumed during the transport of ions and the synthesis of compatible solutes for development (Flowers & Colmer, 2008). However, when seeds are subjected to salt stress conditions, part of the energy from the reserves is consumed during the transport of the Na^+ ions, leading to serious damage to seedling development and eventually preventing their development, a fact observed in the present study. Dal'Maso et al. (2013) also observed reductions in shoot and root development in *S. hispanica* seedlings as saline levels increased.

As to the chloroplast pigments, *S. hispanica* seeds that germinated at temperatures of 20 and 25°C obtained satisfactory levels of chlorophyll *a* and *b* under saline conditions up to 13.5 dS m^{-1}; from this point on, the chlorophyll values started to decrease (Figure 3A). It should be noted that at these temperatures, the response was quadratic, with increases in the contents of chlorophyll *a* and *b* that were much greater in the saline treatments than in the control treatment. Most likely due to the limitations caused by the degradation and absorption of reserves resulting from the increase in the osmotic potential, *S. hispanica* seedlings were stimulated to produce more photosynthesizing pigments to escape the deleterious effects of the salt stress.

For chlorophyll *a*, there were reductions of 35 and 4% at the highest saline level (18.0 dS m^{-1}) compared to the control saline level (0 dS m^{-1}) at 20 and 25°C, respectively. However, at 20 and 25°C, there were 5 and 32% increases in chlorophyll *b*, respectively, between the highest saline level (18.0 dS m^{-1}) and the control (0 dS m^{-1}) (Figure 3A). The increase in the content of chlorophyll *b*, but not of chlorophyll *a*, with the increase in temperature may be related to greater potentiation of photosynthetic activity since chlorophyll *b* is an accessory pigment that captures wavelengths of light energy that are not captured by chlorophyll *a* (Silveira, Silva, Silva, & Viégas, 2010). Nevertheless, at the highest temperature (30°C) and the alternated temperature treatment of 20-30°C, the contents of chlorophyll *a* and *b* showed quadratic behaviors, with an increasing trend up to 4.5 dS m^{-1}. Considering that the increase in temperature potentiated the effect of salinity on *S. hispanica* seedlings, under these temperature conditions, the increase in salinity probably stimulated the synthesis of the enzyme chlorophyllase, which is responsible for the degradation of chlorophyll (Taiz & Zaiger, 2013).

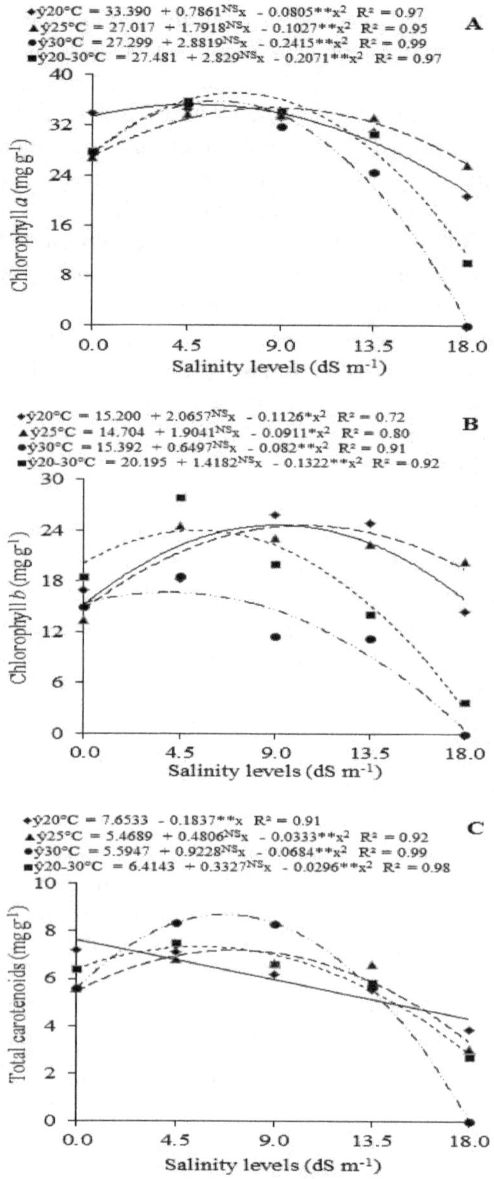

Figure 3. Contents of chlorophyll *a* (A), chlorophyll *b* (B) and total carotenoids (C) in *S. hispanica* seedlings at different levels of salinity and temperature. NS,★★, and ★ = non significant, significant at $p < 0.01$ and $p < 0.05$ propabaility level, respectively.

The content of carotenoids in the *S. hispanica* seeds that germinated at 20°C linearly decreased as a function of the increase in salinity, with a reduction of 0.18 mg g^{-1} per dS m^{-1}. At the other temperatures, there was a quadratic response, with maximum peaks in the content of carotenoids at the saline levels of 7.21, 5.62, and 7.75 dS m^{-1} at 25, 20-30, and 30°C, respectively, and subsequent decreases in carotenoid content with salinity (Figure 3C). The quadratic behavior of the content of carotenoids observed at 25, 20-30, and 30°C agreed with the results found for chlorophyll *a* and *b*, considering that the carotenoids act as protecting pigments working against antioxidant activity activities and protecting the lipid membranes from oxidative stress caused in the plants exposed to saline conditions (Falk & Munné-Bosch, 2010). Thus, the increase in the content of carotenoids may be related to the protection of chlorophyll molecules by the reduction of the activities of enzymes such as chlorophyllase.

The contents of the organic compounds, free amino acids, proline and sugars that were evaluated in the *S. hispanica* seedlings were significantly altered by the salt stress and changing temperatures (Figure 4A, B and C). The contents of free amino acids, proline and sugars in the seedlings in seeds germinated at temperatures of 20, 25, and 20-30°C increased linearly with the increase in salinity, and the highest increases occurred at 25°C. The organic compounds of the seedlings germinated at 30°C showed a quadratic response with the increase in salinity, with the highest contents of free amino acids, proline and sugars occurring at saline levels of 3.40, 7.41, and 7.19 dS m^{-1}, respectively. An increase in the synthesis of organic substances, such as nitrogen compounds and sugars, frequently occurs in plants under salt stress because these substances protect structures and support the osmotic balance of the plant (Esteves & Suzuki, 2008). Based on the results of this study, *S. hispanica* seedlings demonstrated osmotic adjustments to salt stress conditions by the synthesis of organic compounds at temperatures up to 30°C.

The dry matter accumulation in *S. hispanica* seedlings was drastically reduced by the increase in salinity, regardless of the temperature, and seeds germinated mainly at saline levels higher than 9.0 dS m^{-1} (Figure 4D). The inhibition of growth and phytomass accumulation in the presence of salinity is related to the ionic toxicity caused by the salts in the metabolic processes of the cells, which prevents or reduces plant growth (Panuccio, Jacobsen, Akhtar, & Muscolo, 2014; Bernardes et al., 2015). Silva et al. (2008) observed that in wheat seeds, there were alterations in the cell constituents when the seeds were subjected to salt stress with NaCl. According to the results of this study, it is evident that high NaCl concentrations are phytotoxic to *S. hispanica* and cause cell alterations that damage the physiological processes of germination and seedling development and that are potentiated under high temperature conditions.

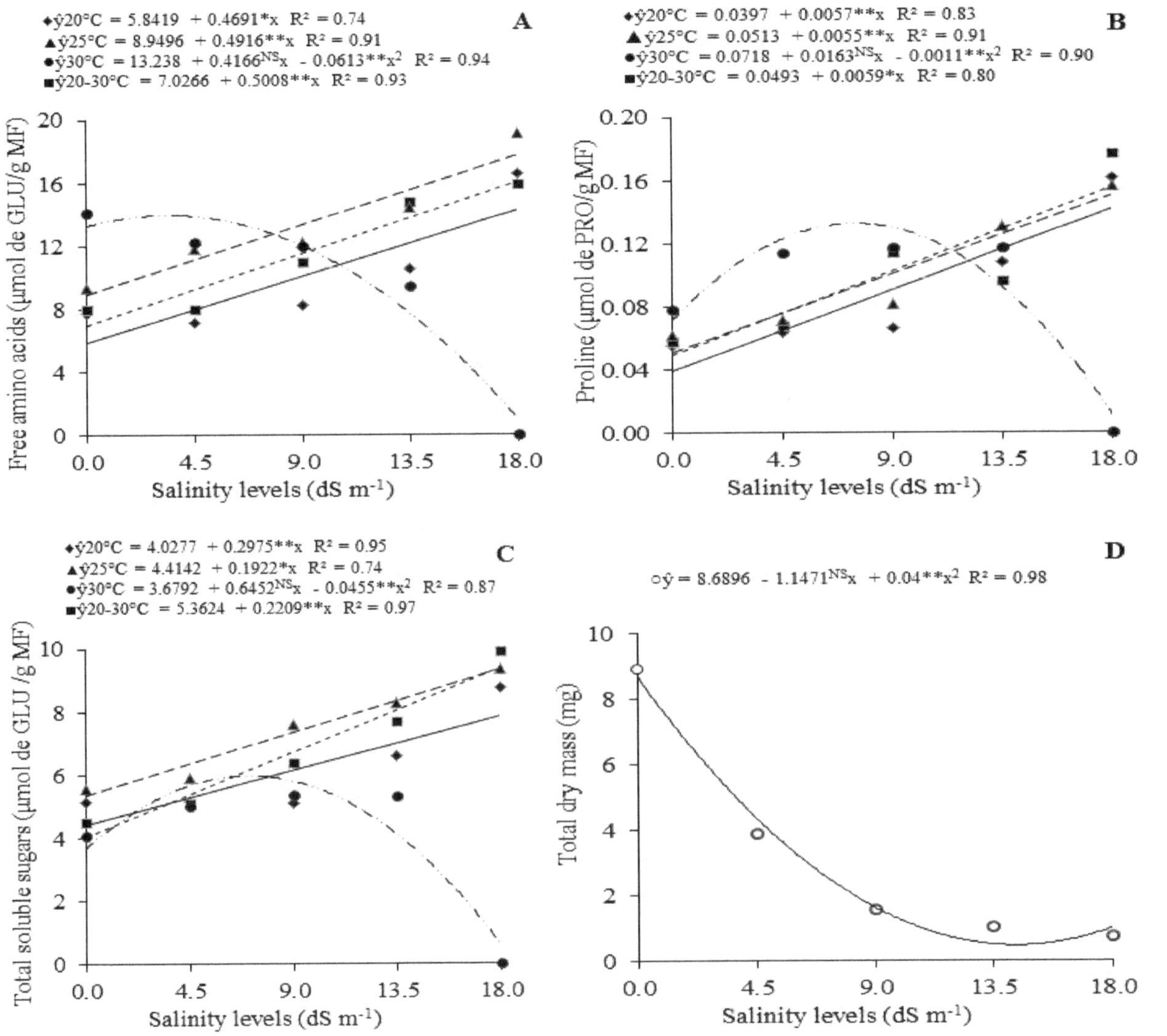

Figure 4. Contents of free amino acids (A), proline (B), total soluble sugars (C) and total dry mass (D) of *S. hispanica* seedlings germinated at different levels of salinity and temperature. ^{NS},★★, and ★ = non significant, significant at p < 0.01 and p < 0.05 propabaility level, respectively.

Conclusion

Salinity levels above 4.5 dS m^{-1} together with temperatures of 30 or 20-30°C negatively affect the germination, vigor, growth and biochemical components of *S. hispanica* seedlings.

The treatment temperature of 25°C promotes the best conditions for the development of *S. hispanica* seedlings up to the saline level of 9.0 dS m^{-1}.

References

Ali, N. M., Yeap, S. K., Ho, W. Y., Beh, B. K., Tan, S. W., & Tan, S. G. (2012). The promising future of chia, *Salvia hispanica* L. *Journal of Biomedicine and Biotechnology*, *12*(1), 1-9. doi: 10.1155/2012/171956

Bates, L. S., Waldren, R. P., & Teare, I. D. (1973). Rapid determination of free proline for water-stress studies. *Plant and Soil*, *39*(1), 205-207.

Bernardes, P. M., Mengarda, L. H. G., Lopes, J. C., Nogueira, M. U., & Rodrigues, L. L. (2015). Qualidade fisiológica de sementes de repolho de alta e baixa viabilidade sob estresse salino. *Nucleus*, *12*(1), 77-86. doi: 10.3738/1982.2278.1105

Brasil. Ministério da Agricultura, Pecuária e Abastecimento. (2009). Secretaria de Defesa Agropecuária. Coordenação Geral de Apoio Laboratorial. *Regras para análise de sementes*. Brasília, DF: MAPA/ACS.

Capitani, M. I., Corzo-Rios, L. J., Chel-Guerrero, L. A., Betancur-Ancona, D. A., Nolasco, S. M., & Tomás, M. C. (2015). Rheological properties of aqueous dispersions of chia (*Salvia hispanica* L.) mucilage. *Journal of Food Engineering*, *149*, 70-77. doi: 10.1016/j.jfoodeng.2014.09.043

Dal'Maso, E. G., Casarin, J., Costa, P. F., Cavalheiro, D. B., Santos, B. S., & Guimarães, V. F. (2013). Salinidade na germinação e desenvolvimento inicial de sementes de chia. *Cultivando o Saber*, *6*(3), 26-39.

Esteves, B. S., & Suzuki, M. S. (2008) Efeito da salinidade sobre as plantas. *Oecologia Brasiliensis*, *12*(4), 662-679.

Falk, J., & Munné-Bosch, S. (2010). Tocochromanol functíons in plants: antioxidation and beyond. *Journal Experimental Botany*, *61*(4), 1549-1566. doi: 10.1093/jxb/erq030

Ferreira, D. F. (2011). Sisvar: A computer statistical analysis system. *Ciência e Agrotecnologia*, *35*(6), 1039-1042. doi: 10.1590/S1413-70542011000600001

Flowers, T. J., & Colmer, T. D. (2008). Salinity tolerance in halophytes. *New Phytologist*, *179*(4), 945-963. doi: 10.1111/j.1469-8137.2008.02531.x

Gorai, M., Gasmi, H., & Neffati, M. (2011). Factors influencing seed germination of medicinal plant *Salvia aegyptiaca* L. (Lamiaceae). *Saudi Journal of Biological Sciences*, *18*(3), 255-60. doi: 10.1016/j.sjbs.2011.01.004

Imran, M., Nadeem, M., Manzoor, M. F., Javed, A., Ali, Z., Akhtar, M. N., & Hussain, Y. (2016). Fatty acids characterization, oxidative perspectives and consumer acceptability of oil extracted from pre-treated chia (*Salvia hispanica* L.) seeds. *BioMed Central*, *15*(162), 1-13. doi: 10.1186/s12944-016-0329-x

Lichthenthaler, H. K. (1987). Chlorophylls and carotenoids: pigments of photosynthetic biomembranes. In I. Packer, & R. Douce (Ed.), *Methods in enzimology* (p. 350-382). London, UK: Academic Press.

Lopes, K. P., Nascimento, M. G. R., Barbosa, R. C. A., & Costa, C. C. (2014). Salinidade na qualidade fisiológica em sementes de *Brassicas oleracea* L. var. Itálica. *Semina: Ciências Agrárias*, *35*(5), 2251-2260. doi: 10.5433/1679-0359.2014v35n5p2251

Maguire, J. D. (1962). Speed of germination-aid in selection and evaluation for seedling emergence and vigor. *Crop Science*, *2*(1), 176-177.

Meiado, M. V., Albuquerque, L. S. C., Rocha, E. A., Rojas-Aréchiga, M., & Leal, I. R. (2010). Seed germination responses of *Cereus jamacaru* DC. ssp. *jamacaru* (Cactaceae) to environmental factors. *Plant Species Biology*, *25*(2), 120-128. doi: 10.1111/j.1442-1984.2010.00274.x

Miranda, R. Q., Correia, R. M., Almeida-Cortez, J. S., & Pompelli, M. F. (2014). Germination of *Prosopis juliflora* (Sw.) D.C. seeds at different osmotic potentials and temperatures. *Plant Species Biology*, *29*(3), 9-20. doi: 10.1111/1442-1984.12025

Munns, R., & Tester, M. (2008). Mechanisms of salinity tolerance. *Annual Review of Plant Biology*, *59*(4), 651–681. doi: 10.1146/annurev.arplant.59.032607.092911

Oliveira, A. K. M., Souza, J. S., Carvalho, J. M. B., & Souza, S. A. (2015). Germinação de sementes de pau-de-espeto (*Casearia gossypiosperma*) em diferentes temperaturas. *Floresta*, *45*(1), 97-106. doi: 10.5380/rf.v45i1.27599

Panuccio, M. R., Jacobsen, S. E., Akhtar, S. S., & Muscolo, A. (2014). Effect of saline water on seed germination and early seedling growth of the halophyte quinoa. *Journal for Plant Science*, *6*(1), 1-18. doi: 10.1093/aobpla/plu047

Sanchez, P. L., Chen, M., Pessarakli, M., Hill, H. J., Gore, M. A., & Jenks, M. A. (2014). Effects of temperature and salinity on germination of non-pelleted and pelleted guayule (*Parthenium argentatum* A. Gray) seeds. *Industrial Crops and Products*, *55*(2), 90-96. doi: 10.1016/j.indcrop.2014.01.050

Silva Júnior, J. F., Klar, A. E., Tanaka, A. A., Freitas-Silva, I. P., Cardoso, A. E. I., & Putti, F. F. (2014). Tomato seeds vigor under water or salt stress. *Brazilian Journal of Biosystems Engineering*, *8*(1), 65-72. doi: 10.18011/bioeng2014v8n1p65-72

Silva, R. N., Duarte, G. L., Lopes, N. F., Moraes, D. M., & Pereira, A. L. A. (2008). Composição química de sementes de trigo (*Triticum aestivum* L.) submetidas a estresse salino na germinação. *Revista Brasileira de Sementes*, *30*(1), 215-220. doi: 10.1590/S0101-31222008000100027

Silveira, J. A. G., Silva, S. L. F., Silva, E. N., & Viégas, R. A. (2010). Mecanismos biomoleculares envolvidos com a resistência ao estresse salino em plantas. In H. R. Gheyi, N. S. Dias, & C. F. Lacerda (Ed.), *Manejo da salinidade na agricultura irrigada: estudos básicos e aplicados* (p. 171-180). Fortaleza, CE: INCTSal.

Stefanello, R., Neves, L. A. S., Abbad, M. A. B., & Viana, B. B. (2015). Resposta fisiológica de sementes de chia (*Salvia hispanica* – Lamiales: Lamiaceae) ao estresse salino. *Biotemas*, *28*(4), 35-39. doi: 10.5007/2175-7925.2015v28n4p35

Taiz, L., & Zeiger, E. (2013). *Fisiologia vegetal*. 6. ed. Porto Alegre, RS: ARTMED.

Yemm, E. W., & Coccking, E. C. (1955). The determination of amino-acids with ninhydrin. *Analyst*, *80*(2), 209-213.

Yemm, E. W., & Willis, A. J. (1954). The estimation of carbohydrates in plant extracts by anthrone. *Biochemical Journal*, *57*(3), 508-514.

Zhang, H., Zhang, G., Lü, X., Zhou, D., & Han, X. (2014). Salt tolerance during seed germination and early seedling stages of 12 halophytes. *Plant and Soil*, *388*(2), 229–241. doi: 10.1007/s11104-014-2322-3

21

Germination and seedling growth of ornamental species of *Passiflora* under artificial shade

Marcel Viana Pires, Alex-Alan Furtado de Almeida*, Abel Louro de Figueiredo, Fábio Pinto Gomes and Margarete Magalhães Souza

Departamento de Ciências Biológicas, Universidade Estadual de Santa Cruz, Rod. Ilhéus-Itabuna, km 16, 45650-000, Ilhéus, Bahia, Brazil.
**Author for correspondence. E-mail: alexalan@uesc.br*

ABSTRACT. *Passiflora morifolia*, *P. suberosa litoralis*, and *P. palmeri* var. *sublanceolata* are three wild species with ornamental potential that occur in Brazil. These species were evaluated with the purpose of determining the effects of different shade levels on seedling emergence and initial growth. Treatment with 50% shade resulted in the highest percentage of seedling emergence and the highest emergence speed index for all three species. The 50% shade level was the best condition for growth of *P. morifolia* and *P. palmeri* var. *sublanceolata* seedlings because in this condition, they presented higher growth in relation to control and the other treatments. However, *P. suberosa litoralis* can be considered a sun species because it grew better in full sunlight and 25% shade, but its growth was stunted at 75% shade. Increased leaf area and specific leaf area were due to changes in leaf dimension and shape in response to an increase in shade levels. These observations related to the initial growth reflect the development and adaptation of adult plants to heterogeneous environments. Therefore, *P. morifolia* and *P. palmeri* var. *sublanceolata* seedlings adapted better to moderate shade conditions, and *P. suberosa litoralis* seedlings should be cultivated in full sunlight.

Keywords: Passifloraceae, passion flowers, emergence speed index, development, irradiance.

Introduction

The *Passiflora* L. genus originates in South America and has its greatest geographical distribution in the Central-North area of Brazil (MELETTI; MAIA, 1999). The species are cultivated for their edible fruits, medicinal properties and, more recently, ornamental characteristics (ABREU et al., 2009). The use of passion flowers as ornamental plants has existed since the 15th century (VANDERPLANK, 2000) and currently continues in European and North American countries in the hybrid seedling market. Its beautiful and exotic flowers and leaves increase the ornamental value of the plant, making passion flowers a prosperous product in the ornamental plant agribusiness.

Some other features place the passion flower on the list of ornamental plants: the peculiar characteristics of the flowers, the capacity for flowering all year round, and the abundance of blooms and exuberant foliage (ULMER; MACDOUGAL, 2004). However, this potential is practically unexplored in the Brazilian plant market (ABREU et al., 2009).

Passiflora hybrids with ornamental potential have been cultivated for this purpose in various countries in the northern hemisphere, where the climate does not favour the breeding of tropical species. Brazil, however, has not yet made use of the wide genetic variability present in the country. The edaphoclimatic conditions in Brazil and the large number of species could potentiate a practically unexploited market. Thus, through competent programs of mass diffusion, the passion flower can be introduced into the Brazilian tropical ornamental plant market. Exportation, especially in Europe, which has cultivated ornamental passion flowers for years, could be a new source of income for the country (ABREU et al., 2009).

The propagation of *Passiflora* is done mainly through seeds at the commercial level, although it may also be done through grafting and cutting (DELANOY et al., 2006; LEONEL; PEDROSO, 2005). In order to obtain plants with the desired standards of quality, it is essential to understand the viable methods of propagation. The formation of seedlings in greenhouses is a very common practice in the cultivation of *P. Edulis* (ZANELLA et al., 2006) and may be employed in the production of seedlings of species with ornamental value (GUISELINI et al., 2004). Many growth variables have been used to evaluate the behaviour of seedlings in relation to shade, such as height and stem diameter, which are used frequently. The ability to grow quickly in height when shaded is an important adaptation mechanism when species are deprived of their full complement of sunlight (BJÖRKMAN, 1981; VERDIAL et al., 2000; ZANELLA et al., 2006). The production of dry matter, the leaf area, and the relationship between the biomass of the shoot and root portions are also used. Leaf area is a feature that is used to analyze the shade tolerance of different species because it correlates directly with the photosynthetic surface area (FARIAS et al., 1997; MATOS et al., 2009; MENZEL; SIMPSON, 1988). However, studies about the seedling growth of ornamental *Passiflora* species concerning the quantity of light absorbed are rare, which makes it difficult to specifically understand the plant's growth and development.

The *Passiflora* wild species are germplasms of interest because they can be used as a source of genes to improve traits in the cultivated species, and, for this purpose, attempts have been made to obtain ornamental hybrids (ABREU et al., 2009; SOUZA et al., 2003). The main objective of this study was to evaluate the effects of different shade levels on seed germination and the initial growth of selected seedlings of wild *Passiflora* species with ornamental potential. Such factors may efficiently subsidize the production of seedlings, indicating which levels of shade are most adequate to optimize cultivation.

Material and methods

The experiment was conducted in Ilhéus, Bahia State, Brazil (39°13'59" W, 14°45'15" S). The three species under study, *P. morifolia* Mast., *P. suberosa litoralis* (Kunth) K. Porter-Utley, and *P. palmeri* var. *sublanceolata* Killip, have been cultivated at UESC's Active Germoplasm Bank. *P. morifolia* occurs widely in Guatemala, Mexico, Venezuela, Bolivia, Colombia, Brazil, Ecuador, Peru, Paraguay, and Argentina. This species has beautiful white flowers with a purplish corona, which gives it, together with its intermediate plant size, favourable characteristics for growing in pots for indoor decoration (MILWARD-DE-AZEVEDO; BAUMGRATZ, 2004; VANDERPLANK, 2000). *P. suberosa litoralis* occurs in almost all of Central and South America, Hawaii, New Guinea, Fiji, and Samoa (VANDERPLANK, 2000). Lobed leaves of *P. suberosa litoralis* are examples of the wide variety of shapes, sizes and shades of green leaves in the genus. Many passion flower species have ornamental value only as a function of foliage (ABREU et al., 2009). *P. palmeri* var. *sublanceolata* is an ornamental tropical species; it is an herbaceous climber of medium size with beautiful flowers with pink and white petals and a purple corona. It is already being used as a parent in interspecific crosses to obtain hybrids for ornamental plants in the USA and Europe, as in the hybrids *P.* 'Aurora' and *P.* 'Pink Jewel' (ULMER; MACDOUGAL, 2004).

Sowing was performed in ten polyethylene bags per treatment, with a capacity of 2 L, with washed sand and matured manure in a 1:1 ratio used as substrate. Ten seeds per bag were sown at a depth of approximately 1.5 cm. At 46 days after emergence, a selection was performed to leave only the most vigorous seedling per plastic bag. After sowing, irrigation was performed daily by a micro-sprinkler system.

Artificial shade was obtained with black plastic screens fixed in wooden frames with dimensions of 5 x 5 x 2 m, under field conditions, which allowed for the reduction of 25% (light shade), 50% (partial shade), and 75% (full shade) of global radiation, along with a control treatment under full sun. The

PPFD, temperature, and relative air humidity values (Figure 1), measured at the upper extremity of seedlings between 8h and 18h, were obtained with a quantum sensor S-LIA-M003 connected to a climatologic station Hobo Micro Station Data Logger (Onset, EUA).

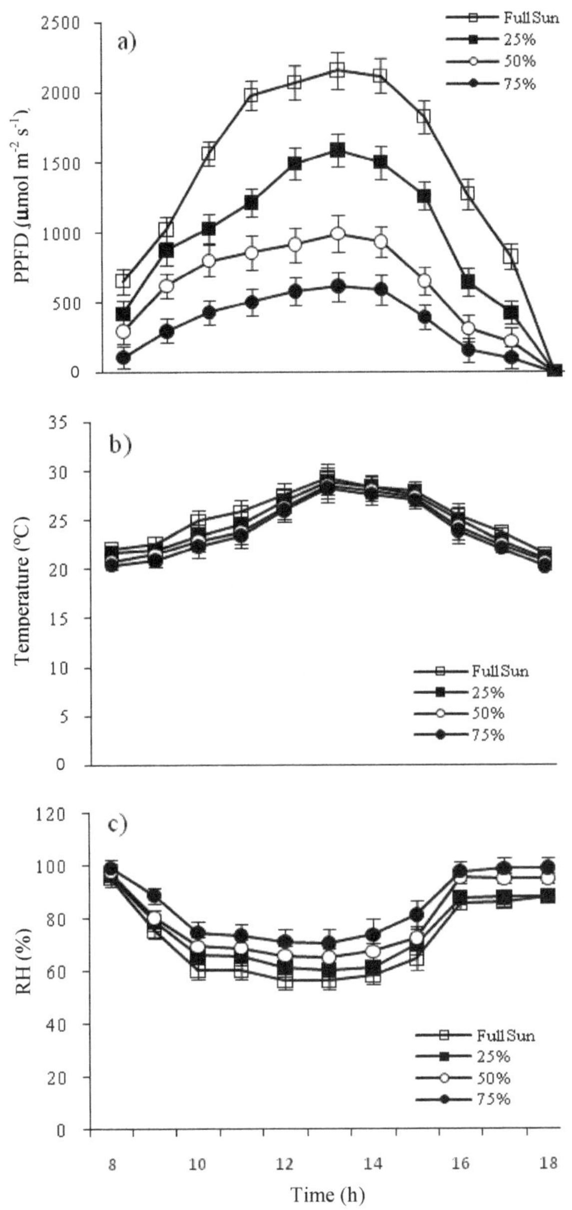

Figure 1. Daily changes in (a) photosynthetic photon flux density (PPFD), (b) temperature, and (c) relative air humidity (RH) in the full sunlight and shaded environments.

Seedling emergence was evaluated from the eighth day after sowing at regular intervals of two days, until the seedling emergence rate proved to be constant at approximately 90 days after sowing, considering the percentage of seedlings emerged with open cotiledonary leaves. The emergence speed index (ESI) was determined using the procedures described by Edmond and Drapala (1958), where counts of the seedlings were performed at regular intervals of five days, starting at the beginning of emergence until stabilization, according the following formula:

$$ESI = \frac{(N_1.G_1)+(N_2.G_2)+...+(N_n.G_n)}{G_1+G_2+...+G_n}$$

where:

N is the number of sowing days at each count, and G is the number of emerged seedlings in each counting.

The initial seedling growth evaluations were performed 46, 60, 74, and 88 days after the emergence, using ten seedlings per treatment. Evaluations were made by measuring the height of the seedlings from the collar to the apex extremity, with the help of a measuring tape and the stem's diameter at the level of the 2nd knot, obtained with the help of a digital pachymeter.

At 94 days after the emergence, five plants per treatment were randomly collected and used to determine the leaf area (LA), ratio shoot/root (S/R), the specific leaf area (SLA), and the leaf area ratio (LAR). Root (RDW), stem (SDW), leaf (LDW), and total (TDW) dry weight were obtained after drying at 70°C until a constant mass was reached. Leaf area was estimated using an automatic Leaf Area Meter LI-3100 (Li-Cor, USA). The S/R ratio was calculated as the quotient between shoot (SDW ± LDW) and RDW. The SLA and LAR were estimated as the quotients between individual mean values of LA and LDW and between total LA and LDW per seedling, respectively (BENINCASA, 1988).

The experiment was conducted with a completely randomized design, using a 4 x 3 factorial arrangement with four light levels (full sunlight and 25, 50, and 75% shade) and three *Passiflora* species, with ten replicates for evaluations of seed germination and initial growth (height and stem diameter) and five replicates for analysis of dry weight, using one seedling per experimental unit. The results were subjected to a variance analysis (ANOVA) followed by Tukey's mean comparison test ($p < 0.05$) and regression analysis ($p < 0.05$ and $p < 0.01$) (STORCK et al., 2006). All statistical procedures were performed using the software Statistica 6.0 (STATSOFT Inc., 2007).

Results and discussion

Germination

The seedling emergence for the three species studied started eight days after sowing. Seedling emergence concluded after 65 days after sowing for *P. morifolia* and *P. suberosa litoralis* and 90 days after sowing for *P. palmeri* var. *sublanceolata*. Seedling emergence and ESI percentage averages differed ($p < 0.05$) between shade levels for the three species (Table 1). *P. morifolia* presented higher percentages of seedling emergence at shade levels of 50 and 75% (49.3 and 44%, respectively). In the same conditions, the lowest ESI values were found (10.83 and 11.27 days), indicating that there was faster initial development of these seedlings under dense shade treatment. However, *P. palmeri* var. *sublanceolata* species presented low percentage rates of seedling emergence in all treatments, of which the highest value was 26.6 at 50% of shade. In the same treatment, this species presented the lowest ESI value, which indicates a higher rate of growth for recently emerged seedlings under moderate shade conditions. In contrast, differences in seedling emergence rates for *P. suberosa litoralis* were not found between shade levels of 50 and 75%, while the lowest ESIs occurred at 50 and 75% of shade (15.4 and 18.7 days, respectively).

Some reports about the seedling emergence of *Passiflora* raise concern over a slow and irregular period of emergence (ALEXANDRE et al., 2004; DELANOY et al., 2006; PASSOS et al., 2004). The treatment using 50% shade presented with the best percentage of seedling emergence and of emergence speed index for all three species studied (Table 1). The low values of observed seedling emergence suggested that the conditions were not fully satisfactory for the germination of seeds of the three wild species of *Passiflora*. Other abiotic factors also affect the germination process, such as temperature and relative humidity (PINHEIRO; BORGHETTI, 2003; SOCOLOWSKI et al., 2010). Some species require light for germination at near optimum temperatures, and some species are indifferent to light under these conditions (BENÍTEZ-RODRÍGUES et al., 2004). However, no difference in temperature was observed between the four light availability levels (Figure 1). The higher relative humidity observed in the more shaded conditions, together with the reduction of irradiance, may have benefited the germination rates by creating a micro-climate that is favourable for seedling emergence.

Studies of the germinative behaviour of *P. molissima* and *P. tricuspis* have shown germination rates close to 27 and 57%, respectively, confirming the low germination for wild *Passiflora* species (DELANOY et al., 2006). In contrast, *P. edulis*, the purple passion fruit, an already domesticated species, presents a seedling emergence capacity of approximately 90% (MELETTI; MAIA, 1999). *Passiflora* seeds demonstrate dormancy due to the impermeability of the integument (MORLEY-BUNKER, 1980; DELANOY et al., 2006). However, nothing is known about the mechanisms of physiological or physical dormancy in the species studied; rather, it is suspected that dormancy exists.

Seedling growth

The estimated regressions (Table 2) showed that the growth of the three species studied was influenced by shade treatments. There was an increase in the height of seedlings with the increase of shade starting from 60 days after emergence, especially at levels of 50 and 75%, which presented the highest average values for the three species (Figure 2a, b and c). In contrast, the increase in the diameter of the stems was proportional to the decrease in the shade for the three species, with the highest average values occurring in full sun and 25% shade (Figure 2d, e and f). Moreover, the relation between shade and the increase in the diameter of the stems became more intense starting at 74 days after emergence.

The low height values of seedlings that grew under full sun indicate that this condition limits these species' initial development (Figure 2a, b and c). In contrast, the increase in height of shaded plants is considered to be a typical morphogenetic response (MATOS et al., 2009). Similarly, shade induced more growth in height in *P. edulis* f. *flavicarpa* plants (ZANELLA et al., 2006). Apical dominance tends to increase when plants are submitted to high shade levels, due to a decrease in the production of photoassimilates and the highest level of auxin at the stem apex bud (VANNESTE; FRIML, 2009; WOODWARD; BARTEL, 2005).

Table 1. Seedling emergence (%) and emergency speed index (ESI) of *P. morifolia*, *P. suberosa litoralis*, and *P. palmeri* var. *sublanceolata*, under different shade conditions.

Shade	*P. morifolia*		*P. suberosa litoralis*		*P. palmeri* var. *sublanceolata*	
	%	ESI	%	ESI	%	ESI
Full Sun	28.0 ± 0.1^{Ab}	15.8 ± 0.2^{Ca}	26.7 ± 0.2^{Ab}	24.1 ± 0.3^{Ba}	2.7 ± 0.8^{Bc}	30.9 ± 0.2^{Aa}
25%	29.3 ± 0.2^{Ab}	16.0 ± 0.1^{Ba}	32.0 ± 0.4^{Aab}	20.6 ± 0.2^{Aa}	6.7 ± 0.5^{Bb}	21.5 ± 0.3^{Ab}
50%	49.3 ± 0.2^{Aa}	10.8 ± 0.2^{Bb}	42.7 ± 0.1^{Aa}	15.4 ± 0.2^{Ab}	26.7 ± 0.5^{Ba}	18.2 ± 0.1^{Ab}
75%	44.0 ± 0.05^{Aa}	11.3 ± 0.3^{Cb}	41.3 ± 0.2^{Aa}	18.7 ± 0.4^{Bb}	12.0 ± 0.3^{Bb}	31.2 ± 0.4^{Aa}

Note: means of ten replicates ± S.E. Means comparison were done using Tukey's test ($p < 0.05$). For each variable lower case letters indicate comparisons between treatments and capital case between species.

The increase in stem diameter is directly related to the exchange activity, which depends on photosynthetic products, such as carbohydrates and hormones, from the apical bud (PAIVA et al., 2003). *P. edulis* f. *flavicarpa* seedlings cultivated under 70% shade seemed very etiolated, with wide internode intervals and a small number of leaves, when compared to seedlings grown under full sunlight (SILVA et al., 2006). This fact was also seen in more prominent shade (75%), where a light etiolation in *P. morifolia*, *P. suberosa litoralis*, and *P. palmeri* var. *sublanceolata* seedlings occurred (Figure 2d, e and f).

At 94 days of exposure, differences ($p < 0.05$) were observed between treatments for RDW, SDW, LDW, and TDW (Table 3). The highest RDW values occurred at 50% shade for *P. morifolia* and *P. palmeri* var. *sublanceolata* and at full sunlight for *P. suberosa litoralis*. A difference in SDW between shade levels for *P. suberosa litoralis* and *P. palmeri* var. *sublanceolata* was not observed. However, for *P. morifolia*, the highest SDW values, observed in the treatment with 50% shade, differed in relation to the other levels of shade. A decrease in the shade level had a diminishing effect on LDW for *P. suberosa litoralis*, while for *P. morifolia* and *P. palmeri* var. *sublanceolata* the LDW value increased along with a decrease in the light intensity. As a result of the increment of shade, there was a reduction of TDW for *P. suberosa litoralis*, while for *P. morifolia* and *P. palmeri* var. *sublanceolata*, higher TDW values occurred at 50% shade (Table 3).

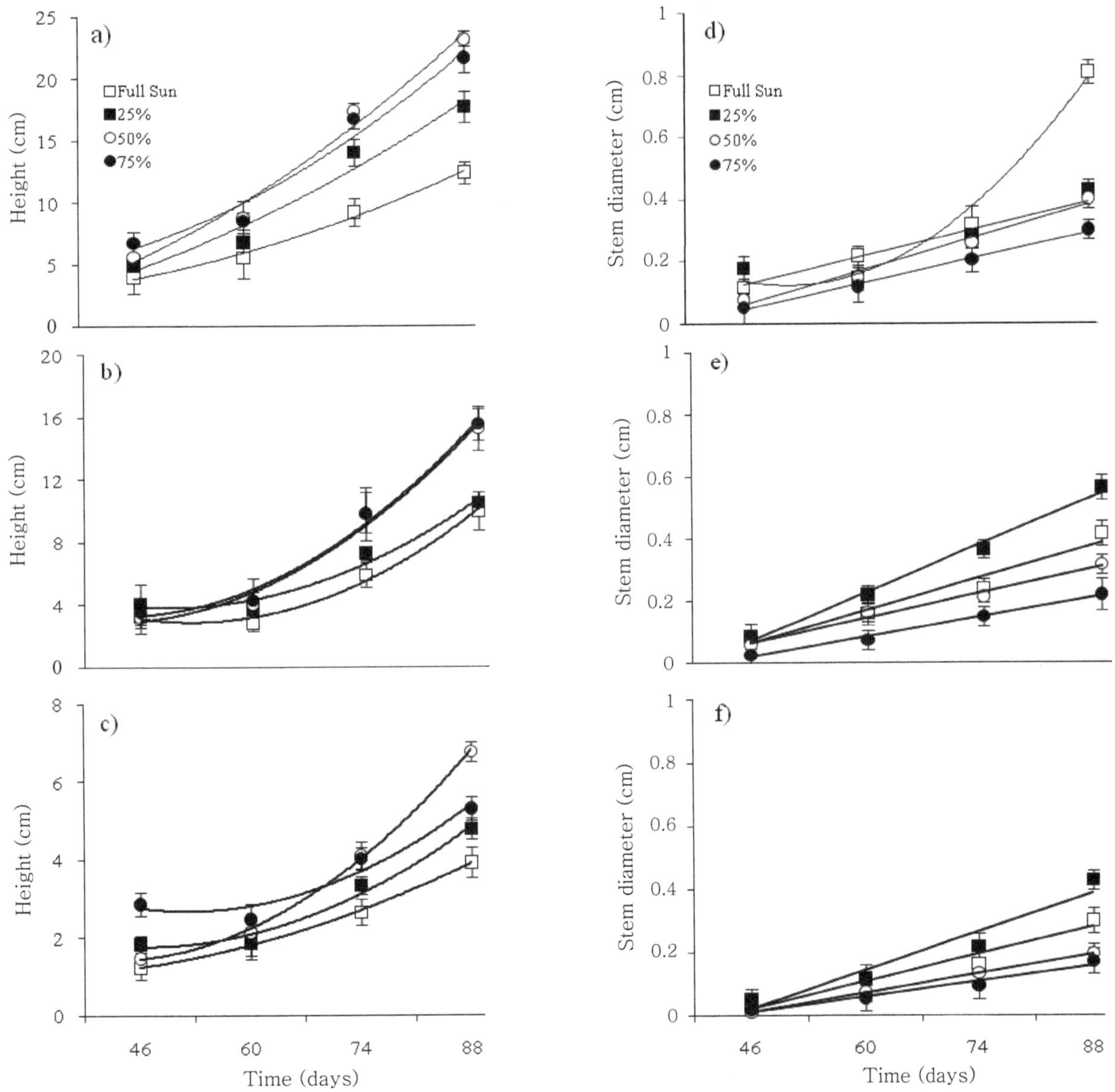

Figure 2. Height (a-c) and stem diameter (d-f) of *P. morifolia* (a, d), *P. suberosa litoralis* (b, e), and *P. palmeri* var. *sublanceolata* (c, f) seedlings under different shade conditions, at 46, 60, 74, and 88 days after emergence. Points are means of ten replications (± S.E).

Table 2. Regressions for the relationship between shade levels and time (days) to height and stem diameter.

Species	Shade	Height	Stem Diameter
P. morifolia	Full Sun	$\hat{y} = 0.40x^2 \pm 0.87x \pm 2.56$, $r^2 = 0.99$**	$\hat{y} = 0.09x^2 - 0.27x \pm 0.31$, $r^2 = 0.97$**
	25%	$\hat{y} = 0.44x^2 \pm 2.32x \pm 1.72$, $r^2 = 0.96$*	$\hat{y} = 0.09x \pm 0.04$, $r^2 = 0.81$*
	50%	$\hat{y} = 0.66x^2 \pm 2.82x \pm 1.73$, $r^2 = 0.98$**	$\hat{y} = 0.11x - 0.05$, $r^2 = 0.98$**
	75%	$\hat{y} = 0.79x^2 \pm 1.34x \pm 4.15$, $r^2 = 0.96$*	$\hat{y} = 0.08x - 0.04$, $r^2 = 0.99$**
P. suberosa litoralis	Full Sun	$\hat{y} = 1.11x^2 - 3.23x \pm 5.21$, $r^2 = 0.98$**	$\hat{y} = 0.10x - 0.04$, $r^2 = 0.95$*
	25%	$\hat{y} = 0.91x^2 - 2.24x \pm 5.21$, $r^2 = 0.96$*	$\hat{y} = 0.16x - 0.09$, $r^2 = 0.99$**
	50%	$\hat{y} = 1.15x^2 - 1.53x \pm 3.29$, $r^2 = 0.98$**	$\hat{y} = 0.08x - 0.02$, $r^2 = 0.98$**
	75%	$\hat{y} = 1.21x^2 - 1.89x \pm 3.96$, $r^2 = 0.98$**	$\hat{y} = 0.06x - 0.05$, $r^2 = 0.99$**
P. palmeri var. *sublanceolata*	Full Sun	$\hat{y} = 0.16x^2 \pm 0.08x \pm 0.99$, $r^2 = 0.99$**	$\hat{y} = 0.08x - 0.06$, $r^2 = 0.96$*
	25%	$\hat{y} = 0.35x^2 - 0.73x \pm 2.14$, $r^2 = 0.98$**	$\hat{y} = 0.12x - 0.09$, $r^2 = 0.93$*
	50%	$\hat{y} = 0.49x^2 - 0.70x \pm 1.65$, $r^2 = 0.99$**	$\hat{y} = 0.06x - 0.05$, $r^2 = 0.99$**
	75%	$\hat{y} = 0.42x^2 - 1.19x \pm 3.54$, $r^2 = 0.95$*	$\hat{y} = 0.05x - 0.04$, $r^2 = 0.96$*

Note: (*) and (**) represent significant difference at $p < 0.05$ and $p < 0.01$, respectively, by the *F*-test.

Table 3. Root (RDW, g), stem (SDW, g), leaf (LDW, g), and total dry weight (TDW, g) of *P. morifolia*, *P. suberosa litoralis*, and *P. palmeri* var. *sublanceolata* seedlings under different shade conditions until 94 days after emergence.

Species	Shade	RDW	SDW	LDW	TDW
P. morifolia	Full Sun	1.84 ± 0.1^{Bd}	1.52 ± 0.1^{Ab}	2.33 ± 0.08^{Bc}	5.69 ± 0.2^{Bd}
	25%	2.57 ± 0.05^{Ab}	1.77 ± 0.1^{Ab}	3.04 ± 0.08^{Ab}	7.38 ± 0.1^{Ab}
	50%	3.04 ± 0.07^{Aa}	2.17 ± 0.03^{Aa}	3.82 ± 0.1^{Aa}	9.03 ± 0.2^{Aa}
	75%	2.27 ± 0.04^{Ac}	1.75 ± 0.09^{Ab}	2.55 ± 0.1^{Ac}	6.57 ± 0.2^{Ac}
P. suberosa litoralis	Full Sun	2.74 ± 0.2^{Aa}	1.58 ± 0.1^{Aa}	2.96 ± 0.1^{Aa}	7.28 ± 0.2^{Aa}
	25%	2.72 ± 0.2^{Aa}	1.49 ± 0.2^{Aa}	2.32 ± 0.2^{Bab}	6.53 ± 0.3^{Ba}
	50%	2.04 ± 0.1^{Bab}	1.28 ± 0.1^{Ba}	2.39 ± 0.3^{Bab}	5.71 ± 0.1^{Ba}
	75%	1.8 ± 0.2^{Bb}	1.16 ± 0.1^{Ba}	1.89 ± 0.1^{Bb}	4.85 ± 0.4^{Ba}
P. palmeri var. *sublanceolata*	Full Sun	1.25 ± 0.1^{Cb}	0.97 ± 0.1^{Ba}	1.62 ± 0.1^{Cb}	3.84 ± 0.1^{Cb}
	25%	1.96 ± 0.1^{Ba}	1.18 ± 0.07^{Ba}	2.18 ± 0.2^{Bab}	5.33 ± 0.2^{Ca}
	50%	2.08 ± 0.1^{Ba}	1.25 ± 0.03^{Ba}	2.53 ± 0.2^{Ba}	5.86 ± 0.2^{Ba}
	75%	1.55 ± 0.07^{Bb}	1.02 ± 0.06^{Ba}	1.84 ± 0.1^{Bb}	4.41 ± 0.1^{Bb}

Note: Means of five replicates ± S.E. Means comparison were done using Tukey's test ($p < 0.05$). For each variable lower case letters indicate comparisons between treatments and capital case comparisons between species.

A study of the development of *P. edulis* f. *flavicarpa* seedlings showed the highest values of RDW in plants grown in full sunlight (SILVA et al., 2006). The same study showed that in 30% shade, there was an increase in SDW in *P. edulis* f. *flavicarpa* seedlings, while the lowest values were found in full sunlight and 70% shade. Low RDW values in *P. suberosa litoralis* that grow in environments with more intense shade may have occurred due to the decrease in the amount of auxin that was carried by roots in shaded plants, resulting in the decrease in the formation of roots (MORELLI; RUBERTI, 2000; VANNESTE; FRIML, 2009).

The highest LDW values for *P. suberosa litoralis* seedlings in full sunlight may have occurred due to an increase in foliar thickness that normally occurs under high amounts of light (SCALON et al., 2001). A similar result was found by Silva et al. (2006), who observed a decrease in the production of leaf biomass in *P. edulis* f. *flavicarpa* accompanying an increase in the shade level. The low light intensity for passion flowers caused branch etiolation and reduced the aerial part dry biomass (CLAUSSEN, 1996; KLUGE, 1998); these data corroborate those obtained in this experiment for *P. suberosa litoralis*. Moreover, the *P. morifolia* and *P. palmeri* var. *sublanceolata* TDW values are similar to those observed for *P. edulis* f. *flavicarpa*, where the accumulation of TDW in plants in full sunlight was lower than that obtained from shade treatments (ZANELLA et al., 2006).

An effect ($p < 0.05$) of shade levels on LA, SLA, and LAR was found for the three species studied, whose highest averages were observed at 50 and 75% shade (Table 4). *P. morifolia* and *P. palmeri* var. *sublanceolata* presented maximum values of LA, SLA, and LAR in 75% shade, while for *P. suberosa litoralis*, maximum values for these parameters occurred in 50% shade. The S/R value did not differ between shade levels (Table 4).

The increase in LA and SLA is caused by changes in leaf dimensions and shape as a response to increasing shade levels. The expansion of the leaf under low irradiance is frequently reported and indicates the way that the plant compensates for the decrease in light, making better use of this resource by increasing the surface area (CAMPOS; UCHIDA, 2002). Increased SLA may improve light harvesting per unit of resources invested in construction of photosynthetic tissues (LUSK et al., 2008); it represents an adaptive mechanism, demonstrating the most efficient utilization of photoassimilates, as a larger photosynthetic area is produced per unit of accumulated dry matter (CHAVES et al., 2008; DAMATTA, 2004).

Table 4. Leaf area (LA, cm^2), specific leaf area (SLA, $cm^2\ g^{-1}$), leaf area ratio (LAR, $cm^2\ g^{-1}$), and ratio shoot $root^{-1}$ (S/R) of *P. morifolia*, *P. suberosa litoralis*, and *P. palmeri* var. *sublanceolata* seedlings under different shade conditions until 94 days after emergence.

Species	Shade	LA	SLA	LAR	S/R
P. morifolia	Full Sun	137.97 ± 4.6^{Ac}	36.41 ± 2.2^{Ad}	15.33 ± 0.8^{Ad}	1.96 ± 0.1^{Aa}
	25%	185.01 ± 8.0^{Ab}	61.27 ± 4.1^{Ac}	25.09 ± 1.3^{Ac}	1.87 ± 0.1^{Aa}
	50%	256.55 ± 11.5^{Aa}	101.53 ± 6.5^{Ab}	39.22 ± 2.2^{Ab}	1.89 ± 0.1^{Aa}
	75%	292.42 ± 9.8^{Aa}	126.12 ± 6.5^{Aa}	51.66 ± 2.6^{Aa}	2.12 ± 0.1^{Aa}
P. suberosa litoralis	Full Sun	83.51 ± 5.2^{Bbc}	28.35 ± 1.9^{Ab}	11.48 ± 0.7^{Ab}	1.69 ± 0.1^{Aa}
	25%	76.31 ± 5.0^{Bc}	33.39 ± 2.7^{Bb}	11.85 ± 1.2^{Bb}	1.43 ± 0.1^{Aa}
	50%	129.49 ± 7.0^{Ba}	53.42 ± 6.2^{Ba}	22.64 ± 1.0^{Ba}	1.85 ± 0.2^{Aa}
	75%	100.51 ± 5.6^{Bb}	57.18 ± 2.0^{Ba}	20.91 ± 1.2^{Ba}	1.75 ± 0.2^{Ba}
P. palmeri var. *sublanceolata*	Full Sun	34.1 ± 2.4^{Cb}	13.83 ± 1.6^{Bb}	5.89 ± 0.6^{Bb}	1.82 ± 0.1^{Aa}
	25%	42.29 ± 2.5^{Cb}	19.93 ± 2.1^{Cb}	8.04 ± 0.7^{Bb}	1.75 ± 0.2^{Aa}
	50%	61.24 ± 2.7^{Ca}	34.06 ± 3.3^{Ca}	13.91 ± 0.6^{Ca}	1.87 ± 0.1^{Aa}
	75%	61.32 ± 3.7^{Ca}	38.3 ± 3.3^{Ca}	16.04 ± 1.2^{Ba}	2.14 ± 0.2^{Aa}

Note: Means of five replicates ± S.E. Means comparison were done using Tukey's test ($p < 0.05$). For each variable lower case letters indicate comparisons between treatments and capital case comparisons between species.

Therefore, the low LAR values observed in passion flowers at full sunlight may have benefited these plants, decreasing the exposure of plant tissues to the sun and reducing water loss and self-shade (MATOS et al., 2009). An increase in the S/R ratio was found with an increase in shade for *P. edulis* f. *flavicarpa* (SILVA et al., 2006). Shading favored the growth of the aerial part; however, the values obtained at full sunlight revealed a balance in the production of dry weight for both parts (Table 2). Low S/R values in plants from environments with higher light indicate that the biomass was distributed more to the roots than to photosynthesizing organs, allowing higher absorption of water and nutrients, a strategy that guarantees higher rates of photosynthesis and transpiration in these environments.

Conclusion

The shade treatments were most effective for seedling emergence of *P. morifolia*, *P. suberosa litoralis* and *P. palmeri* var. *sublanceolata*. Moreover, it can be indicated for the formation of *P. morifolia* and *P. palmeri* var. *sublanceolata* seedlings because in these conditions, the shade-tolerant species presented higher vegetative growth in relation to the other treatments. *P. suberosa litoralis* was revealed to be a sun-tolerant species, as it grew and developed better under full sunlight. The functional meaning of these morphological characteristics and structural adjustments were related to differential development under the imposed shade conditions, at a whole plant level. The development of these species in each environment was best demonstrated by growth rates and biomass accumulation because environment interferes in the production of biomass. Notwithstanding, relevant importance should be given to the adoption of handling techniques. These results may be useful for selecting sun and shade-tolerant species for landscaping projects, and they provide support for introducing the cultivation of *Passiflora* into the Brazilian tropical ornamental plant market. *P. suberosa litoralis* can be used as ornamental plants in outdoor gardens, green live fencing or pergola green cover, whereas *P. morifolia* and *P. palmeri* var. *sublanceolata* can be better used as indoor vase plants in shaded environments.

Acknowledgements

Thanks are due to CAPES (Coordination for Scientific Support for Post-Graduate Level Training) for the scholarships granted to M. V. Pires, to CNPq (Brazilian Council for Advancement of Science and Technology) for the scholarships awarded to A-A. F. Almeida, and to FAPESB (Foundation for Research Support of the State of Bahia) for the scholarships granted to A. L. Figueiredo and financial and technical assistance.

References

ABREU, P. P.; SOUZA, M. M.; SANTOS, E. A.; PIRES, M. V.; PIRES, M. M.; ALMEIDA, A.-A. F. Passion flower hybrids and their use in the ornamental plant market: perspectives for sustainable development with emphasis on Brazil. **Euphytica**, v. 166, n. 3, p. 307-315, 2009.

ALEXANDRE, R. S.; WAGNER JÚNIOR, A.; NEGREIROS, J. R. S.; PARIZZOTTO, A.; BRUCKNER, C. H. Germinação de sementes de genótipos de maracujazeiro. **Pesquisa Agropecuária Brasileira**, v. 39, n. 12, p. 1239-1245, 2004.

BENINCASA, M. M. P. **Análise do crescimento em plantas**: noções básicas. Jaboticabal: Funep, 1988.

BENÍTEZ-RODRÍGUES, J.; OROZCO-SEGOVIA, A.; ROJASARÉCHIGA, M. Light effect on seed germination of four *Mammillari* species from the Tehuacán-cuicatlán valley, central Mexico. **The Southwestern Naturalist**, v. 49, n. 1, p. 11-17, 2004.

BJÖRKMAN, O. Responses to different quantum flux densities. In: LANGE, O. L.; NOBEL, P. S.; OSMOND, C. B.; ZIEGLER, H. (Ed.). **Encyclopedia**

of Plant Physiology. Berlin: Springer-Verlag, 1981. p. 57-107.

CAMPOS, M. M.; UCHIDA, T. Influência do sombreamento no crescimento de mudas de três espécies amazônicas. **Pesquisa Agropecuária Brasileira**, v. 37, n. 3, p. 281-288, 2002.

CHAVES, A. R. M.; TEN-CATEN, A.; PINHEIRO, H. A.; RIBEIRO, A.; DAMATTA, F. M. Seasonal changes in photoprotective mechanisms of leaves from shaded and unshaded field-grown coffee (*Coffea arabica* L.) trees. **Trees**, v. 22, n. 3, p. 351-361, 2008.

CLAUSSEN, J. W. Acclimation abilities of three tropical rainforest seedlings to an increase in light intensity. **Forest Ecology and Management**, v. 80, n. 1-3, p. 245-255, 1996.

DAMATTA, F. Ecophysiological constraints on the production of shaded and unshaded coffee: a review. **Field Crops Research**, v. 86, n. 2-3, p. 99-114, 2004.

DELANOY, M.; VAN DAMME, P.; SCHELDEMAN, X.; BELTRAN, J. Germination of *Passiflora mollissima* (Kunth) L.H.Bailey, *Passiflora tricuspis* Mast. and *Passiflora nov* sp. seeds. **Scientia Horticulturae**, v. 110, n. 2, p. 198-203, 2006.

EDMOND, J. B.; DRAPALA, W. J. The effects of temperature, sand and soil, and acetone on germination of okra seed. **Proceedings of the American Society for Horticultural Science**, v. 71, n. 1, p. 428-434, 1958.

FARIAS, V. C.; SILVA, D. E.; ALMEIDA, A. J. Análise de crescimento de mudas de cedrorana (*Cedrelinga catenaeformis* (Ducke) Ducke) cultivadas em condições de viveiro. **Revista Brasileira de Sementes**, v. 19, n. 2, p. 193-200, 1997.

GUISELINI, C.; SENTELHAS, P. C.; OLIVEIRA, R. C.; PRELA, A. Uso de malhas de sombreamento em ambiente protegido III: efeito sobre o crescimento e a produção comercial da *Gerbera jamesonii*. **Revista Brasileira de Agrometeorologia**, v. 12, n. 1, p. 27-34, 2004.

KLUGE, R. A. Maracujazeiro (*Passiflora* sp.). In: CASTRO, P. R.; KLUGE, R. A. (Ed.). **Ecofisiologia de fruteiras tropicais**. São Paulo: Nobel, 1998. p. 32-47.

LEONEL, S.; PEDROSO, C. J. Produção de mudas de maracujazeiro-doce com o uso de bioregulador. **Revista Brasileira de Fruticultura**, v. 27, n. 1, p. 107-109, 2005.

LUSK, C.; REICH, P. B.; MONTGOMERY, R. A.; ACKERLY, D. D.; CAVENDER-BARES, J. Why are evergreen leaves so contrary about shade? **Trends in Ecology and Evolution**, v. 23, n. 6, p. 299-303, 2008.

MATOS, F. S.; WOLFGRAMM, R.; CAVATTE, P. C.; VILLELA, F. G.; VENTRELLA, M. C.; DAMATTA, F. M. Phenotypic plasticity in response to light in the coffee tree. **Environmental and Experimental Botany**, v. 67, n. 2, p. 421-427, 2009.

MELETTI, L. M.; MAIA, M. L. **Maracujá**: produção e comercialização. Campinas: Instituto Agronômico, 1999.

MENZEL, C. M.; SIMPSON, D. R. Effect of continuous shade on growth, flowering and nutrient uptake of passion fruit. **Scientia Horticulturae**, v. 35, n. 1-2, p. 77-88, 1988.

MILWARD-DE-AZEVEDO, M. A.; BAUMGRATZ, J. F. A. *Passiflora* L. subgênero Decaloba (DC.) Rchb. (Passifloraceae) na Região Sudeste do Brasil. **Rodriguésia**, v. 55, n. 85, p. 17-54, 2004.

MORELLI, G.; RUBERTI, I. Shade avoidance responses: driving auxin along lateral routes. **Plant Physiology**, v. 122, n. 3, p. 621-626, 2000.

MORLEY-BUNKER, M. J. S. Seed coat dormancy in *Passiflora* species. **Annual Journal of the Royal New Zealand Institute of Horticulture**, v. 8, n. 1, p. 72-84, 1980.

PAIVA, C. L.; GUIMARÃES, R. J.; SOUZA, C. A. Influência de diferentes níveis de sombreamento sobre o crescimento de mudas de cafeeiro (*Coffea arabica* L.). **Ciência e Agrotecnologia**, v. 27, n. 1, p. 134-140, 2003.

PASSOS, I. R. S.; MATOS, G. V. C.; MELETTI, L. M. M.; SCOTT, M. D. S.; BERNACCI, L. C.; VIEIRAS, M. A. R. Utilização do ácido giberélico para a quebra de dormência de sementes de *Passiflora nitida* Kunth germinadas *in vitro*. **Revista Brasileira Fruticultura**, v. 26, n. 2, p. 380-381, 2004.

PINHEIRO, F.; BORGHETTI, F. Light and temperature requirements for germination of seeds of *Aechmea nudicaulis* (L.) Griesebach and *Streptocalyx floribundus* (Martius ex Schultes F.) Mez (Bromeliaceae). **Acta Botanica Brasilica**, v. 17, n. 1, p. 27-35, 2003.

SCALON, S. P.; SCALON FILHO, H.; RIGONI, M. R.; VERALDO, F. Germinação e crescimento de mudas de pitangueira (*Eugenia uniflora* L.) sob condições de sombreamento. **Revista Brasileira de Fruticultura**, v. 23, n. 3, p. 652-655, 2001.

SILVA, M. L.; VIANA, A. E.; SÃO JOSÉ, A. R.; AMARAL, C. L.; MATSUMOTO, S. N.; PELACANI, C. R. Desenvolvimento de mudas de maracujazeiro (*Passiflora edulis* Sims f. *flavicarpa* Deg.) sob diferentes níveis de sombreamento. **Acta Scientiarum. Agronomy**, v. 28, n. 4, p. 513-521, 2006.

SOCOLOWSKI, F.; VIEIRA, D. C. M.; SIMÃO, E.; TAKAKI, M. Influence of light and temperature on seed germination of *Cereus pernambucensis* Lemaire (Cactaceae). **Biota Neotropica**, v. 10, n. 2, p. 53-56, 2010.

SOUZA, M. M.; PEREIRA, T. N.; SILVA, L. C.; REIS, D. S.; SUDRÉ, C. P. Karyotype of six *Passiflora* species collected in the state of Rio de Janeiro. **Cytologia**, v. 68, n. 2, p. 165-171, 2003.

STATSOFT Inc. **Statistica 6.0**. Tulsa, 2007.

STORCK, L.; GARCIA, D. C.; LOPES, S. J.; ESTEFANEL, V. **Experimentação vegetal**. Santa Maria: UFSM, 2006.

ULMER, T.; MACDOUGAL, J. M. ***Passiflora***: passionflowers of the world. Portland: Timber Press, 2004.

VANDERPLANK, J. **Passion flowers**. Cambridge: The MIT Press, 2000.

VANNESTE, S.; FRIML, J. Auxin: a trigger for change in plant development. **Cell**, v. 136, n. 6, p. 1005-1016, 2009.

VERDIAL, M. F.; LIMA, M. S.; TESSARIOLI NETO, J.; DIAS, C. T.; BARBANO, M. T. Métodos de formação de mudas de maracujazeiro amarelo. **Scientia Agricola**, v. 57, n. 4, p. 795-798, 2000.

WOODWARD, A. W.; BARTEL, B. Auxin: regulation, action, and interaction. **Annals of Botany**, v. 95, n. 5, p. 707-735, 2005.

ZANELLA, F.; SONCELA, R.; LIMA, A. L. Formação de mudas de maracujazeiro 'amarelo' sob níveis de sombreamento em Ji-Paraná/RO. **Ciência e Agrotecnologia**, v. 30, n. 5, p. 880-884, 2006.

22

Di-nitrogen fixation at the early and late growth stages of soybean

Raphael Lemes Hamawaki[1,2*] and Stella K Kantartzi[2]

*[1]Coordenação de Aperfeiçoamento de Pessoal de Nível Superior, Ministério da Educação, Setor Bancário Norte, Quadra 2, Bloco L, Lote 06, Asa Norte, 70040-020, Brasília, Distrito Federal, Brazil. [2]Department of Plant, Soil and Agricultural Systems, Southern Illinois University Carbondale, 1205 Lincoln Dr., Carbondale, Illinois, 62901, USA. *Author for correspondence. E-mail: hamawaki@siu.edu*

ABSTRACT. Soybean derives a significant portion of the required nitrogen (N) from the symbiosis with rhizobia bacteria. However, information on the available genetic variation for N_2 fixation capacity in different growth stages of soybean is limited. The objective of this study was to investigate the N_2 fixation capacity of 22 soybean lines compared with that of non-nodulating and supernodulating checks at the early and late growth stages and identify the most informative traits for selection. Two cycles of greenhouse experiments were carried out to estimate the percentage of N derived from the atmosphere (%Ndfa) as well as 10 different traits related to N_2 fixation. The results showed that %Ndfa was significantly different among the lines at the early and late growth stages. SPAD readings showed the highest correlation with the early N_2 fixation, whereas shoot dry weight with the late N_2 fixation. Early and late %Ndfa could be used to select superior lines for N_2 fixation and study the underlying physiological and molecular mechanism.

Keywords: *Glycine max*; %Ndfa; ^{15}N dilution technique; nodulation; symbiosis.

Introduction

The nitrogen (N) requirements of soybean are probably the highest among all major crops, since it demands 29 mg of N g^{-1} photosynthate compared with 11 mg of N g^{-1} photosynthate in corn and 26 mg of N g^{-1} photosynthate in cowpea (Sinclair & Wit, 1975). The crop derives 36–69% of its total N through its symbiotic relationship with rhizobia bacteria (Salvagiotti et al., 2008) and the rest from the available soil N.

Yield improvement in soybean is relatively high and averages 31.2 kg ha^{-1} $year^{-1}$. Thus, some concerns have been raised regarding the capacity of the new high-yielding cultivars to reach their yield potential without the input of nitrogenous fertilizers (Specht, Hume, & Kumudini, 1999; Salvagiotti et al., 2008). Moreover, previous studies have suggested a reduction in the N_2 fixation capacity of new cultivars compared with that of old ones (Nicolás, Arias, & Hungria, 2002; Van Kessel & Hartly, 2000). Therefore, the evaluation of lines and traits, aiming to improve the N_2 fixation capacity, is considered crucial (Nicolás et al., 2002).

In soybean, N_2 fixation varies among different growth stages, reaching the highest rate at the R2–R4 (Keyser & Li, 1992; Imsande, 1989; Latimore, Giddens, & Ashley, 1977; Lawn & Brun, 1974; Thibodeau & Jaworski, 1975). However, N accumulation follows a different pattern during the soybean growth cycle, suggesting a limitation in the N supplied through symbiosis (Keyser & Li, 1992; Phillips & DeJong, 1984).

Multiple evaluations of the N_2 fixation activity of the nodulating line Chippewa and its non-nodulating isoline were performed and unveiled significant variation in the proportion of N derived from fixation at different growth stages (Zapata, Danso, Hardarson, & Fried, 1987). Using the acetylene reduction activity assay for evaluating nitrogenase activity, Pazdernik, Graham, Vance, and Orf (1996) screened 20 soybean lines and identified considerable variation in early nodulation traits and N_2 fixation activity, whereas Fabre and Planchon (2000) concluded that the studied lines differed in N_2 fixation activity at the R5 and R6 growth stages, but not at the R2 growth stage. Herridge, Bergersen, and Peoples, (1990) used the ureide abundance and ^{15}N natural abundance methods, collecting data at the V8–R6.5 growth stages of seven soybean lines and demonstrated an overall increase in N_2 fixation with the advance of growth cycle; however, the pattern of fixation was genotype- and environment-dependent.

Variations in the N_2 fixation activity throughout the growth cycle of soybean have been attributed to differences in the expression of genes that control nodule development (Kaewsuralikhit, Yokoyama, Kouchi, & Arima, 2005). However, information on the underlying genetic mechanism of symbiosis in soybean is still limited mainly due to the difficulties (Santos et al., 2013) and relatively high cost of accurately measuring the N_2 fixation rate. In this context, a limited number of studies have focused on the study of traits related to N_2 fixation (Nicolás et al., 2002). In addition to nodulation, some other traits, such as photosynthesis parameters, seed weight, seed protein, plant height, and time to maturity, have been used to discriminate soybean genotypes regarding their N_2 fixation capacity (Vollmann, Walter, Sato, & Schweiger, 2011).

Therefore, information on measurements that can help to identify soybean lines with enhanced N_2 fixation capacity throughout the soybean growth cycle is needed. Furthermore, evaluations of different soybean lines at different growth stages are critical for identifying genetic variation for high N_2 fixation. The objectives of this study were to: (1) assess the value of early and late evaluations of soybean lines for high N_2 fixation capacity; (2) test the association of measurements with traits directly and indirectly related to N_2 fixation; and (3) select lines that can be used for studying the underlying mechanisms involved in the genetic control and inheritance of traits responsible for high N_2 fixation capacity.

Material and methods

Plant material

Twenty-two soybean lines of different maturity groups (II–IX) and genetic backgrounds were selected for this study based on their potential N_2 fixation, nodulation, and higher yield in order to investigate the association of growth stages and N_2 fixation (Table 1). Additionally, three checks were included; the non-nodulating soybean cultivars Nitrasoy (Burton, Israel, & Bishop, 2006) and D68-099 (Hartwig, 1994) for estimating the amount of N derived from fixation as well as the supernodulating mutant SS2-2 (Youn et al., 2008; Youn et al., 2009) for evaluating the levels of N_2 fixation.

Greenhouse Experiments

The study was conducted under greenhouse conditions at Southern Illinois University, Carbondale, IL, USA. All the lines were sown in plastic pots (15 cm in diameter; 14 cm in depth) filled with Fafard Growing Mix 2 (Conrad Fafard Inc., Agawam, MA, USA), consisting of 70% peat moss, 20% perlite, and 10% vermiculite. A solution, containing 14 mg of P and 18 mg of K, was applied to each pot prior to sowing. The first cycle of the study was sown on February 6, 2015 and the second on December 12, 2015; plants were then grown until they reached the R7 growth stage (Fehr, Caviness, Burmood, & Pennington, 1971). Three seeds were planted per pot, and two weeks later thinned to a single plant per pot. The lines were assigned to the experimental units following a randomized complete block design with four blocks and four replications within each block. Plants were kept throughout the cycle in a 16-h photoperiod at 26/22°C day/night. All the pots were watered once a day with a volume of water to prevent leaching of the ^{15}N fertilizer.

Inoculation with *Bradyrhizobium japonicum*

The *B. japonicum* strain USDA 110 was selected to inoculate soybean seeds due to its vastly and long term use in N_2 fixation soybean studies, and no additional strains were applied to prevent any interaction effects. USDA 110 was kindly provided by the USDA Soybean Genomics and Improvement Lab in Beltsville, MD, USA. The supplied sample was grown in liquid modified arabinose gluconate medium, pH 6.6 (Van Berkum, 1990) for 7 days in a shaking incubator until reaching the stationary phase. The inoculation was performed on the seed prior to sowing at a rate of 2×10^5 cells seed^{-1}.

^{15}N labelling

A solution of ^{15}N labeled urea with 10% atom ^{15}N excess was carefully applied to the growing medium of each pot prior to sowing. The medium was removed individually from each pot and mixed with the labeled solution using a plastic bag to achieve uniform distribution, and then, the labeled medium was returned to each pot. The total amount of labeled N was 7.82 mg N kg^{-1} of potting medium. A plastic tray was placed underneath each pot to collect any leaching and prevent the ^{15}N labeled material to spread on the greenhouse benches.

Sampling

Sampling was performed at 35 d after emergence (DAE; early growth stage) and R7 (late growth stage). The upper most fully developed leaf was collected at the early growth stage and pod walls at the late growth stage. These samples were then oven dried at 70°C for 72h in a convection oven (Isotemp 500; Fisher Scientific, Waltham, MA, USA) and ball milled in steel vials using a 5100 Mixer/Mill (SPEX SamplePrep LLC, Metuchen, NJ, USA) for 15 min. After weighing and packing in tin capsules, the samples were placed in 96-well plates and sent to the Mass Spectrometry Facility of Southern Illinois University for the analysis of ^{15}N enrichment. Prior to the collection of leaf samples at the early growth stage, three measurements of chlorophyll concentration were obtained from the center of the leaflets using a Minolta SPAD-502 chlorophyll meter (Konica Minolta Sensing Inc., Osaka, Japan), and the mean value was recorded as SPAD meter readings (SPD). Shoot, nodule, and seed samples were collected at the late growth stage and oven dried at 70°C for 72h prior to weighing and recording of shoot dry weight (SDW), nodule number (NN), nodule dry weight (NDW), total seed number (TSN), total seed weight (TSW), and mean seed weight (MSW). The number of days to maturity was recorded as DTM.

^{15}N Analysis and Total %N

For estimating ^{15}N enrichment and total %N, leaf and pod wall samples were analyzed using continuous flow elemental analysis isotope ratio mass spectrometry (CF-EA-IRMS) with a Thermo-Scientific Delta V Plus isotope-ratio mass spectrometer (Bremen, Germany) connected to a Costech 4010 Elemental Combustion System (Costech Analytical Technologies Inc., Valencia, CA, USA) via a Conflo IV unit. The measured ^{15}N abundance of each sample represented by the parameter atom% ^{15}N was subtracted from the natural abundance of ^{15}N in the atmosphere (0.3663 atom% ^{15}N) to obtain the atom% ^{15}N excess (Unkovich et al., 2008) in the leaf (LAT%) and pod wall samples (PAT%); the analysis also provided the percentage of N in the leaf (%LN) and pod wall (%PN) samples. The percentage of N derived from the atmosphere (%Ndfa) was estimated as follows (Unkovich et al., 2008):

$$\%\text{Ndfa} = \left(1 - \frac{\text{atom}\%^{15}\text{N excess N}_2 - \text{fixing plant}}{\text{atom}\%^{15}\text{N excess reference plant}}\right) \times 100$$

The non-nodulating lines Nitrasoy and D68-099 were used as references.

Statistical Analysis

All statistical analyses were performed using JMP 13 (SAS Institute, Cary, NC, USA). Analysis of variance (ANOVA) was performed in conjunction with Tukey's test to identify differences among lines or Dunnett's multiple comparison test for comparing the %Ndfa of each line at the early and late growth stages with that of the supernodulating check SS2-2. Evaluation cycle (EC) and the interaction between genotype and EC (G × EC) were considered as fixed effects, whereas block as a random effect.

Pearson's correlation was used to determine the relations among measurements and traits. A correlation coefficient was estimated for each pair-wise combination of traits related to N_2 fixation (Table 3) using JMP 13. To analyze the association of these traits with early and late fixation activity, the estimates of leaf (-LAT%) and pod (-PAT%) atom% ^{15}N excess were included. For these two parameters, the data were transformed using the negative value of each data point. This approach allowed to have an estimation of N_2 fixation activity for all genotypes, including the non-nodulating checks. Since the negative values of atom% ^{15}N excess were used, all the traits associated with the fixation activity yielded a positive correlation with -LAT% and -PAT%.

The Ward's minimum variance method (Ward, 1963) was used for clustering analysis. ANOVA was performed using the minimum variance criterion in the sum of squares to separate the lines in different clusters. These analyses were performed separately for the early and late growth stage and consequently, two dendrograms were constructed.

Results and discussion

Assessment of Di-nitrogen Fixation

Measurements of di-nitrogen fixation parameters at the early and late growth stages are presented in Table 1. The estimates of %Ndfa using D68-099 as a check were significantly different from those obtained using Nitrasoy as a check at the early (t = -9.5; $p < 0.01$) and late growth stages (t = 5.5; $p < 0.01$).

Table 1. Leaf (LAT%) and pod (PAT%) enrichment of ^{15}N and least squares means for early and late percentage of nitrogen derived from atmosphere (%Ndfa) for 22 lines, one supernodulating line, and two non-nodulating lines grown under greenhouse conditions for two growing seasons in ^{15}N-labelled soil.

Soybean line	Early (35 DAE) %Ndfa			Late (R7) %Ndfa		
	LAT%	D68-099	Nitrasoy	PAT%	D68-099	Nitrasoy
JTN-4307	0.1642	20	18	0.0121	85	88
JTN-5203	0.1759	14	12	0.0157	82	85
Osage	0.1469	27	25	0.0207	74	81
Ozark	0.1517	27	24	0.0138	84	87
R05-3239	0.1671	20	18	0.0216	74	79
Jake	0.1727	16	14	0.0149	83	86
Saluki 4910	0.1837	12	9	0.0148	83	85
Bragg	0.1784	14	12	0.0091	90	91
Davis	0.1398	31	29	0.0098	89	91
Enrei	0.1498	26	25	0.0679	20	32
Williams	0.1512	27	25	0.0180	80	82
PI 471938	0.1793	10	9	0.0139	82	86
Clark	0.1771	12	10	0.0118	87	88
Bossier	0.1537	25	24	0.0091	89	91
Centennial	0.1618	22	20	0.0124	85	88
Hardee	0.1546	26	24	0.0109	87	90
Jackson	0.1822	11	9	0.0267	67	73
S.J.2	0.1968	4	2	0.0105	88	90
J-200	0.1724	14	13	0.0161	83	84
R01-416F	0.1529	25	23	0.0128	85	87
PI 96169B	0.1890	7	6	0.1322	0	0
PI 96171	0.1469	27	25	0.0632	26	34
Supernodulating check						
SS2-2	0.1255	38	38	0.0129	86	88
Non-nodulating checks						
Nitrasoy	0.2023			0.1071		
D68-099	0.2080			0.0930		
Analysis of variance						
Genotype (G)	< 0.0001	< 0.0001	< 0.0001	< 0.0001	< 0.0001	< 0.0001
Evaluation Cycle (EC)	< 0.0001	< 0.0001	< 0.0001	ns	ns	ns
G x EC	< 0.0001	< 0.0001	< 0.0001	< 0.0001	< 0.0001	< 0.0001
Multiple comparison test						
Tukey HSD 95%	0.0443	22	21	0.0277	32	27

However, differences in ranking were minor, and the estimates for %Ndfa obtained from the two checks were highly correlated at both the early (r = 0.97; $p < 0.0001$) and late growth stages (r = 0.91; $p < 0.0001$). The LSmean of atom% ^{15}N excess was used as an auxiliary criterion for distinguishing the capacity of N_2 fixation of each line (Table 1).

All the estimates of %Ndfa were significantly affected by G ($p < 0.0001$), EC ($p < 0.0001$), and G×EC ($p < 0.0001$), except for that at the late growth stage that was not affected by EC. These results showed that the host control of N_2 fixation in soybean has an important contribution to the overall symbiotic N_2 fixation, but also that this trait is strongly affected by environmental factors.

At the early growth stage, S.J.2 and Davis showed the lowest and highest %Ndfa, respectively, and the estimates varied from 4–31% and 2–29% when using D68-099 and Nitrasoy as a check, respectively. The proportion of N derived from fixation at the early growth stage was in accordance to that reported by George and Singleton (1992), in which the average was 28% for the total N fixed during all the vegetative growth stages (George and Singleton, 1992). The limitation of soybean to supply N through N_2 fixation during the early growth stages has been well documented and can be attributed to the inability of the symbiotic rhizobia to supply all the required N by the crop (Phillips & DeJong, 1984; Keyser & Li, 1992).

At the late growth stage, PI96169B and Bragg showed the lowest and highest %Ndfa, respectively, and the estimates varied from 0–90% and 0–91% when using D68-099 and Nitrasoy as a check, respectively. Except for PI96169B that did not form any nodules with USDA 110, %Ndfa values estimated in the present study were in accordance to those reported by Salvagiotti et al. (2008) that in a review paper of 61 studies on soybean N_2 fixation reported a range of 58–98%.

Our results showed that atom% ^{15}N excess in leaf (LAT%) and pod walls (PAT%) were significantly influenced by G ($p < 0.0001$), EC ($p < 0.0001$), and G × EC ($p < 0.0001$), except for PAT% that was not affected by EC. Of these parameters, the high values for atom% ^{15}N excess indicated a reliance of the line on the N derived from the soil instead of that obtained from N_2 fixation. This pattern was confirmed by the higher enrichment of leaf and pod wall samples in both non-nodulating lines. The only exception was LAT% and PAT% of PI96169B, probably due to an incompatibility between the line and USDA 110; however, further studies need to be conducted to confirm this assumption. Davis and S.J.2 showed the lowest and highest LAT%, whereas Bossier and PI96169B showed the lowest and highest PAT%, respectively.

The overall high proportions of N derived from fixation identified at the late growth stage were consistent with those found by Harper (1987) that studied N_2 fixation under low soil-N conditions and reported that N derived from fixation was 80–94%. Although the average contribution of N_2 fixation at the late growth stage ($mean_{\%Ndfa}$ = 66.3; data not shown) was markedly higher than that at the early growth stage ($mean_{\%Ndfa}$ = 17.3; data not shown), the level of genetic variation followed the opposite trend. These results were in agreement with those reported by Pazdernik et al. (1996) that analyzed 20 soybean lines for early N_2 fixation and nodulation efficiency and suggested that early nodule formation can markedly improve N_2 fixation capacity.

We used the Dunnett's test to compare the N_2 fixation capacity of each line with that of the supernodulating check SS2-2. At the early growth stage, the %Ndfa of nine lines (Davis, Enrei, R01-416F, Williams, PI96171, Osage, Hardee, Bossier, and Ozark) did not show any significant differences from that of SS2-2 (Figure 1), indicating their high N_2 fixation capacity at the early growth stage. The remaining 13 lines yielded a significantly lower %Ndfa compared with that of SS2-2, demonstrating a limited N_2 fixation capacity at the early growth stage.

At the late growth stage, the %Ndfa of 19 lines (Bossier, S.J.2, Davis, Bragg, Hardee, Centennial, JTN-4307, Clark, Ozark, R01-416F, Jake, Saluki 4910, PI471938, JTN-5203, J-200, Williams, Osage, R05-3239, and Jackson) did not show any significant differences from that of SS2-2 (Figure 2). PI96169B, PI96171, and Enrei showed %Ndfa significantly different from that of the supernodulating check, indicating their poor performance in N_2 fixation at the late growth stage.

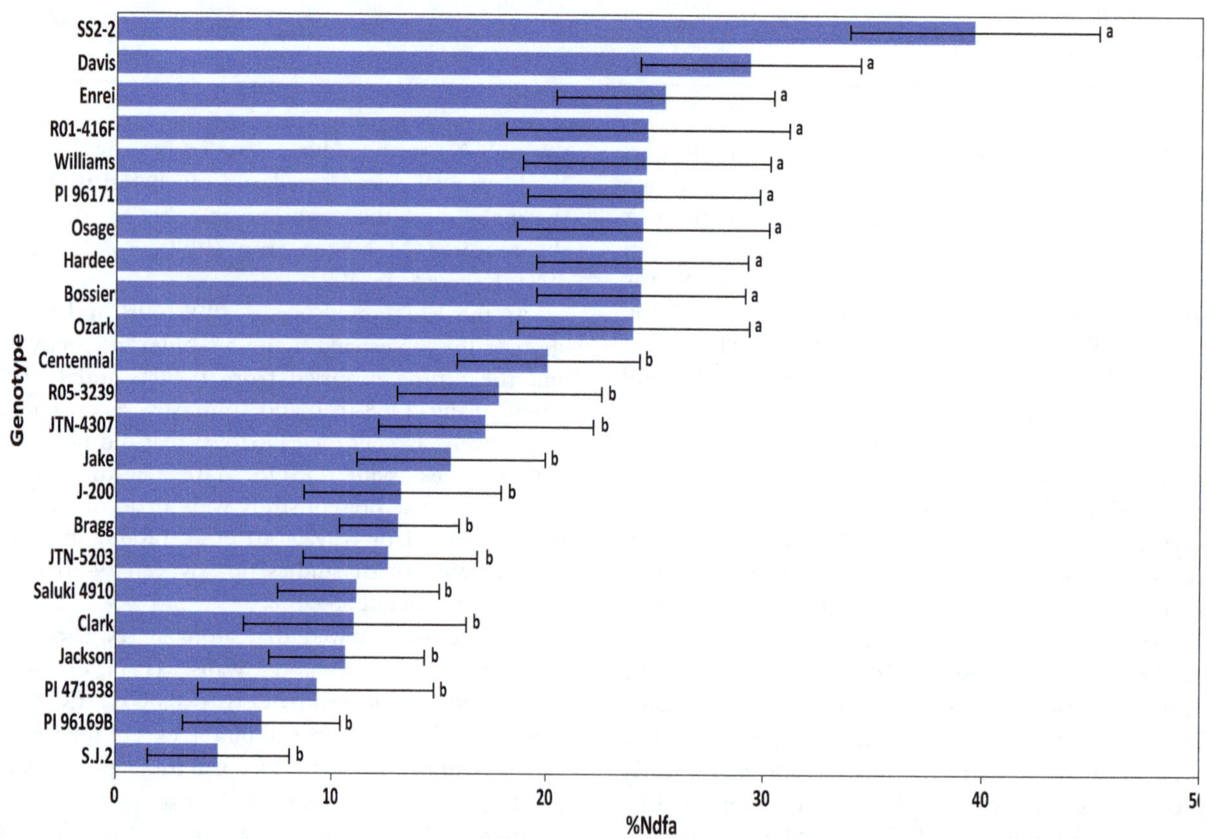

Figure 1. Comparison of %Ndfa at the early growth stage (35 d after emergence) between each of the 22 lines and the supernodulating line, SS2-2, using the Dunnett's multiple comparison test. Different letters indicate significant differences between the line and the check. Error bars indicate one standard error of the mean.

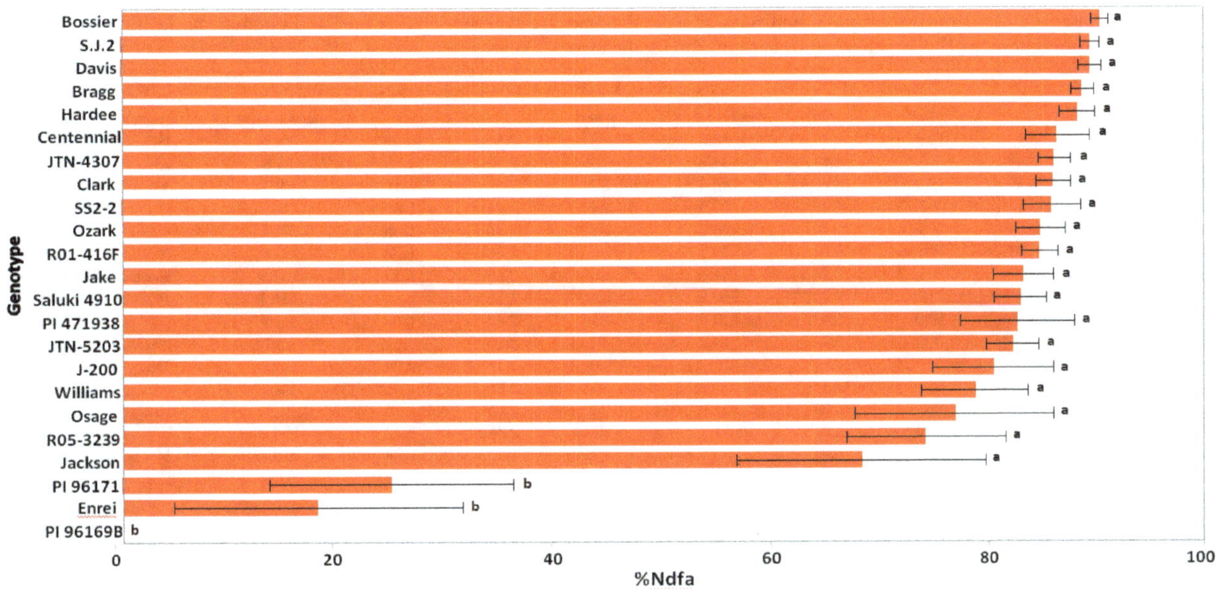

Figure 2. Comparison of %Ndfa at the late growth stage (R7) between each of the 22 lines and the supernodulating line, SS2-2, using the Dunnett's multiple comparison test. Different letters indicate significant differences between the line and the check. Error bars indicate one standard error of the mean.

The contribution of N_2 fixation to the total N accumulated during the vegetative period is estimated to be 21%, markedly lower than that at the R3–R7 growth stages that is approximately 56% (Zapata et al., 1987). In the present study, given the substantial variability found among the lines in the measurement of early N_2 fixation, we compared the mean of the two groups of lines discriminated using the Dunnett's test, and a difference of 14% in %Ndfa was found (data not shown); this difference in early N_2 fixation capacity might help to increment the accumulation of N derived from fixation during the vegetative period.

Further studies are needed to investigate the identified restriction of N_2 fixation in the vegetative phase. This limitation is markedly higher in the beginning of growth cycle, and thus, various studies have reported the benefits of N application at the early growth stages of soybean, a practice that is commonly known as 'starter N'. Osborne and Riedell (2006) obtained a 6% yield increase when 16 kg N ha^{-1} was applied at planting.

Traits related to N_2 fixation

The mean values of the 10 N_2 fixation-related traits (SPD, SDW, TSW, TSN, MSW, L%N, P%N, NN, NDW, and DTM) assessed in the 22 studied lines as well as the non-nodulating and supernodulating checks are presented in Table 2. The results showed that all the 10 N_2 fixation-related traits assessed in this study were significantly ($p < 0.001$) affected by G, EC, and G×EC, except for MSW that was not affected by G×EC or L%N that was not affected by EC. Among the 10 N_2 fixation-related traits, NN and NDW were the most directly related to nodulation; and the supernodulating check was significantly different from all the studied lines ($p < 0.0001$) for these two nodulation traits. This indicates that the studied lines had an adequate nodulation, but none of them reached the nodulation pattern of the supernodulating check. Although a superior nodulation ability is desirable for breeding lines with higher N_2 fixation capacity, the level of nodulation displayed by supernodulating mutants can restrict root growth and lead to a 30–40% yield reduction (Day, Lambers, Bateman, Carroll, & Gresshoff, 1986; Gremaud & Harper, 1989; Wu & Harper, 1991; Herridge, 2003). The MSW of Enrei was significantly ($p < 0.0001$) higher than that of the supernodulating check and all the other lines. The non-nodulating checks Nitrasoy and D68-099 performed poorly or showed the lowest values for P%N, SDW, MSW, L%N, TSW, and TSN. Additionally, Bossier showed the highest SDW, Enrei the highest SPD, and R01-416F the highest P%N.

A previous study that assessed soybean plants under controlled conditions demonstrated that the mean seed weight, total biomass, net photosynthetic output, and total plant N are related to both N_2 fixation and seed yield and that these effects were more pronounced at the pod-fill stage (Imsande, 1989). Analogous traits studied herein also accompanied the N_2 fixation activity among the studied lines; the supernodulating check showed the highest values of TSW, TSN, and L%N, whereas the non-nodulating checks the lowest values, indicating an overall robust relationship of these traits with the N_2 fixation activity.

Table 2. Comparison of nitrogen fixation-related traits among 22 lines, one supernodulating line, and two non-nodulating lines grown under greenhouse conditions for two growing seasons in ^{15}N-labelled soil.

	SPD	SDW	TSW	TSN	MSW	L%N	P%N	NN	NDW	DTM
		g plant^{-1}	g plant^{-1}	number plant^{-1}	mg seed^{-1}	%	%	number plant^{-1}	mg plant^{-1}	number of days
JTN-4307	28.3	10.3	4.6	48.8	98	2.59	0.99	107.5	192	170
JTN-5203	26.6	7.2	2.8	34.8	76	2.51	1.58	55.8	139	164
Osage	25.5	12.9	3.6	42.4	82	2.86	1.10	89.7	205	176
Ozark	25.1	7.2	2.1	24.3	87	2.21	1.65	54.5	101	166
R05-3239	28.0	7.5	2.4	28.2	85	2.45	1.48	70.3	128	162
Jake	26.6	8.6	3.0	29.2	96	2.46	1.28	60.4	105	163
Saluki 4910	26.2	9.1	3.2	34.6	88	2.60	1.39	103.9	159	171
Bragg	20.1	14.0	2.7	30.8	108	1.88	1.16	136.1	245	180
Davis	24.8	10.9	1.9	20.9	73	2.63	1.64	56.4	100	175
Enrei	34.3	3.6	3.7	14.2	250	2.85	0.61	43.7	190	97
Williams	22.4	7.9	4.6	36.7	125	2.18	0.94	119.0	235	152
PI 471938	23.7	9.8	3.1	31.9	85	2.05	1.51	68.5	107	161
Clark	21.2	9.8	3.8	37.9	96	2.01	1.26	71.7	162	161
Bossier	23.4	15.6	2.6	31.7	77	2.38	1.32	154.0	294	186
Centennial	24.3	12.2	3.4	35.5	90	1.93	1.24	128.0	249	177
Hardee	25.5	12.2	2.1	26.9	68	2.25	1.45	81.4	156	189
Jackson	22.6	13.3	3.2	31.3	95	2.13	0.81	95.1	184	169
S.J.2	23.1	12.7	1.2	17.6	45	2.11	1.31	94.5	213	196
J-200	23.7	14.6	1.3	15.5	70	1.96	1.69	88.6	132	196
R01-416F	26.2	8.0	2.5	27.1	80	2.44	1.84	92.9	167	159
PI 96169B	27.7	3.0	1.6	10.3	163	1.85	0.47	2.9	32	107
PI 96171	30.9	3.5	1.9	16.4	117	2.47	0.85	40.1	139	108
Supernodulating check										
SS2-2	31.3	9.8	4.6	48.8	112	3.49	1.58	294.1	541	157
Non-nodulating checks										
Nitrasoy	21.7	4.0	0.4	7.3	54	1.58	0.73	0	0	179
D68-099	23.9	2.9	0.5	8.5	50	1.92	0.78	0	0	184
Analysis of variance										
Genotype (G)	< 0.0001	< 0.0001	< 0.0001	< 0.0001	< 0.0001	< 0.0001	< 0.0001	< 0.0001	< 0.0001	< 0.0001
Evaluation Cycle (EC)	< 0.0001	< 0.0001	< 0.0001	<0.0002	< 0.0001	ns	< 0.0001	< 0.0001	< 0.0001	< 0.0001
G x EC	< 0.0001	< 0.0001	< 0.0001	< 0.0001	ns	0.0437	< 0.0001	< 0.0001	< 0.0001	< 0.0001
Multiple comparison test										
Test mean	25.5	9.2	2.9	28.8	100	2.32	1.19	84.8	168	165
Tukey 95%	5.4	3.2	1.8	18.2	70	0.86	0.60	79.9	151	21

SPD, SPAD readings (mg chlorophyll m^{-2}); SDW, shoot dry weight (g plant^{-1}); TSW, total seed weight (g plant^{-1}); TSN, total seed number per plant; MSW, mean seed weight (mg seed^{-1}); L%N, leaf % nitrogen; P%N, pod % nitrogen, NN, number of nodules per plant; NDW, nodule dry weight (mg plant^{-1}); DTM, d to maturity.

Table 3. Correlation matrix of ^{15}N enrichment and N fixation related traits for 22 lines, one supernodulating line, and two non-nodulating lines grown under greenhouse conditions for two growing seasons in ^{15}N-labelled soil.

	NN	NDW	TSN	TSW	MSW	DTM	SPD	SDW	L%N	P%N	-LAT%	-PAT%
NN	1.00	0.88**	0.33**	0.20**	-0.10**	-0.12**	-0.18**	0.39**	0.07^{ns}	0.20**	0.07^{ns}	0.27**
NDW		1.00	0.39**	0.30**	-0.01^{ns}	-0.16**	-0.11**	0.39**	0.08*	0.11**	0.09*	0.26**
TSN			1.00	0.78**	-0.02^{ns}	0.06^{ns}	-0.01^{ns}	0.31**	0.10*	-0.13**	0.20**	0.34**
TSW				1.00	0.40**	-0.06^{ns}	0.10*	0.12**	0.09*	-0.29**	0.24**	0.26**
MSW					1.00	-0.31**	0.23**	-0.27**	0.09*	-0.32**	0.08*	-0.18**
DTM						1.00	0.16**	0.26**	0.01^{ns}	-0.04^{ns}	0.16**	0.35**
SPD							1.00	-0.28**	0.44**	-0.23**	0.43**	-0.10**
SDW								1.00	0.00^{ns}	0.21**	0.03^{ns}	0.47**
L%N									1.00	0.10*	0.12**	0.09*
P%N										1.00	-0.18**	0.35**
-LAT%											1.00	0.15**
-PAT%												1.00

*, **. Significance at $p < 0.05$ and $p < 0.01$, respectively. NN, number of nodules per plant; NDW, nodule dry weight (mg plant^{-1}); TSW, total seed weight (g plant^{-1}); TSN, total seed number per plant; MSW, mean seed weight (mg seed^{-1}); DTM, d to maturity; SPD, SPAD readings (mg chlorophyll m^{-2}); SDW, shoot dry weight (g plant^{-1}); L%N, leaf % nitrogen; P%N, pod % nitrogen; -LAT%, negative leaf atom% ^{15}N excess; -PAT%, negative pod atom% ^{15}N excess.

Correlation analysis

A weak, but significant, correlation between the early -LAT% and the late -PAT% ($r = 0.15$; $p < 0.01$) suggested significant differences in N_2 fixation profile among the lines, showing that the superior N_2 fixation during the early growth stages did not last until the late growth stages. The strongest correlation was identified between -LAT% and SPD as well between -PAT% and SDW (Table 3). Since N_2 fixation was related with different traits at the early and late growth stages this confirmed the independence of the two evaluations. The variability in the N_2 fixation pattern of each line and the traits related to each of the measurements might be a promising tool for selecting lines with improved N_2 fixation capacity in breeding programs.

The nodulation parameters NN and NDW were found to be highly correlated ($r = 0.88$; $p < 0.01$), indicating that only one of those needs to be used for screening lines for N_2 fixation capacity. Additionally, both NN and NDW were moderately correlated with SDW (r =0.39; p < 0.01). This result might reflect the influence of the N derived from the nodules in the biomass accumulation. Mastrodomenico and Purcell (2012) reported that the N derived from fixation is largely accumulated into the biomass; however, a large portion of this N is not remobilized to the seed.

At the late growth stage, -PAT% was moderately correlated with SDW (r = 0.47; p < 0.01), similarly as reported by Herridge et al. (1990), in which P that represented the proportion of N derived from fixation was significantly correlated with the crop dry matter (r = 0.41; p < 0.05); however, the coefficient of correlation between NDW and -PAT% (r = 0.26; p < 0.01) found in the present study was markedly lower than that between nodule weight and P (r = 0.79; p < 0.001).

The coefficient of correlation between NN and -PAT% (r = 0.27; p < 0.01) was similar to that reported by Houngnandan et al. (2008) between nodule number and $\delta^{15}N$ (r = 0.325*) in a diverse group of soybean cultivars. Additionally, Pazdernik, Graham, and Orf (1997) studied a soybean population to investigate early nodulation traits and found that %Ndfa at R5 was significantly correlated with nodule fresh weight (r = 0.33; p < 0.01).

Correlation analysis also revealed associations with traits that might be used to reduce the cost of population screening for N_2 fixation capacity. SDW was moderately correlated with NN (r = 0.39; p < 0.01) and NDW (r = 0.39; p < 0.01), which are commonly used to improve N_2 fixation in soybean, but can impose practical and economical limitations in the screening of large populations. The correlation between SPD and L%N (r = 0.44; p < 0.01) might also be useful due to the lower cost to assess the former. Additionally, the correlation of TSN with NN (r = 0.33; p < 0.01) and NDW (r = 0.39; p < 0.01) could be advantageous, since the screening of the former is less laborious than that of the latter two.

Cluster analysis

Considerable variation was identified among the 22 soybean lines and the supernodulating check for N_2 fixation at the early and late growth stages, which was also demonstrated by the weak correlation between %Ndfa at the early and late growth stage (r = 0.14; p < 0.001) using both checks. These results confirmed the overall different performance of lines regarding the N_2 fixation capacity at different growth stages.

At the early growth stage, cluster analysis classified the 22 lines and SS2-2 into three different clusters (Figure 3): Cluster 1 included 10 lines with a high N_2 fixation capacity; Cluster 2 included 12 lines with a low N_2 fixation capacity; and Cluster 3 included only SS2-2 that had the highest N_2 fixation activity.

At the late growth stage, the 22 lines and the check SS2-2 were also classified into three different clusters: Cluster 1 included 20 lines with a high N_2 fixation capacity; Cluster 2 included Enrei and PI96171 with a low N_2 fixation capacity; and Cluster 3 included PI96169B that had the lowest N_2 fixation activity.

Soybean is estimated to spend 5.2–18.8 g of C g^{-1} of fixed N_2 (Minchin & Witty, 2005). To meet both the requirements of photosynthate from the plant and the N_2-fixing rhizobia, the elevated N_2 fixation activity is likely to be accompanied by a higher photosynthetic rate (Kaschuk, Hungria, Leffelaar, Giller, & Kuyper, 2010). Thus, the identification of this pattern may allow to develop new lines without compromising the photosynthate required for other physiological processes.

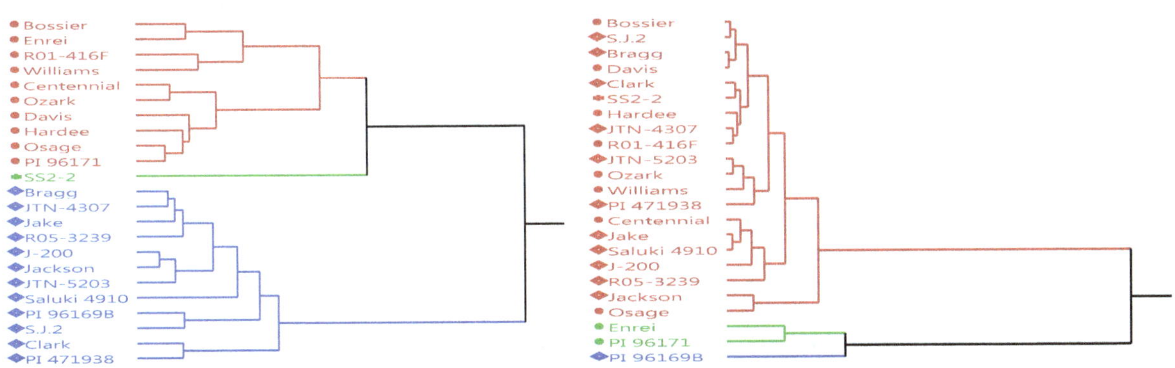

Figure 3. Clustering of 23 soybean lines and the two non-nodulating lines, Nitrasoy and D68-099, based on %Ndfa at the early (35 days after emergence; left) and late (R7; right) growth stages using the Ward's method.

In the present study, the identified variation in N_2 fixation activity during the plant cycle might be associated with the time and pattern of nodule formation on soybean roots. Nodules that are formed during the early growth stages are usually located at the top of the main root and are known to last on average 65 days. Later in the growth cycle, the N_2 fixation activity is maintained by a secondary set of nodules that is usually located on deep and lateral roots (Keyser & Li, 1992; Zapata et al., 1987). In the nodule aging process, the peak of N_2 fixation activity occurs at 4–5 weeks after the infection of plant cells by rhizobia; then, the nodules start to senesce and the rate of N_2 fixation is decreased (Dupont et al., 2012). Nonetheless, further research is needed on the formation and aging of nodules in soybean and its effect on N_2 fixation activity.

Conclusion

Overall, our results revealed significant variation in N_2 fixation capacity among the studied lines and also, between and within the early and late growth stages. The traits that were found closely related to N_2 fixation as well as the lines with contrasting N_2 fixation capacity at the early and late growth stages could be used in future studies to better understand the underlying physiological and molecular mechanisms of N_2 fixation in soybean.

Acknowledgements

This study was supported in part by the Brazilian Coordination for the Improvement of Higher Education Personnel (CAPES-Process BEX 11900/13-0). We would like to thank Dr. Chiun-Kang Hsu for his assistance and guidance.

References

Burton, J. W., Israel, D. W., & Bishop, P. E. (2006). Registration of 'Nitrasoy' soybean. *Crop Science*, *46*(6), 2709-2710. doi: 10.2135/cropsci2005.12.0503

Day, D. A., Lambers, H., Bateman, J., Carroll, B. J., & Gresshoff, P. M. (1986). Growth comparisons of a supernodulating soybean (*Glycine max*) mutant and its wildtype parent. *Physiologia Plantarum*, *68*(3), 375-382. doi: 10.1111/j.1399-3054.1986.tb03368.x

Dupont, L., Alloing, G., Pierre, O., Msehli, S. E., Hopkins, J., Hérouart, D., & Frendo, P. (2012). The Legume Root Nodule: From Symbiotic Nitrogen Fixation to Senescence. In N. Tetsuji (Ed.), *Senescence* (p. 137-168). Rejeka, CR: InTech. doi: 10.5772/34438

Fabre, F., & Planchon, C. (2000). Nitrogen nutrition, yield and protein content in soybean. *Plant Science*, *152*(1), 51-58. doi: 10.1016/S0168-9452(99)00221-6

Fehr, W. R., Caviness, C. E., Burmood, D. T., & Pennington, J. S. (1971). Stage of development descriptions for soybeans, *Glycine max* (L.) Merril. *Crop Science*, *11*(6), 929-931. doi: 10.2135/cropsci1971.0011183X001100060051x

George, T., & Singleton, P. W. (1992). Nitrogen assimilation traits and dinitrogen fixation traits in soybean and common bean. *Agronomy Journal*, *84*(6), 1020-1028. doi: 10.2134/agronj1992.00021962008400060022x

Gremaud, M. F., & Harper, J. E. (1989). Selection and initial characterization of partially nitrate tolerant nodulation mutants of soybean. *Plant Physiology*, *89*(1), 169-173. doi: 10.1104/pp.89.1.169

Harper, J. E. (1987). Nitrogen metabolism. In Wilcox Jr. (Ed.), *Soybeans: improvement, production, and uses* (p. 497-533). Madison, WI: American Society of Agronomy.

Hartwig, E. E. (1994). Registration of near-isogenic soybean germplasm lines D68-099 and D68-0102, differing in ability to form nodules. *Crop Science*, *34*(3), 822. doi: 10.2135/cropsci1994.0011183X003400030046x

Herridge, D. F. (2003). Breeding for traits supportive of nitrogen fixation in legumes. In G. Hardarson, & W. J. Broughton (Eds.), *Maximizing the use of biological nitrogen fixation in agriculture* (p. 199-212). Hingham, MA: Kluwer Academic Publishers.

Herridge, D. F., Bergersen, F. J., & Peoples, M. B. (1990). Measurement of nitrogen fixation by soybean in the field using the ureide and natural ^{15}N abundance methods. *Plant Physiology*, *93*(2), 708-716. doi: 10.1104/pp.93.2.708

Houngnandan, P., Yemadje, R. G. H., Oikeh, S. O., Djidohokpin, C. F., Boeckx, P., & Cleemput, O. V. (2008). Improved estimation of biological nitrogen fixation of soybean cultivars (*Glycine max* L. Merril) using ^{15}N natural abundance technique. *Biology and Fertility of Soils*, *45*(2), 175-183. doi: 10.1007/s00374-008-0311-5

Imsande, J. (1989) Rapid dinitrogen fixation during soybean pod fill enhances net photosynthetic output and seed yield: A new perspective. *Agronomy Journal*, *81*(4), 549-556. doi: 10.2134/agronj1989.00021962008100040001x

Kaewsuralikhit, S., Yokoyama, T., Kouchi, H., & Arima, Y. (2005). Comprehensive analysis of plant gene expression in soybean root nodules at different growth stages. *Soil Science and Plant Nutrition*, *51*(4), 535-547. doi: 10.1111/j.1747-0765.2005.tb00062.x

Kaschuk, G., Hungria, M., Leffelaar, P. A., Giller, K. E., & Kuyper, T. W. (2010). Differences in photosynthetic behaviour and leaf senescence of soybean (*Glycine max* [L.] Merril) dependent on N_2 fixation or nitrate supply. *Plant Biology*, *12*(1), 60-69. doi: 10.1111/j.1438-8677.2009.00211.x

Keyser, H. H., & Li, F. (1992). Potential for increasing biological nitrogen fixation in soybean. *Plant and Soil*, *141*(1), 119-135. doi: 10.1007/BF00011313

Latimore, M., Giddens, J., & Ashley, D. A. (1977). Effect of ammonium and nitrate nitrogen upon photosynthate supply and nitrogen fixation by soybeans. *Crop Science*, *17*(3), 399-404. doi: 10.2135/cropsci1977.0011183X001700030015x

Lawn, R. J., & Brun, W. A. (1974). Symbiotic nitrogen fixation in soybeans. 1. Effect of photosynthetic source-sink manipulations. *Crop Science*, *14*(1), 11-16. doi: 10.2135/cropsci1974.0011183X001400010004x

Mastrodomenico, A. T., & Purcell, L. C. (2012). Soybean nitrogen fixation and nitrogen remobilization during reproductive development. *Crop Science*, *52*(3), 1281-1289. doi: 10.2135/cropsci2011.08.0414

Minchin, F. R., & Witty, J. F. (2005). Respiratory/Carbon costs of symbiotic nitrogen fixation in legumes. In H. Lambers, & M. Ribas-Carbo (Eds.), *Plant Respiration* (p. 195-205). Dordrecht, GY: Springer. doi: 10.1007/1-4020-3589-6_11

Nicolás, M. F., Arias, C. A. A., & Hungria, M. (2002). Genetics of nodulation and nitrogen fixation in Brazilian soybean cultivars. *Biology and Fertility of Soils*, *36*(2), 109-117. doi: 10.1007/s00374-002-0511-3

Osborne, S. L., & Riedell, W. F. (2006). Starter nitrogen fertilizer impact on soybean yield and quality in the northern great plains. *Agronomy Journal*, *98*(6), 1569-1574. doi: 10.2134/agronj2006.0089

Pazdernik, D. L., Graham, P. H., & Orf, J. H. (1997). Heritability in the early nodulation of F_3 and F_4 soybean lines. *Canadian Journal of Plant Sciences*, 77(2), 201-205. doi: 10.4141/P96-098

Pazdernik, D. L., Graham, P. H., Vance, C. P., & Orf, J. H. (1996). Host genetic variation in the early nodulation and dinitrogen fixation of soybean. *Crop Science*, *36*(5), 1102-1107. doi: 10.2135/cropsci1996.0011183X003600050005x

Phillips, D. A., & DeJong, T. M. (1984). Dinitrogen fixation in leguminous crop plants. In R. D. Hauck (Ed.), *Nitrogen in crop production* (p. 121-132). Madison, WI: American Society of Agronomy.

Salvagiotti, F., Cassman, K. G., Specht, J. E., Walters, D. T., Weiss, A., & Dobermann, A. (2008). Nitrogen uptake, fixation and response to fertilizer N in soybeans: A review. *Field Crops Research*, *108*(1), 1-13. doi: 10.1016/j.fcr.2008.03.001

Santos, M. A., Geraldi, I. O., Garcia, A. A. F., Bortolatto, N., Schiavon, A., & Hungria, A. (2013). Mapping of QTLs associated with biological nitrogen fixation traits in soybean. *Hereditas*, *150*(2-3), 17-25. doi: 10.1111/j.1601-5223.2013.02275.x

Sinclair, T. R., & Wit, C. T. (1975). Photosynthate and nitrogen requirements for seed production by various crops. *Science*, *189*(4202), 565-567. doi: 10.1126/science.189.4202.565

Specht, J. E., Hume, D. J., & Kumudini, S. V. (1999). Soybean yield potential a genetic and physiological perspective. *Crop Science*, *39*(6), 1560-1570. doi: 10.2135/cropsci1999.3961560x

Thibodeau, P. S., & Jaworski, E. G. (1975). Patterns of nitrogen utilization in the soybean. *Planta*, *127*(2), 133-147. doi: 10.1007/BF00388374

Unkovich, M., Herridge, D., Peoples, M., Cadisch, G., Boddey, R., Giller, K., Alves, B., & Chalk, P. (2008). *Measuring plant-associated nitrogen fixation in agricultural systems*. Canberra, AU: ACIAR.

Van Berkum, P. (1990). Evidence for a third uptake hydrogenase phenotype among the soybean bradyrhizobia. *Applied and Environmental Microbiology*, *56*(12), 3835-3841.

Van Kessel, C., & Hartley, C. (2000). Agricultural management of grain legumes: has it led to an increase in nitrogen fixation. *Field Crops Research*, *65*(2-3), 165-181. doi: 10.1016/S0378-4290(99)00085-4

Vollmann, J., Walter, H., Sato, T., & Schweiger P. (2011). Digital image analysis and chlorophyll metering for phenotyping the effects of nodulation in soybean. *Computers and Electronics in Agriculture*, *75*(1), 190-195. doi: 10.1016/j.compag.2010.11.003

Ward, J. H. (1963). Hierarchical grouping to optimize an objective function. *American Statistical Association Journal*, *58*(301), 236-244. doi: 10.1080/01621459.1963.10500845

Wu, S., & Harper, J. E. (1991). Dinitrogen fixation potential and yield of hypernodulating soybean mutants: a field evaluation. *Crop Science*, *31*(5), 1233-1240. doi: 10.2135/cropsci1991.0011183X003100050031x

Youn, J. T., Van, K. J., Lee, J. E., Kim, S. K., Song, J., Kim, W. H., & Lee, S. H. (2009). Effect of N fertilizer top-dressing on N accumulation and N_2 fixation of supernodulating soybean mutant. *Journal of Crop Science and Biotechnology*, *12*(3), 153-159. doi: 10.1007/s12892-009-0125-5

Youn, J. T., Van, K. J., Ryu, Y. H., Shinji, S., Lee, C. K., Kim, S. L., Seo, S. J., & Lee, S. H. (2008). Evaluation of nitrogen fixation between supernodulating soybean mutants and their wild-types using ^{15}N in field conditions. *Journal of Crop Science and Biotechnology*, *11*(1), 17-23. doi: 10.1038/hdy.2016.27

Zapata, F., Danso, S. K. A., Hardarson, G., & Fried, M. (1987). Time course of nitrogen fixation in field-grown soybean using nitrogen-15 methodology. *Agronomy Journal*, *79*(1), 172-176. doi: 10.2134/agronj1987.00021962007900010035x

23

Plant regeneration and histological study of the somatic embryogenesis of sugarcane (*Saccharum* spp.) cultivars RB855156 and RB72454

Giovana Bomfim de Alcantara*, Roberson Dibax, Ricardo Augusto de Oliveira, João Carlos Bespalhok Filho and Edelclaiton Daros

*Laboratório de Micropropagação de Plantas, Setor de Ciências Agrárias, Departamento de Fitotecnia e Fitossanitarismo, Universidade Federal do Paraná, Rua dos Funcionários, 1540, Cabral, Curitiba, Paraná, Brazil. *Author for correspondence. E-mail: giobomfim@ufpr.br*

ABSTRACT. The development of an efficient regeneration system is crucial for the genetic transformation of sugarcane. The aims of this study were to regenerate plants from somatic embryos and analyze the origin and division pattern of cells during the different embryonic developmental stages of the sugarcane cultivars RB855156 and RB72454. For both cultivars, the best results for embryogenic callus induction were obtained with explant cultures on culture medium containing 13.5 μM 2,4-dichlorophenoxyacetic acid (2,4-D) followed by callus culture on solid medium with 4.5 μM 2,4-D. A higher rate of shoot induction was observed in RB855156 with the addition of 8.9 and 17.8 μM benzylaminopurine (BAP); this higher rate was also observed in RB72454 with the addition of 17.8 μM BAP. Murashige and Skoog (1962) (MS) medium containing 2.5 and 5.0 μM indolebutyric acid (IBA) for RB855156 and RB72454, respectively, was suitable for rooting. The plantlets were successfully acclimatized and the plant survival rate was 100% for both cultivars. Histological analysis revealed that the shoots in both cultivars originated from somatic embryogenesis.

Keywords: tissue culture, somatic embryos, anatomy, 2,4-D, Poaceae.

Introduction

Sugarcane is a renewable energy source for the production of ethanol that is grown in the tropical and subtropical regions of more than 70 countries. Brazil is emerging as a world leader in the use of sugarcane as a source of renewable energy and sugar exports. There is a growing interest in developing breeding programs for this species. Genetic transformation is a potential alternative to sugarcane breeding programs, and somatic embryogenesis is a fundamental method for the *in vitro* regeneration of plants.

The RB855156 and RB72454 sugarcane cultivars were developed by "Rede Interuniversitária para o Desenvolvimento do Setor Sucroalcooleiro" (RIDESA). The RB855156 cultivar exhibits early maturation and has the ability to sprout from stumps, while RB72454 has a high sucrose content, exhibits late maturation and high agricultural productivity and has a long, useful period of industrialization (DAROS et al., 2010). Many of the stages of somatic embryogenesis in these cultivars have not yet been studied. Furthermore, some sugarcane cultivars are recalcitrant to plant regeneration; for example, the RB72454 cultivar does not respond to established protocols (LIMA et al., 2001).

The induction of somatic embryogenesis has been achieved in sugarcane explant cultures on medium supplemented with various types of auxins (BRISIBE et al., 1994; FITCH; MOORE, 1990; HO; VASIL, 1983). Plant regeneration from an embryogenic callus obtained from immature leaves was observed by Nieves et al. (2008), Blanco et al. (1997) and Vickers et al. (2005) using Murashige and Skoog (1962) (MS) medium supplemented with 2,4-dichlorophenoxyacetic acid (2,4-D). Although various studies of callus induction in sugarcane had been reported, few studies have analyzed all stages of somatic embryogenesis.

This study is important because of the need to identify the cells associated with induction processes and the formation of structures capable of organized growth and eventual development into seedlings. For plant regeneration from somatic embryogenesis, histological analyses must be carried out to determine the route of regeneration and discover whether the plant originated from somatic embryogenesis or organogenesis. Additionally, it is important to determine whether somatic embryogenesis is a reliable technique for use in the genetic transformation of sugarcane.

The aims of this study were to regenerate plants via somatic embryogenesis and to analyze the origin and division pattern of cells during the various stages of somatic embryo development in the sugarcane cultivars RB72454 and RB855156.

Material and methods

Plants of the sugarcane cultivars RB855156 and RB72454 at 6 to 8 months of cultivation were used for the experiments. Immature leaves (cylinders of approximately 2 × 10 cm) were surface sterilized for 1 min. in 70% ethanol (v/v), rinsed three times with sterile distilled water, incubated for 20 min. in a 2.0% (v/v) sodium hypochlorite solution and rinsed again three times in sterile distilled water. Meristematic regions of transverse segments were removed, transversely cut into segments of 5-6 mm and used as explants.

The explants were cultured for 8 weeks on callus induction medium. Embryogenic calli were then separated, transferred to callus multiplication medium and cultured for 4 weeks. Subsequently, embryogenic calli were transferred to regeneration medium (embryo conversion into plantlets), they were rooted, acclimatized and transferred to a greenhouse. The material obtained with the best treatment at each phase was selected to proceed to other phases.

Induction of somatic embryogenesis

The explants were cultured in flasks (6 cm in diameter and 9 cm in height) with 30 mL of induction medium (MSC3), which was composed of MS medium supplemented with 50 mg L^{-1} arginine, 50 mg L^{-1} cysteine, 0.15 mg L^{-1} citric acid, 500 mg L^{-1} hydrolyzed casein, 20 g L^{-1} sucrose and 0.6% agar (Vetec®).

Mixtures containing a standard concentration of of 2,4-D and various concentrations of benzylaminopurine (BAP) or kinetin (KIN) (13.5 μM 2.4-D (control); 13.5 μM 2.4-D + 8.9 μM BAP; 13.5 μM 2.4-D + 17.8 μM BAP; 13.5 μM 2.4-D + 26.7 μM BAP; 13.5 μM 2.4-D + 9.3 μM KIN; 13.5 μM 2.4-D + 18.6 μM KIN; and 13.5 μM 2.4-D + 27.9 μM KIN) were compared for callus induction.

Additionally, various types and concentrations of auxins (6.8, 13.5 and 20.4 μM 2.4-D; 6.7, 13.4 and 20.1 μM trichlorophenoxyacetic acid; and 6.6, 13.2 and 19.8 μM picloran) were compared. The explants were maintained for a culture period of 8 weeks at 25 ± 2°C in the dark; following this, the percentage growth of embryogenic callus was evaluated.

Growth of embryogenic callus

Embryogenic callus was grown in liquid and solid media composed of MSC3 supplemented with 30 g L^{-1} sucrose (MSC1). Petri dishes were used for the solid medium and flasks were used for the liquid medium, and both contained 30 cm^3 of culture medium. Four concentrations of 2,4-D (0, 4.5, 9.0 and 13.5 μM) were compared in the liquid and solid media. Initially, 14 mg of callus was introduced for a subculture period of 15 days. The callus in the liquid medium was maintained under intermittent agitation at 120 rpm. The fresh weight of the callus (mg) was measured after 30 days of culture.

The cultures were maintained in a growth chamber at 25 ± 2°C under a 16/8h light/dark cycle under cold white fluorescent light with a photon flux density of 30 μmol m^{-2} s for 30 days.

Conversion of somatic embryos into plantlets

For the conversion of somatic embryos into plantlets, the treatments were composed of clusters of callus that formed embryogenic masses at the same stage of development. The callus clusters were transferred into Petri dishes containing 30 cm^3 of MSC1 medium and subjected to the same *in vitro* conditions as previously mentioned. Four BAP concentrations (0, 8.9, 17.8, 26.7 and 35.6 μM) were compared. The percentage of shoot induction was evaluated after 30 days of culture.

Rooting and plant acclimatization

The rooting sugarcane shoots obtained in the previous experiment were transferred to MS/2 medium containing three concentrations (0, 2.5 and 5.0 μM) of indolebutyric acid (IBA). After 30 days of culture, the percentage of rooting plants, number of roots, root length (cm) and percentage of normal plant development were evaluated. The cultures were maintained in a growth chamber at 25 ± 2°C under cold white fluorescent light with a photon flux density of 30 μmol m^{-2} s under a 16/8h light/dark cycle for 30 days.

For acclimatization, the rooted plants were transferred to plastic cups (7 cm in diameter and 10 cm in height) containing Plantmax® substrate. The percentage of surviving plants was evaluated after 30 days; these plants were then transferred to a greenhouse in pots (20 cm diameter and 17 cm height) containing Plantmax®.

Histology

Samples of plant material at different stages of cell differentiation were collected at various points during callus and shoot induction. The samples were fixed in 50% formaldehyde/acetic acid/alcohol (FAA) (v/v) for 48h in accordance with the protocol of Johansen (1940). Next, they were transferred into 70% ethanol and dehydrated in an ethanol series. The samples were embedded in glycol methacrylate (GMA) following the procedure described by Feder and O'Brien (1968) and the manufacturer's instructions (JB4 - PolyScience®). Longitudinal sections were cut using a rotary microtome with a steel razor (type C) to a section thickness of 5 mm. The cuttings were placed onto slides and stained with 0.05% (v/v) toluidine blue for 10 min. (O'BRIEN et al., 1965). The permanent slides were mounted in synthetic resin (Entelan®). Images were captured with a Sony Cyber-shot® P200 digital camera attached to a Zeiss Axiolab® microscope. The scales were projected in the same optical conditions.

Experimental design

The experimental design was a completely randomized factorial scheme with five replications and ten explants per experimental unit. For each phase, the factors included the treatments with regulators and the cultivars (RB855156 and RB72454). The regulators of somatic embryogenesis induction were 2,4-D, BAP and KIN. Growth of the embryogenic callus occurred in the presence of 2,4-D, conversion of somatic embryos into plantlets occurred in the presence of BAP and IBA was used for rooting and plant acclimatization.

The homogeneity of means was tested using Bartlett's test. For the callus induction and plant rooting experiments, the treatment effect was analyzed by ANOVA and the means were compared with Tukey's multiple range test (α = 0.05) using Sisvar® software. The results of callus growth and plantlet conversion experiments were subjected to a regression analysis using the MSTAT-C® program version 2.10 (Michigan State University, USA).

Results and discussion

Callus induction

The greatest percentage of embryogenic callus induction in both cultivars was achieved using 13.5 μM 2,4-D, with 70% induction observed for RB855156 and 65% observed for RB72454. No callus induction occurred with the treatments that lacked 2,4-D (Table 1). Lima et al. (2001) reported that embryogenic callus induction of the sugarcane cultivar RB72454 occurred on MS medium with 9.0 μM 2,4-D; however, this callus did not regenerate into plants.

Table 1. Percentage of embryogenic callus induction in two sugarcane cultivars cultured on different concentrations of BAP (benzylaminopurine) and KIN (kinetin) in the presence of 2,4-D.

Treatments	RB855156			RB72454		
Control	0.0	d	A	0.0	d	A
13.5 μM 2,4-D	70.0	a	A	65.0	a	B
13.5 μM 2,4-D + 8.9 μM BAP	20.0	b	A	21.7	b	A
13.5 μM 2,4-D + 17.8 μM BAP	10.0	c	A	5.0	c	B
13.5 μM 2,4-D + 26.7 μM BAP	0.0	d	A	0.0	d	A
13.5 μM 2,4-D + 9.3 μM KIN	0.0	d	A	0.0	d	A
13.5 μM 2,4-D + 18.6 μM KIN	0.0	d	A	0.0	d	A
13.5 μM 2,4-D + 27.9 μM KIN	0.0	d	A	0.0	d	A
CV (%)			11.9			

Means followed by same lowercase letter (vertically) and capital letter (horizontally) do not differ according to Tukey's multiple range test ($p < 0.05$).

The induction of somatic embryogenesis in Poaceae was positively affected by the association between auxins and cytokinin (BHASKARAN; SMITH, 1990). However, the association of 2,4-D with BAP at concentrations ≤ 17.8 μM BAP led to a low percentage of embryogenic callus induction in this study. No embryogenic callus induction was observed with 26.7 μM BAP or treatments containing 2,4-D combined with KIN (Table 1).

The response of both cultivars to the treatments was similar, and significant differences were only observed for the treatments containing 13.5 μM 2,4-D and 13.5 μM 2,4-D with 17.8 μM BAP. The RB855156 cultivar exhibited superior somatic embryogenesis induction under these treatments (Table 1). A similar result was observed for the Q165^{A} sugarcane cultivar cultured on medium with auxin (LAKSHMANAN et al., 2006).

The highest percentage of embryogenic callus induction was observed with 13.5 μM 2,4-D. The use of trichlorophenoxyacetic acid and picloran resulted in a lower percentage of embryogenic callus induction in the RB855156 cultivar, whereas no embryogenic callus formation was observed in the RB72454 cultivar, with the exception of explants cultured on medium containing 13.2 μM picloran. Comparing the two cultivars, RB855156 exhibited a higher percentage of embryogenic callus induction (Table 2). Similar results were observed by Taylor et al. (1992), who studied embryogenic callus induction in 14 sugarcane cultivars. The best results for all the cultivars were observed when the explants were cultured on MS medium containing 13.5 μM 2,4-D.

Table 2. The percentage of embryogenic callus induction of two sugarcane cultivars cultured on different concentrations of auxins, 2,4-dichlorophenoxyacetic acid, (2,4-D), triclorophenoxyacetic acid and picloran.

Treatments	RB855156			RB72454		
6.8 μM 2,4-D	24.0	b	A	16.0	b	B
13.5 μM 2,4-D	68.0	a	A	52.0	a	B
20.4 μM 2,4-D	8.0	cd	A	10.0	c	B
6.7 μM triclorophenoxyacetic	12.0	c	A	0.0	c	B
13.4 μM triclorophenoxyacetic	4.0	d	A	0.0	c	B
20.1 μM triclorophenoxyacetic	4.0	d	A	0.0	c	B
6.6 μM picloran	12.0	c	A	0.0	c	B
13.2 μM picloran	8.0	cd	A	4.0	bc	B
19.8 μM picloran	15.0	b	A	0.0	c	B
CV (%)	34.8					

Means followed by same lowercase letter (vertically) and capital letter (horizontally) do not differ according to Tukey's multiple range test ($p < 0.05$).

The use of 2,4-D for the induction of somatic embryogenesis in sugarcane has been reported in various studies (AHLOOWALIA; MARETZKI, 1983; CHEN et al., 1987; CHENGALRAYAN; GALLO-MEAGHER, 2001; FITCH; MOORE, 1990; GALLO-MEAGHER et al., 2000; HO; VASIL, 1983; LIU, 1993; OROPEZA; DE-GARCIA, 1996). However, the results observed here differ from those reported by Ho and Vasil (1983). These authors reported that low concentrations of 2,4-D were sufficient for embryogenic callus induction. In another study, the percentage of embryogenic callus induction in the sugarcane cultivar CP-841198 gradually decreased when the concentration of 2,4-D increased from 4.5 to 45.2 μM after 12 weeks of culture; the same results were observed when the explants were cultured on medium containing picloran (CHENGALRAYAN et al., 2005). In a study of the sugarcane cultivar CoJ88t, an increase in the concentration of 2,4-D from 18.1 to 36.2 μM reduced callus induction from 86.4 to 76.0% (SINGH et al., 2008). In another study, dicamba (3,6-dichloro-2 hydroxybenzoic acid) was found to be more effective at inducing callus formation in the Nco310 sugarcane cultivar compared to 2,4-D, naphthaleneacetic acid (NAA) and 4-fluorophenoxyacetic acid (4-FPA) (BRISIBE et al., 1994). However, most studies showed that 2,4-D or 4-chlorophenoxyacetic acid (pCPA) were superior to other regulators in terms of the induction of somatic embryogenesis in sugarcane (LAKSHMANAN, 2006). Shoot regeneration from somatic embryogenesis of the sugarcane cultivar Q165^{A} was found to occur in MS medium supplemented with NAA and pCPA (LAKSHMANAN et al., 2006).

Callus growth

The greatest amount of callus growth in liquid medium was observed in the RB855156 cultivar in the presence of 13.5 μM 2,4-D; under these conditions, the callus weight was over 1,000 mg. Because the initial weight of each callus was approximately 14 mg, there was an average increase of 986 mg per flask. The optimal concentration of 2,4-D for callus induction in RB72454 was 4.5 mM.

Callus growth occurred on solid medium at all concentrations of 2,4-D (4.5, 9.0 and 13.5 μM); however, no significant differences were observed between the treatments (Figure 1). The growth rate decreased in the absence of 2,4-D. The 4.5 μM concentration of 2,4-D was chosen because it resulted in the same growth rate as the other concentrations. In another study, an increase in the concentration of 2,4-D was found to promote somaclonal variation in plants (SHOEMAKER et al., 1991).

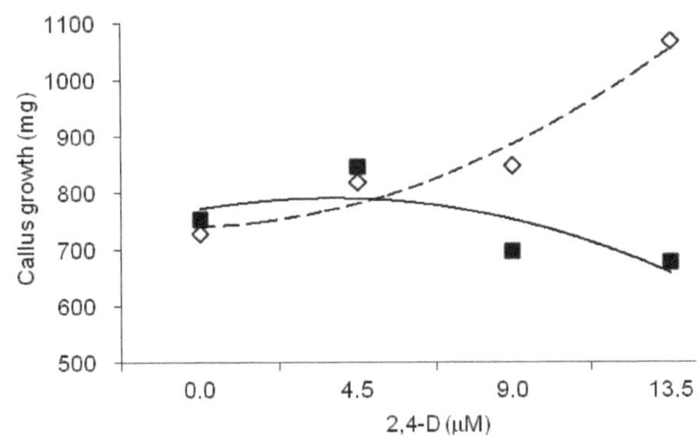

Figure 1. Callus growth of the sugarcane cultivars RB855156 and RB72454 in the presence of various concentrations of 2,4-dichlorophenoxyacetic acid (2,4-D).

Conversion of somatic embryos into plantlets

A higher shoot induction rate was observed in the presence of 8.9 and 17.8 μM BAP for the RB855156 cultivar and 17.8 μM BAP for the RB72454 cultivar. Increasing the concentration of BAP to 35.5 μM decreased the rate of shoot induction in both cultivars (Figure 2). However, when the Q165[A] sugarcane cultivar was cultured on medium containing KIN, BAP or NAA, no shoots were produced, although the authors observed some shoot induction when auxins and cytokinin were added to the MS medium at a minimum concentration of 4 μM. Shoots were induced in nearly half of the explants cultured on MS medium with 4 μM BAP and 10 μM NAA after 8 weeks of culture. However, the proportion of explants that induced shoots was significantly lower, especially the proportion that produced more than 20 shoots per explant when BAP was combined with KIN (LAKSHMANAN et al., 2006).

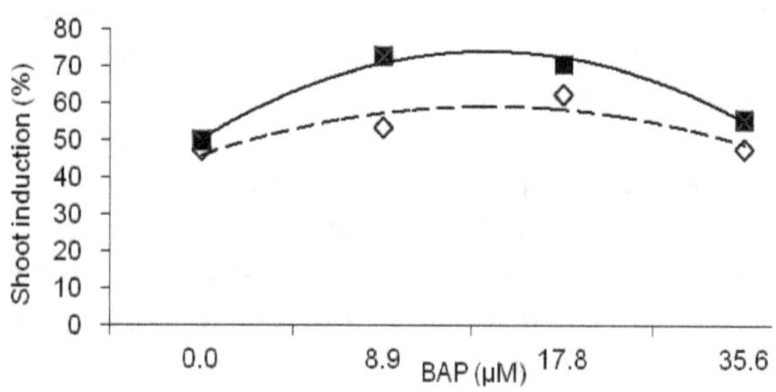

Figure 2. The effect of benzylaminopurine (BAP) on shoot induction in the sugarcane cultivars RB855156 and RB72454.

In the second experiment, the highest rates of shoot induction in the RB855156 cultivar were observed with 17.8 μM and 26.7 μM BAP. In the RB72454 cultivar, the addition of 26.7 μM BAP resulted in the highest rate of shoot induction (Figure 3). The results presented here confirm the effect of cytokines on shoot induction as described by Ali et al. (2008). These authors studied the effect of a combination of BAP and KIN on shoot induction from sugarcane somatic embryos. The shoot induction rate was 60% in the presence of 1.11 μM BAP and 1.16 μM KIN. In another study, the addition of 4.44 μM BAP to MS medium resulted in a shoot induction rate of 31.5% from embryogenic masses of sugarcane after 4 weeks of culture (ATHER et al., 2009).

Similar results were observed for sugarcane calli induced on medium containing 2,4-D. Shoot induction from somatic embryos occurred after the calli were transferred to MS medium containing BAP and KIN (FALCO et al., 1996). In another study, shoot induction from embryogenic masses of sugarcane occurred in the presence of cytokines (FRANKLIN et al., 2006), confirming the results reported herein. Plant regeneration from an embryogenic callus obtained from immature inflorescence explants was observed on media containing 2,4-D (BLANCO et al., 1997) and various combinations of different concentrations of 2,4-D, KIN and BAP (LIU, 1993).

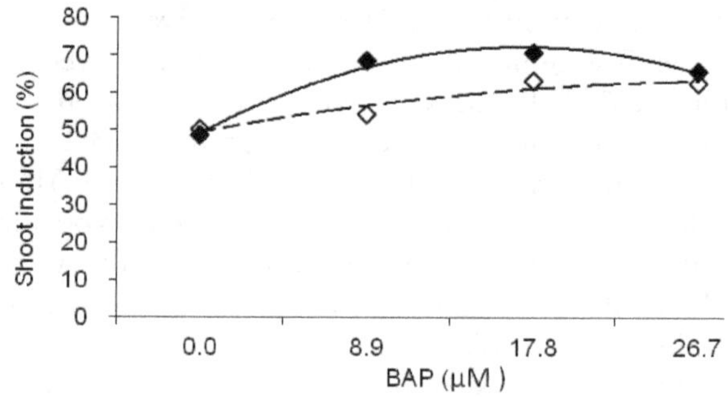

Figure 3. Effect of benzylaminopurine (BAP) concentration on shoot induction in the sugarcane cultivars RB855156 and RB72454.

Rooting and plant acclimatization

Culture of the RB855156 cultivar on MS medium resulted in a higher percentage of rooting plants, greater number of roots, longer root length and greater percentage of normal shoot development compared to culture on MS/2 medium. Culture on MS medium containing 5.0 μM IBA resulted in the highest percentage of rooted plants (98%) (Figure 4E); however, no statistically significant differences were observed between treatments (0; 2.5 and 5.0 μM IBA) (Table 3).

Table 3. Effects of MS and MS/2 media and indolebutyric acid (IBA) concentration on rooting in the sugarcane cultivar RB855156.

IBA	Rooting plants (%)		Number of roots		Root length (cm)		Normal shoot development (%)	
(μM)	MS	MS/2	MS	MS/2	MS	MS/2	MS	MS/2
0	80.0 b A	70.0 a A	4.0 a A	3.0 a B	3.4 b A	2.0 a B	100.0 a A	30.2 a B
2.5	90.0 ab A	75.0 a B	4.5 a A	3.5 a B	4.8 a A	2.0 a B	98.0 a A	35.0 a B
5.0	98.0 a A	78.0 a B	4.2 a A	3.5 a B	5.0 a A	2.2 a B	100.0 a A	32.0 a B
CV (%)	10.5		9.6		6.7		9.9	

Means followed by same lowercase letter (vertically) and capital letter (horizontally) do not differ according to Tukey's multiple range test ($p < 0.05$).

When the RB72454 cultivar was treated with IBA, culture on MS medium resulted in a significantly higher rate of rooting compared to MS/2 medium. No significant differences were observed between the two culture media in terms of the number of roots and root length, although treatment with 5.0 μM IBA and culture on MS medium resulted in longer root lengths. A significantly higher percentage of normal shoot development was observed with MS medium compared to the MS/2 medium. Regarding the IBA treatments, the longest root lengths were observed on MS medium containing 5.0 μM IBA and MS/2 medium containing 2.5 and 5.0 μM IBA (Table 4). Different results were observed for rooting in the sugarcane cultivars CP 77.400 and BL-4. In another study, the use of MS medium supplemented with 10.74 μM NAA increased rooting by 90% after 6 days of culture (ALI et al., 2008).

Table 4. Effect of MS and MS/2 media and indolebutyric acid (IBA) concentration on rooting of the sugarcane cultivar RB72454.

IBA	Rooting plants (%)		Number of roots		Root length (cm)		Normal shoot development (%)	
(μM)	MS	MS/2	MS	MS/2	MS	MS/2	MS	MS/2
0	78.0 b A	70.0 a A	3.0 a A	3.4 a A	1.8 b A	2.0 b A	99.0 a A	39.5 a B
2.5	85.0 ab A	74.0 a B	3.5 a A	3.2 a A	2.1 b A	2.3 ab A	100.0 a A	32.0 a B
5.0	91.0 a A	72.0 a B	3.2 a A	2.9 a A	4.7 a A	2.6 a B	100.0 a A	35.5 a B
CV (%)	10.5		9.8		9.7		8.8	

Means followed by same lowercase letter (vertically) and capital letter (horizontally) do not differ according to Tukey's multiple range test ($p < 0.05$).

According to the plant acclimatization results, the survival rate of plants regenerated *in vitro* was 100%, and no phenotypic changes were observed (Figure 4F).

Figure 4. Plant regeneration of the sugarcane cultivar RB855156 via somatic embryogenesis. A) nec = non-embryogenic callus; ec = embryogenic callus. B, C and D) Somatic embryo development and formation of shoots; sh = shoots. E) Rooted plants cultured on MS medium containing 5.0 μM IBA. F) Acclimatized plants.

Histological study

According to the results of the histological study, there were two types of callus induction: mucilaginous and nodular. The mucilaginous callus had a soft consistency, became necrotic, did not exhibit an embryogenic nature and did not regenerate into plants (Figure 4A). The nodular callus was whitish in color, composed of cells with a dense cytoplasm and a nucleus with an evident nucleolus and had a high nucleus/cytoplasm ratio, indicating the presence of meristematic zones that could lead to the development of somatic embryos (Figure 4A) and the formation of shoots (Figure 4B, C and D).

No significant differences were observed between the histological structures of RB72454 and RB855156. The anatomy of both sugarcane cultivars was found to be similar to that previously reported for other monocotyledonous species (GEORGE, 1993) (Figure 5), including grasses (GUIDERDONI; DEMARLY, 1988; VASIL et al., 1984) such as *Saccharum* spp. (NADAR et al., 1976).

In some cases, non-embryogenic calli were found (Figure 5A), which had larger cells with generally obscured nuclei. However, various developmental stages of somatic embryogenesis were observed. Initially, we observed the formation of pro-embryos with six or more cells (Figure 5B). Generally, clusters of somatic embryos are formed by meristematic cells with a dense cytoplasm, large nucleus and many starch grains (GEORGE, 1993).

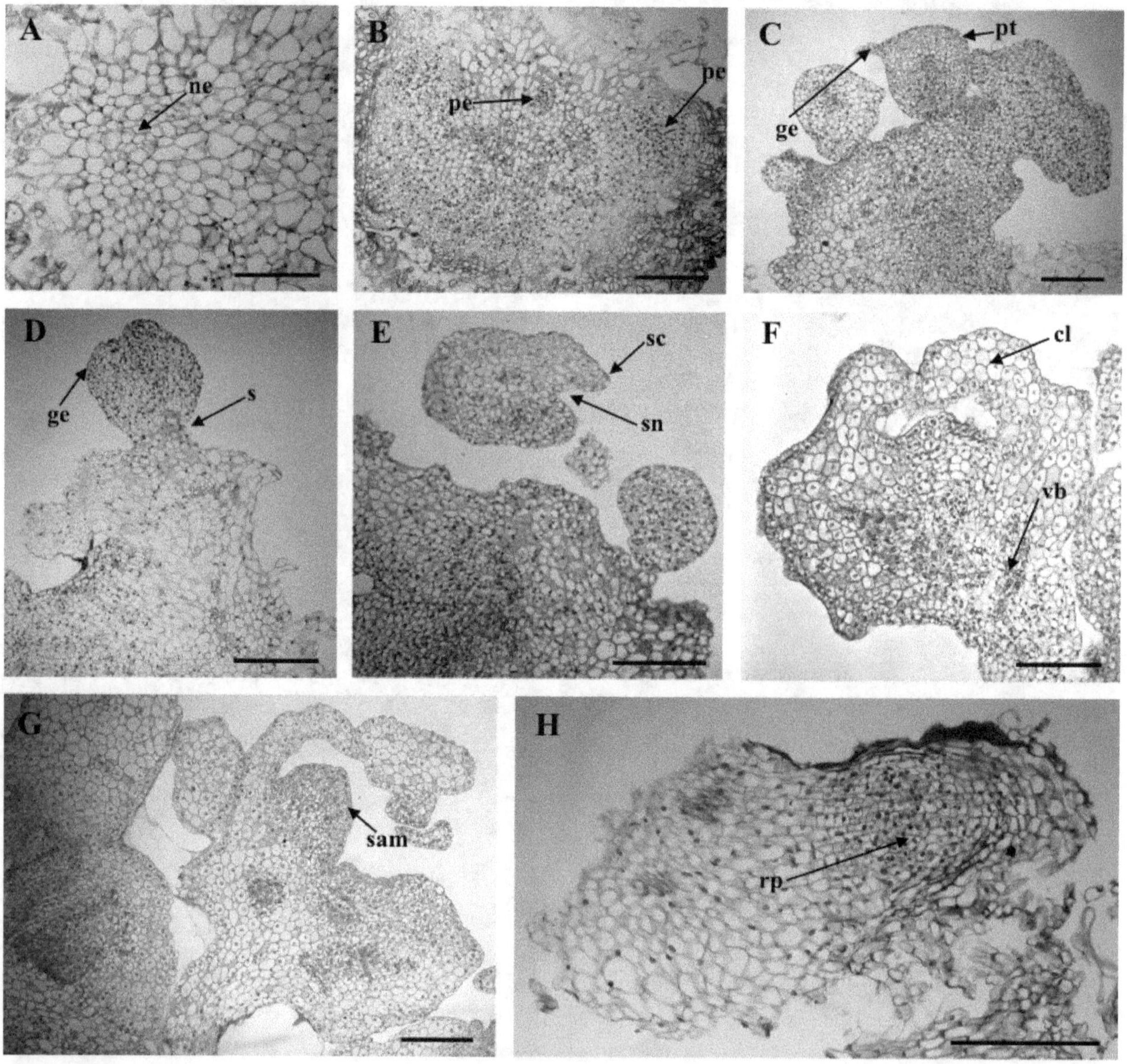

Figure 5. Histological study of somatic embryogenesis of the sugarcane cultivar RB72454. A) ne = non-embryogenic callus. B) pe = pro-embryos. C) ge = globular embryos, pt = protoderm. D) ge = globular embryos, s = suspensor. E) sn = node-scutellar, sc = scutellum. F) cl = coleoptile, vb = vascular bundle. G) sam = shoot apical meristema. H) rp =root pole. Bar = 200 μm.

Globular embryos formed after cell division, and a protoderm was evident (Figure 5C). In some cases, globular embryos attached to the callus surface by a suspensor were observed (Figure 5D). However, in many cases, the suspensor was not clearly distinguishable or was not observed as a multicellular structure. When the suspensor was present, it ranged from a uniserial or biserial structure to a large multicellular structure, as previously observed by Ho and Vasil (1983).

The further development of globular somatic embryos was characterized by the formation of a terminal leaf node, called a scutellar node, which indicated the beginning of the formation of a scutellum (Figure 5E). The scutellum consists of cells that are richly cytoplasmic and irregularly shaped and contains tracheids in procambium that is bordered by typical epithelial cells; this was previously observed in sugarcane by Guiderdoni and Demarly (1988). An increased number of cell layers in the scutellum and other stages of embryonic development were observed when the coleoptile emerged (Figure 5F). The somatic embryos of monocotyledons initially go through a globular developmental stage; however, unlike dicotyledons, somatic embryos develop in the scutellum and coleoptile (GEORGE, 1993).

After several days of culture, an increased number of cell layers were observed in the scutellum and coleoptile; the apical shoot was also evident in the region of the scutellar node (Figure 5G). This showed continuity with the root vascular system (Figure 5F), as observed by Vasil et al. (1985). Later, a more advanced stage of somatic embryo development was observed (Figure 5G) with a shoot apical meristem, leaf primordia and a more developed coleoptile and scutellum aligned with the root meristems.

The cross-sections illustrate the differentiation of a somatic embryo with a bipolar organization formed by the shoot and root axis (Figure 5H), which was observed previously by other authors (HO; VASIL, 1983). According to the histological analysis of the two cultivars, plant regeneration in the present study occurred via somatic embryogenesis. The somatic embryos exhibited typical bipolar orientation of the apical and root poles and an absence of vascular connections between these structures and the underlying callus. The results described herein agree with the morphological characteristics of somatic embryos described by Reinert (1977) and Guiderdoni and Demarly (1988). These authors observed the same characteristics when they studied the somatic embryogenesis of sugarcane. Both studies demonstrated the typical bipolar orientation and the absence of vascular connections with the original tissue of the explant.

Conclusion

The sugarcane explants were responsive to shoot regeneration, and acclimatized plants were obtained for both cultivars. A histological investigation revealed that the adventitious shoots arose via somatic embryogenesis. All stages of development during somatic embryogenesis were observed in both cultivars. The information described herein could be used for regeneration protocol optimization for other sugarcane cultivars and for the production of plants through genetic transformation.

Acknowledgements

The authors thank Coordenação de Aperfeiçoamento de Pessoal de Nível Superior (CAPES, Brazil) for providing a grant to Alcantara and Rede Interuniversitária para o Desenvolvimento do Setor Sucroalcooleiro, Brazil (RIDESA) for funding the research.

References

AHLOOWALIA, B. S.; MARETZKI, A. Plant regeneration via somatic embryogenesis in sugarcane. **Plant Cell Reports**, v. 2, n. 1, p. 21-25, 1983.

ALI, A.; NAZ, S.; SIDDIQUI, F. A.; IQBAL, J. Rapid clonal multiplication of sugarcane (*Saccharum officinarum*) through callogenesis and organogenesis. **Pakistan Journal of Botany**, v. 40, n. 1, p. 123-138, 2008.

ATHER, A.; KHAN, S.; REHMAN, A.; NAZIR, M. Optimization of the protocols for callus induction, regeneration and acclimatization of sugarcane cv. thatta-10. **Pakistan Journal of Botany**, v. 41, n. 2, p. 815-820, 2009.

BHASKARAN, S.; SMITH, R. H. Regeneration in cereal tissue culture: a review. **Crop Science**, v. 30, n. 6, p. 1328-1336, 1990.

BLANCO, M. A.; NIEVES, N.; SÁNCHEZ, M.; BORROTO, C. G.; CASTILLO, R.; GONZÁLEZ, J. L.; ESCALONA, M.; BÁEZ, E.; HERNANDÉZ, Z. Protein changes associated with plant regeneration in embryogenic calli of sugarcane (*Saccharum* sp.). **Plant Cell Tissue and Organ Culture**, v. 51, n. 3, p.153-158, 1997.

BRISIBE, E. A.; MIYAKE, H.; TANIGUCHI, T.; MAEDA, E. Regulation of somatic embryogenesis in long-terms callus cultures of sugarcane (*Saccharum officinarum* L.). **New Phytologist**, v. 126, n. 2, p. 301-307, 1994.

CHEN, W. H.; GARTLAND, K. M. A.; DAVEY, M. R.; SOTAK, R.; GARTLAND, J. S.; MULLIGAN, B. J.;

POWER, J. B.; COCKING, E. C. Transformation of sugarcane protoplasts by direct uptake of a selectable chimeric gene. **Plant Cell Reports**, v. 6, n. 4, p. 297-301, 1987.

CHENGALRAYAN, K.; GALLO-MEAGHER, M. Effect of various growth regulators on shoot regeneration of sugarcane. ***In vitro* Cellular and Development Biology - Plant**, v. 37, n. 4, p. 434-439, 2001.

CHENGALRAYAN, K.; ABOUZID, A.; GALLO-MEAGHER, M. *In vitro* regeneration of plants from sugarcane seed-derived callus. ***In Vitro* Cellular and Development Biology - Plant**, v. 41, n. 4, p. 477-482, 2005.

DAROS, E.; OLIVEIRA, R. A.; ZAMBON, J. L. C.; BESPALHOK FILHO, J. C. **National Catalogue of Sugarcane Varieties "RB"**. Curitiba: Agir, 2010.

FALCO, M. C.; MENDES, B. M. J.; TULMANN NETO, A.; GLÓRIA, B. A. Histological characterization of *in vitro* regeneration of *Saccharum* sp. **Revista Brasileira de Fisiologia Vegetal**, v. 8, n. 2, p. 93-97, 1996.

FEDER, N.; O'BRIEN, T. P. Plant microtechnique: some principles and new methods. **American Journal of Botany**, v. 55, n. 1, p. 123-142, 1968.

FITCH, M. M. M.; MOORE, P. H. Comparison of 2,4-D and picloram for selection of long-term totipotent green callus of sugarcane. **Plant Cell Tissue and Organ Culture**, v. 20, n. 3, p. 157-163, 1990.

FRANKLIN, G.; ARVINTH, S.; SHEEBA, C. J.; KANCHANA, M.; SUBRAMONIAN, N. Auxin pretreatment promotes regeneration of sugarcane (*Saccharum* spp. hybrids) midrib segment explants. **Plant Growth Regulation**, v. 50, n. 2-3, p. 111-119, 2006.

GALLO-MEAGHER, M.; ENGLISH, R. G.; ABOUZID, A. Thidiazuron stimulates shoot regeneration of sugarcane embryogenic callus. ***In vitro* Cellular and Developmental Biology – Plant**, v. 36, n. 1, p. 37-40, 2000.

GEORGE, E. F. **Plant propagation by tissue culture**. Part 1. The technology. Basingstoke: Exegetics, 1993.

GUIDERDONI, E.; DEMARLY, Y. Histology of somatic embryogenesis in cultured leaf segments of sugarcane plantlets. **Plant Cell, Tissue and Organ Culture**, v. 14, n. 2, p. 71-88, 1988.

HO, J.; VASIL, I. K. Somatic embryogenesis in sugarcane (*Saccharum officinarum*). The morphology and physiology of callus formation and the ontogeny somatic embryos. **Protoplasma**, v. 118, n. 3, p. 169-180, 1983.

JOHANSEN, D. A. **Plant microtechnique**. New York: McGraw-Hill Book Company, 1940.

LAKSHMANAN, P. Somatic embryogenesis in sugarcane – an addendum to the invited review 'sugarcane biotechnology: the challenges and opportunities. ***In vitro* Cellular and Developmental Biology – Plant**, v. 42, n. 3, p. 201-205, 2006.

LAKSHMANAN, P.; GEIJSKES, R. J.; WANG, L.; ELLIOTT, A.; GROF, C. P. L.; BERDING, N.; SMITH, G. R. Developmental and hormonal regulation of direct shoot organogenesis and somatic embryogenesis in sugarcane (*Saccharum* spp. interspecific hybrids) leaf culture. **Plant Cell Reports**, v. 25, n. 10, p. 1007-1015, 2006.

LIMA, M. A. C.; GARCIA, R. O.; MARTINS, G. S.; MANSUR, E. *In vitro* morphogenesis of Brazilian sugarcane (*Saccharum officinarum* L.) varieties and callus susceptibility to selective markers used in genetic transformation systems. **Revista Brasileira de Botânica**, v. 24, n. 1, p. 73-77, 2001.

LIU, M. C. Factors affecting induction, somatic embryogenesis and plant regeneration of callus from cultured immature inflorescence of sugarcane. **Journal of Plant Physiology**, v. 141, n. 6, p. 714-720, 1993.

MURASHIGE, T.; SKOOG, F. A revised medium for rapid growth and bioassays with tobacco tissue cultures. **Physiologia Plantarum**, v. 15, n. 3, p. 473-497, 1962.

NADAR, H. M.; SOEPRAPTOPO, S.; HEINZ, D. J.; LADD, S. L. Fine structure of sugarcane (*Saccharum* sp.) callus and the role of auxin in embryogenesis. **Crop Science**, v. 18, n. 2, p. 210-216, 1976.

NIEVES, N.; SAGARRA, F.; GONZÁLEZ, R.; LEZCANO, Y.; CID, M.; BLANCO, M. A.; CASTILLO, R. Effect of exogenous arginine on sugarcane (*Saccharum* sp.) somatic embryogenesis, free polyamines and the contents of the soluble proteins and proline. **Plant Cell, Tissue and Organ Culture**, v. 95, n. 3, p. 313-320, 2008.

O'BRIEN, T. P.; FEDER, N.; MCCULLY, M. E. Polychromatic staining of plant cell walls by toluidine blue O. **Protoplasma**, v. 59, n. 2, p. 368-373, 1965.

OROPEZA, M.; DE-GARCIA, E. Somaclonal variants resistant to sugarcane mosaic virus and their agronomic characterization. ***In vitro* Cellular and Developmental Biology – Plant**, v. 32, n. 1, p. 26-30, 1996.

REINERT, H. Aspects of organization - Organogenesis and embryogenesis. In: STREET, H. E. (Ed.). **Plant tissue and cell culture**. Oxford: Blackwell Scientific Publications, 1977. p. 338-355.

SHOEMAKER, R. C.; AMBERGER, L. A.; PALMER, R. G.; OGLESBY, L.; RANCH, J. P. Effect of 2,4-dichlorophenoxyacetic acid concentration on somatic embryogenesis and heriyable variation in soybean (*Glycine max* (L) Merr.). ***In vitro* Cellular and Developmental Biology – Plant**, v. 27, n. 2, p. 84-88, 1991.

SINGH, G.; SANDHU, S. K.; MEETA, M.; SINGH, K.; GILL, R.; GOSAL, S. S. *In vitro* induction and characterization of somaclonal variation for red rot and other agronomic traits in sugarcane. **Euphytica**, v. 160, n. 1, p. 35-47, 2008.

TAYLOR, P. W. J.; KO, H. L.; ADKINS, S. W.; RATHUS, C.; BIRCH, R. G. Establisment of embryogenic callus and high protoplast yielding suspension cultures of sugarcane (*Saccharum* spp. Hybrids). **Plant Cell, Tissue and Organ Culture**, v. 28, n. 1, p. 69-78, 1992.

VASIL, V.; VASIL, I. K.; LU, C. Somatic embryogenesis in long-term callus cultures *of Zea mays* L. (Gramineae).

American Journal of Botany, v. 71, n. 1, p. 158-161, 1984.

VASIL, V.; LU, C.; VASIL, I. K. Histology of somatic embryogenesis in cultured immature embryos of maize (*Zea mays* L.). **Protoplasma**, v. 127, n. 1-2, p. 1-8, 1985.

VICKERS, J. E.; GROF, C. P. L.; BONNETT, G. D.; JACKSON, P. A.; MORGAN, T. E. Effects of tissue culture, biolistic transformation, and introduction of PPO and SPS gene constructs on performance of sugarcane clones in the field. **Australian Journal of Agricultural Research**, v. 56, n. 1, p. 57-68, 2005.

24

Growth and development of yacon in different periods of planting and growing regions

Diego Mathias Natal da Silva*, Fábio Luiz de Oliveira, Paulo Cezar Cavatte, Mateus Augusto Lima Quaresma and Bruno Fardim Christo

*Departamento de Agronomia, Centro de Ciências Agrárias e Engenharias, Universidade Federal do Espírito Santo, Alto Universitário, s/n° Guararema, 29500-000, Alegre, Espírito Santo, Brazil. *Author for correspondence. E-mail: diegoufvjm@yahoo.com.br*

ABSTRACT. The increasing interest in the commercial exploitation of yacon has demanded adequate technological knowledge for the implantation and management of the crop, including planting periods and cultivation regions that contribute to the satisfactory growth and development of the species. This study was conducted in two different locations, one in a mountainous region and another in a lowland region. In each location, yacon was planted in four different seasons, fall (April), winter (July), spring (September) and summer (December). We collected primary dry matter accumulation data for different parts of the plant and the leaf area at 30 day intervals. We used these values to calculate the following morphophysiological indexes: specific leaf area, leaf area index, tillage growth rate and net assimilation rate. We concluded that planting in the fall in mountainous or lowland sites favoured the growth and development of yacon. Cultivation in mountainous regions, regardless of the planting season, promoted a greater mass accumulation of tuberous roots, which may be reflected in higher productivity. Both fall and winter planting in the mountainous region allows root harvesting at the same time.

Keywords: *Smallanthus sonchifolius*; growth analysis; production.

Introduction

Yacon (*Smallanthus sonchifolius*) is a member of the Asteraceae family and sequesters large amounts of carbohydrates in the roots, many of which are fructans, primarily fructo-oligosaccharides (FOS), which are not metabolizable carbohydrates in the human digestive tract (Vilhena, Câmara, & Kakihara, 2000).

The presence of FOS contributes to several human health benefits that have been attributed to yacon consumption, such as an immunostimulatory effect (Vaz-Tostes et al., 2014), prebiotic effect (Campos et al., 2012) protection against colon cancer (Moura et al., 2012), a role in the reduction of serum lipids (Habib, Honoré, Genta, & Sánchez, 2011), intestinal regularization (Moura et al., 2012), modulation of fasting insulin (Genta et al., 2009), control of chronic diseases such as diabetes, and blood pressure and cholesterol level control (Oliveira, Braga, & Fernandes, 2013).

Yacon consumption has increased because of these benefits, which has led to expectations that it will be cultivated as a new product for commercial exploitation (Santana & Cardoso, 2008).

However, technological knowledge about the management of the commercial exploitation of yacon is necessary, including the best seasons for planting, growth and development under different cultivation conditions.

Yacon has been cultivated in its region of origin (the Andes) mainly during the rainy season; however, it has also been planted outside this season with irrigation. Although the plant occurs in Pacific Coast regions up to 3,600 metres, the best tuberous root production has been noted at medium altitudes between 1500-2000 metres (Seminario, Valderrama, & Manrique, 2003).

In the Andean region, yacon shoots emerge between 30 and 50 days after planting and grow slowly during the first four months such that the diameter of the stem and leaf production reach a maximum at 170 days. Then, root formation and coarsening accelerate (Seminario et al., 2003).

This plant enters the reproductive period between six and ten months after planting, when it blooms, and the onset of flowering depends on the region where the plant is grown - it matures earlier at lower altitudes (Santana & Cardoso, 2008). This stage is followed by a period in which the carbohydrate content in the rhizophores and the tuberous roots increases, leading to a senescence phase of the shoots and the numbness of the underground parts (Vilhena et al., 2000). According to Oliveira and Nishimoto (2004), the tuberous roots and rhizophores should be harvested between the beginning of shoot flowering and senescence.

Despite widespread information about yacon, we have noted that its growth and development varies depending on its growing conditions, including the ability to change the ideal point for harvesting the roots, thus changing the commercial returns. In this context, scientific knowledge could be used to optimize farming practices (tillage).

Therefore, the aim of this study was to evaluate the growth and development of yacon when planted on different dates in two farming regions.

Material and method

The experiment was conducted in two locations in the south of Espírito Santo, Brazil, from April 2013 to April 2014, one in a mountainous region, in the municipality of Ibatiba, at 20° 17 ' south latitude and 41° 37 ' west longitude at an altitude of 837 m; and the other in the lowland region in the municipality of Alegre, 20° 45' south latitude and 41° 29' west longitude at an altitude of 113 m.

The weather in the south of Espírito Santo has two well defined seasons during the year as follows: a hot and rainy season between October and March and a cold and dry season between April and September. The annual rainfall is approximately 1,200 mm. However, the southern region has micro-regions due to the relief, which mainly affects the temperature. The Microrregião Sul-Serrana region includes Ibatiba and is characterized by a mountain climate (tropical altitude) with milder temperatures, and the Vale do Rio Itapemirim region includes Alegre and has a warm tropical micro-region (lowland) with higher temperatures (Pezzopane, Castro, Pezzopane, & Cecílio, 2012).

The monthly average maximum and minimum temperatures were obtained from the automatic weather stations closest to the experimental sites from INMET, Alegre, Espírito Santo State (Itapemirim Valley) and INCAPER, in Iúna, Espírito Santo State (Southern Mountain) (20.357° South, 41.557° West longitude and 758 metres above sea level) (Figure 1).

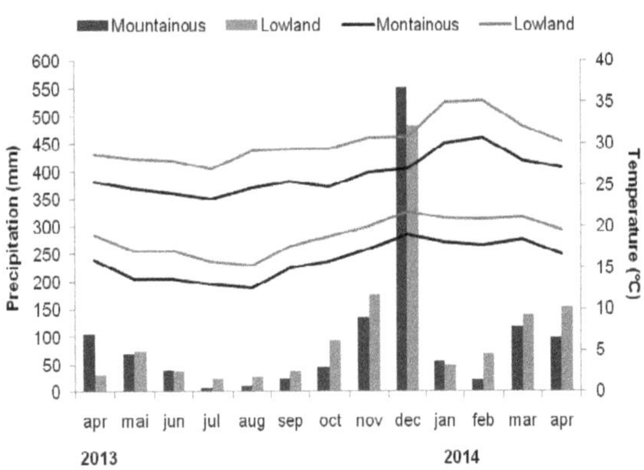

Figure 1. Precipitation and monthly temperature average, maximum and minimum observed during the assessments in mountainous and lowland regions in the state of Espirito Santo, Brazil. Source: INCAPER (2014).

The soil at both sites was classified as Red Yellow Latosol, with a medium texture (sand 68%, Silt 4%, and clay 28%) in the mountain region, and a clayey texture (sand 60%, Silt 5%, and clay 35%) in the lowland (Embrapa, 2014). A sample collected from 0-20 cm, was analysed and showed the following characteristics. Mountain region: pH (water) - 6.20; Phosphorus Mehlich 1 - 53.99 mg dm^{-3}; Potassium - 80.00 mg dm^{-3}; Calcium - 2.12 $cmol_c$ dm^{-3}; Magnesium - 0.87 $cmol_c$ dm^{-3}; Aluminium - 0.0 $cmol_c$ dm^{-3}; Sum of bases - 3.24 $cmol_c$ dm^{-3}; CTC effective - 3.24 $cmol_c$ dm^{-3}; Total organic carbon - 1,83%; total nitrogen - 0.15%; Lowland region: pH (water) - 6.37; Phosphorus Mehlich 1 - 36.99 mg dm^{-3}; Potassium - 57.00 mg dm^{-3}; Calcium - 1.50 cmol$_c$ dm^{-3}; Magnesium - 0.71 cmol$_c$ dm^{-3}; Aluminium - 0.0 cmol$_c$ dm^{-3}; Sum of bases - 2.36 cmol$_c$ dm^{-3}; CTC effective - 2.36 cmol$_c$ dm^{-3}; total organic Carbon - 1.08%; total Nitrogen - 0.10%.

At each site was conducted by planting yacon in four dates of the year 2013: fall (April), winter (July), spring (September) and summer (December). At each planting date, the experiment was set up according to the randomized block design, with four replications. The treatments were constituted by collection times within each planting date. In each planting date, four experimental units were established, consisting of 8 m long with five lines of plants, with a total of 16 plants per line (Figure 2).

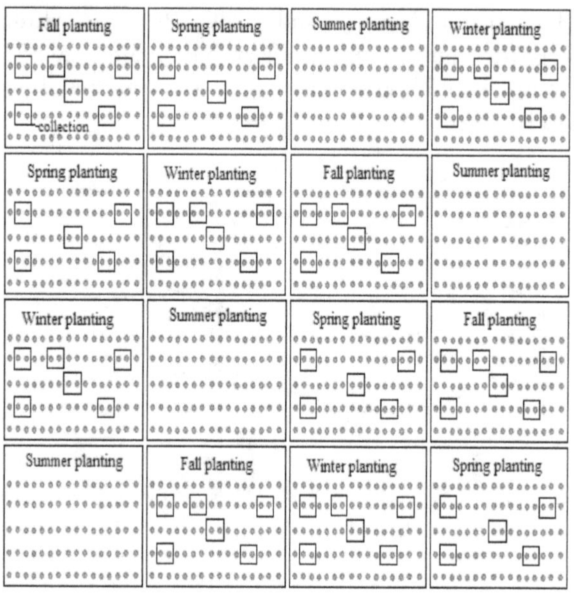

Figure 2. Sketch of the experimental area, in both locations, with the distribution of the experimental units.

Every 30 days, two plants per experimental unit, randomly chosen within the central lines, except for the borders, were collected for evaluation. The leaves, stems, rhizophores and tuberous roots were separated. The collections began 30 days after bud emergence and were discontinued at flowering on 80% of the standing plants. Thus, the number of treatments (collection times) varied according to the site and the planting date, being: 30, 60, 90, 120, 150, 180, and 210 days after emergence (DAE) for the fall planting in mountain; 30, 60, 90, 120, 150, and 180 DAE for the winter plantings in mountain and fall and spring in lowland; and 30, 60, 90, 120, and 150 DAE for spring planting in mountain and winter planting in lowland.

Rhizophores weighing approximately 35 g were planted in grooves at a depth of 10 cm with 1.0 m spacing between the rows and 0.5 m between the plants. The grooves were fertilized with 180 g of tanned bovine manure per linear metre. The manure had the following nutritional composition: 14.21 g kg^{-1} N, 4.75 g kg^{-1} P, 5.28 g kg^{-1} K, 4.29 g kg^{-1} Ca, and 1.92 g kg^{-1} Mg. Throughout the growing season, weeds were controlled by hoeing. Irrigation was performed by applying 30 mm weekly water blades.

The leaf area was obtained by means of a photoelectric meter Licor Area Meter 3100 (LiCor, EUA), and each plant part was oven-dried with forced air at 70 ± 5°C to constant weight, and the dry mass was subsequently obtained.

The leaf mass fraction was the ratio of the leaf dry mass to the total dry mass, the stem mass fraction was the ratio of the stem dry mass to the total dry mass, the rhizophore mass fraction was the ratio of the rhizophore dry mass to the total dry mass, and the tuber root mass fraction was the ratio of the tuber root dry mass to the total dry mass. The specific leaf area was the ratio of leaf area to the leaf dry mass. The leaf area index was the ratio of the leaf area in m^2 contained in 1 m^2 of occupied land by planting.

The dry matter and leaf area data were submitted to analysis of variance by the 5% probability F test, using the statistical program SISVAR (Ferreira, 2014). The natural logarithms of the averages of the original data were adjusted by multiple linear regression to different growth models, with time as an independent variable. The growth rate was estimated by the functional method of growth analysis (Hunt, 1982). To derive a model adjusted to the total dry mass data (W) and the leaf area (A), instantaneous values of the culture growth rate were estimated (CBT = dW / dT), where dW / dT is the derivative of the dry weight accumulation (W) with time and the liquid absorption rate [TAL = (1 / A) x (dW / dT)] at time T.

Result

The satisfactory emergence and growth of the yacon shoots did not occur at either cultivation area during the summer planting; thus, the evaluations presented here are based on the other planting seasons.

The emergence time of the yacon shoots varied with cultivation area and planting season. In the mountainous region, shoot emergence was observed at 60 days after the fall and winter planting (DAP) and 30 DAP during the spring. In the lowland, the emergence of shoots occurred at 60 DAP in the fall and 30 DAP in the winter and spring.

The yacon cultivation cycle also varied by area and planting season. In the mountainous region, flowering, which determined the end of the cultivation cycle, occurred 210 days after emergence (DAE) for the fall planting, at 180 DAE for the winter planting and at 150 DAE for the spring planting. The cycle duration shortened as planting was done closer to the hottest time of the year (summer). In the lowland region, the behaviour was similar. However, the time duration of each cycle was shorter for the fall planting (180 DAE) and the winter planting (150 DAE), but the same for the spring planting (180 DAE).

In the adjustment of dry mass and leaf area data, the exponential polynomial model of the 3rd degree was chosen because it had the highest coefficient of determination (R^2), and all the regression coefficients were significant (Table 1).

The highest total dry mass accumulation (TDM) occurred for the fall planting in both cultivation regions, but in the mountainous region, the cumulative maximum (1,898.38 g m^{-2}) occurred later (210 DAE) than in the lowland region (1,823.75 g m^{-2} at 150 DAE) (Figure 3A).

Another interesting point was related to the spring plantings. In the lowland region, a lower TDM accumulation (599.23 g m^{-2}) was noted than in the mountainous region (1,399.80 g m^{-2}), but the cycle was longer in the mountains (Figure 3A).

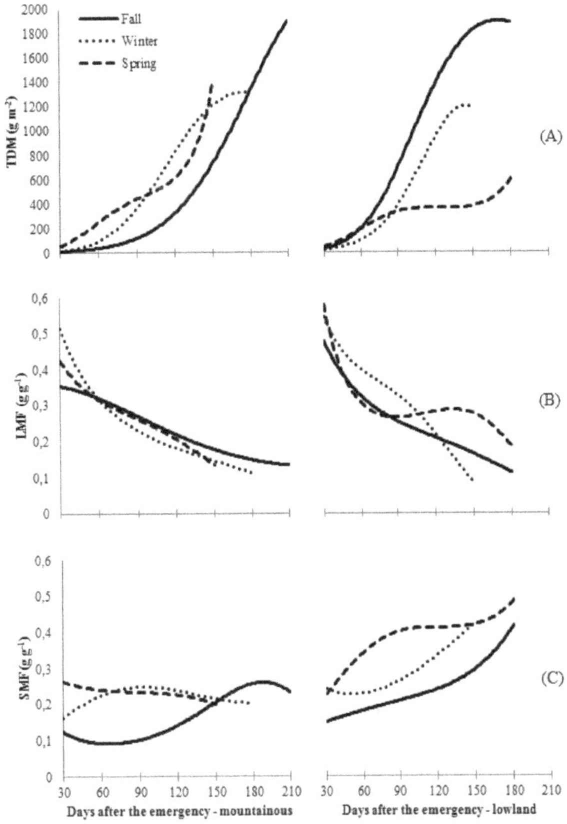

Figure 3. Total dry mass (TDM), leaf mass fraction (LMF) and stem mass fraction (SMF) along the ontogeny of yacon in different planting dates and two growing regions.

In general, the leaf mass fraction (LMF) decreased throughout the yacon cultivation cycle in both locations and planting seasons, except for the spring planting in the lowland region, where the LMF remained somewhat stable between 60 and 150 DAE (Figure 3B).

Table 1. Coefficients of the exponential model polynomial of 3° degree, $Y = \exp(a+bt+ct^2+dt^3)$, fitted to the data of yacon plants grown under different planting dates and two growing regions.

Characteristic	Regions	Planting dates	a	b	c	d	R^2
Total dry mass	Mountainous	Fall	1.01**	0.05**	-2.82e-5**	-2.08e-7*	0.99
		Winter	-0.14**	0.10**	-4.81e-4**	6.84e-7**	1.00
		Spring	0.78**	0.14**	-1.34e-3**	4.49e-6**	0.99
	Lowland	Fall	0.57**	0.11**	-5.61e-4**	9.41e-7**	1.00
		Winter	1.72**	0.04**	1.66e-4**	-1.48e-6**	1.00
		Spring	1.00**	0.12**	-9.95e-4**	2.72e-6**	0.98
Leaf area	Mountainous	Fall	5.11*	0.06**	-1.80e-4**	1.41e-7**	0.99
		Winter	4.88*	0.11*	-7.13e-4*	1.60e-6*	0.98
		Spring	6.59*	0.08*	-7.09e-4*	2.27e-6*	1.00
	Lowland	Fall	6.15**	0.06**	-1.96e-4**	-1.88e-7**	0.98
		Winter	8.46**	-0.04**	1.16e-3**	-5.46e-6**	0.99
		Spring	7.37*	0.04*	-3.17e-4*	8.59e-7*	0.89

** Significant at 1%.* Significant at 5%.

In general, the stem mass fraction (SMF) was lower in plants grown in the mountainous region, independent of the planting time. The SMF reached a later maximum in the fall planting, close to 180 DAE (Figure 3C).

However, in the lowland region, SMF was larger in both planting seasons, and it increased until the end of the cycle. Importantly, the spring planting showed the highest SMF (Figure 3C).

The rhizophores mass fraction (RiMF) always decreased according to the cultivation cycle in the mountainous region, independent of planting season. In contrast to the lowland region, higher values were observed during the initial stage of the cycle, which decreased during the intermediate phase, and increased again at the end of the cycle (Figure 4A).

The tuberous roots mass fraction (TBMF) increased during the yacon cultivation cycle in the mountainous region, except for the fall planting in which a decrease from 120 to 180 DAE occurred, and the plants growth was retaken later (Figure 4B).

In general, in the lowland region, an initial increase in TBMF followed by subsequent decrease was observed. In the fall and winter planting, the maximum was at approximately 120 DAE and 105 DAE for spring planting (Figure 4B). However, a decrease in TBMF between 30 and 60 DAE was observed in the winter, which might have been related to the death of the roots.

Under both cultivation conditions, the specific leaf area (SLA) decreased throughout the yacon cultivation cycle for the fall and the winter plantings. An SLA decrease at the beginning of the cultivation cycle was seen in the spring plantings, followed by an increase beginning at 75 DAE (mountainous region) and 90 DAE (lowland region) (Figure 5A).

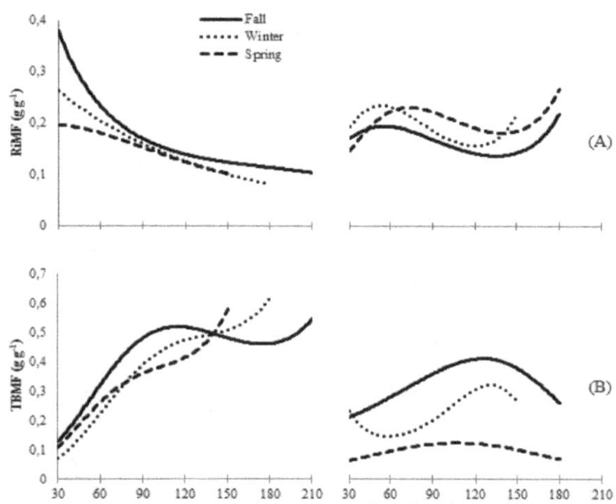

Figure 4. Rhizophores mass fraction (RiMF) and tuberous roots mass fraction (TBMF) along the ontogeny of yacon in different planting dates and two growing regions.

The net assimilation rate (NAR) varied depending on the planting time and growing conditions. In the mountainous region in the fall and winter plantings, the NAR increased during a considerable part of the cycle, at approximately 165 and 105 DAE, respectively. It decreased from that point on, even reaching negative values in the winter planting. For the spring planting the NAR decreased nearly to zero at 90 DAE and began to increase from 120 DAE until the end of the cycle when it had its highest values for the spring planting (Figure 5B).

In general, in the lowland region, the NAR slightly increased at the beginning of the cycle (approximately up to 60 DAE), then decreased until the end of the cycle. The spring planting was an exception in which the decreasing values reached zero. Later, it increased again, from 120 DAE until the end of the cycle, when it had its highest values for the spring planting (Figure 5B).

The leaf area rate (LAI) always increased throughout the cycle in the mountainous region for the fall planting (maximum 3.26 m^2 m^{-2} at 210 days) and the spring planting (maximum 3.10 m^2 m^{-2} at 150 days). For the winter planting, the maximum (2.23 m^2 m^{-2}) occurred earlier (120 days), and this rate persisted until the end of the cycle (approximately 180 days) (Figure 5C).

In contrast to the lowland region, the maximum peaks occurred earlier in the fall plantings (maximum 5.10 m^2 m^{-2} to 138 days) and in the winter (maximum 3.72 m^2 m^{-2} to 119 days), followed by a decrease until the end of the cycle. However, the spring planting showed an increasing LAI, with the maximum (2.11 m^2 m^{-2}) observed at the end of the cultivation cycle (Figure 5C).

The tillage growth rate (CGR) followed a specific pattern in the fall and winter plantings in both growing conditions increasing up to a specific point, reaching a maximum and decreasing until the end of the cycle (Figure 5D).

The maximum point varied, and in the lowland region, the maximum CGR had similar values and occurred at similar times, at 104 DAE in the fall planting (23.42 g m^{-2} day^{-1}) and 112 DAE in the winter planting (19.99 g m^{-2} day^{-1}). In the mountainous region, the maximum (21.40 g m^{-2} day^{-1}) for the fall planting occurred later (177 DAE) than the maximum (16.16 g m^{-2} day^{-1}) in the winter planting (114 DAE) (Figure 5D).

Different behaviour was observed for the spring planting. In both regions, the CGR showed a short decrease in the middle of the cycle and a subsequent increase until the end of the cycle. In the lowland region, the CGR reached zero (120 DAE), then increased and peaked (16.19 g m^{-2} day^{-1}) at 180 DAE.

In the mountainous region, it did not reach zero, and the maximum was much higher (61.18 44 g m^{-2} day^{-1}) and occurred earlier, at 150 DAE (Figure 5D).

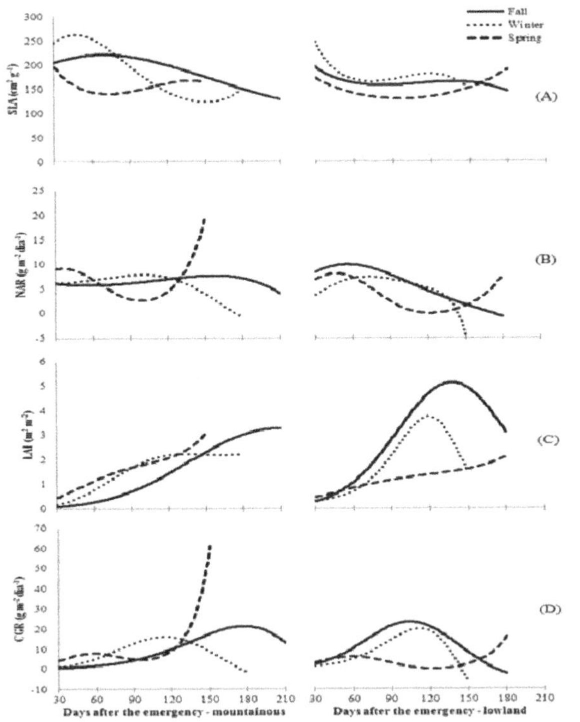

Figure 5. Especific leaf area (SLA), net assimilation rate (NAR), leaf area index (LAI) and crop growth rate (CGR) along the ontogeny of yacon in different planting dates and two growing regions.

Discussion

The absence of emergent and satisfactory shoots in the summer planting at both locations was related to the adverse conditions to which the rhizophores were exposed. A heavy rainfall (551 mm provided mountain and 483 mm in the lowland) occurred during the planting month (December/2013), which made the soil excessively wet. Combined with high temperatures (30°C in the mountainous region and 35°C in the lowland region), it caused the death of the rhizophores. This result indicated that the planting yacon in high temperature and humidity conditions (e.g., heavy rain) is risky for the initial establishment of the tillage.

In mild temperatures, the emergence and growth of shoots occur more slowly, as seen in the mountainous region, or when the crop is planted in the fall, which directly affects the cycle duration, which occurs late. This behaviour has been observed in Peru, where the crop cycle can vary between four and ten months, and, at higher altitudes, ripening also occurs late (Seminario, Valderrama, & Manrique, 2003).

According to Finger, Fontes, and Pauiatti (2005), in the rest and dormancy phases of vegetative propagules, there are changes in the concentrations of phytorregulators (inhibitor / stimulator) that control sprouting. These changes are influenced by environmental factors, with emphasis on temperature. The high temperature in pre-dormancy can provide a return to growth, and after the dormancy period, there is a transition period when growth is resumed, and under high temperature the duration of each stage is shortened in comparison with low temperature.

Fall planting in the mountainous or lowland region favoured the accumulation of biomass by the yacon plants because it allowed the plants to grow in mild temperatures during most of the cycle. In general, the closer to the time of the year when higher temperatures occurred (summer) that planting was carried out, the plants accumulated less biomass.

This behaviour was most apparent in the lowland region, particularly for the spring planting. The plants virtually ceased growing between 90 and 150 days. At that time, high temperatures occurred in the region (approximately 35°C in January and February). At this point, the NAR and CGR were close to zero, which indicates that the photosynthetic activity of the plants was nearly equal to the respiratory activity. For this reason, no increase of vegetal biomass per unit of leaf area occurred. There is almost a cessation of the plant growth, which an extension of the cycle and a decreased biomass accumulation by the plants.

A plant changes its ontogeny after a thermal stress, increasing the SLA after a period of high temperature, investing in the expansion of the leafy limbs to optimize light capture. An increased production of photoassimilates increases the NAR and CGR for the plant to continue to grow when the SLA decreased with a shorter investment in the leaf limb expansion. Moreover, thermal stress may have reduced or inhibited the translocation of photoassimilates from the leaf to other parts of the plant (Finger, Fontes, & Pauiatti, 2005), and thus, being accumulated in the leaf, which has its increased weight, reducing the values of SLA (Cancellier et al., 2010).

This behaviour shows that high temperature is detrimental to CO_2 assimilation by the yacon leaves, which reinforces the argument that yacon is sensitive to thermal stress, and if it is planted under high temperature conditions, the rate of photosynthesis per leaf area unit might be considerably reduced. Similar behaviour was also observed in other plants with a reserve habit in subterranean organs, such as taro (Oliveira, Araujo, & Guerra, 2011).

Plants that grow in mild temperatures (as in the fall plantings or the mountainous region) invest less in the stem fraction, mainly the rhizophores, which are a stem portion, and more in the tuberous roots, which can directly impact the harvest yields because the root is the main commercial part.

It is known that the tuberset process is affected by light, mineral nutrition, temperature and endogenous growth regulates level (Finger, Fontes, & Pauiatti, 2005). Yacon has been described as neutral regarding the effect of the photoperiod on stem and root formation (Kerbauy, 2013; Seminario, Valderrama, & Manrique, 2003). The results of this study suggest that temperature process is an environmental factor that affects tuberset and has consequences on the behaviour of yacon during the cultivation cycle because high temperatures harm the formation of the reserve organ, while mild temperatures favour its growth.

It has been noticed in other plants of Andean origin, such as the potato (*Solanum tuberosum*), for example. High temperature reduces the growth of the plant and the assimilated partition for the tubers. The range of 18-22°C is acceptable as an optimum for potato production. However, depending on the cultivar, in the environment of greater luminous density, the optimum moves to larger values. Above 30°C respiration rapidly increases, resulting in a decrease in net photosynthesis and practically no tubercle production (Finger, Fontes, & Pauiatti, 2005; Rodríguez-Falcón, Bou, & Prat, 2006). Among plant hormones, gibberellins have been indicated as controlling for tuberization, since environmental conditions that promote this process cause a decrease in gibberellin activity in stems. High temperatures stimulate the production of gibberellins in stem buds rather than leaves, which may be related to the inhibition of the tubercle caused by high temperatures (Kerbauy, 2013).

In the fall planting in the mountainous region, temperatures of approximately 13 to 25°C were in a good range for yacon development, according to Seminario, Valderrama, and Manrique (2003). Even in the spring planting in the mountainous region, the temperatures during the cycle were approximately 15 to 28°C, which accounted for the best growth of mass accumulation (TDM) and the best development, mainly of the reserve roots (TBMF), as compared to the lowland region.

An important fact related to the cultivation area and, consequently, to temperature is that in fall and winter planting in the lowland region, TDM decreases when the LAI decreases. This can be related to the leaf senescence rate, which would be countered by the rate of the production of new leaves. Additionally, a decrease in the NAR, which indicates an increase in the average age of a leaf, consequently reduces the photosynthetic rate, which in turn imitates a reduction in the TBMF and RiMF. This indicates that the accumulation of mass that the plant is still demonstrating is being invested in a non-marketable organ. This could be an indication that the optimum harvest time for the roots could be before flowering, which was the final evaluation time.

This observation can be corroborated by observing the yacon growth rate (CGR), which also peaked with the LAI. In a similar manner to the CGR maximum, a greater translocation of reserve substances proceeds to the tuberose roots, consequently resulting in less leaf emission and greater leaf senescence, indicating the cycle closure, as has been observed in sweet potatoes (Conceição, Lopes, & Fortes, 2005). According to Hunt (1982) after the CGR peak, a change in the otogenic development stage of the species also occurs, which normally indicates the beginning of the reproductive period and progressing to greater translocation to reserve organs.

Additionally, this behaviour indicates that the yacon has a rhizophore and a reserve organ, which is the organ that will originate a new plant and has the potential for species perpetuation (Seminario, Valderrama, & Manrique, 2003). This organ also contains the reservoir of simple carbohydrates and fructooligosaccharides (Vitali, Sancho, & Katinas, 2015).

However, these signs of cycle closure were not as clear in the plants in the mountainous region because the LAI kept increasing following the TDM and the TBMF until flowering, which was the end of the evaluation. At this time, the plant was still growing and accumulating root reserves. Although the beginning of a decrease in the NAR was noticed after the 120 days, which might have occurred due to possible leaf auto shading, and which contributed to the reduction of this rate (Pohl et al., 2009). With an LAI increase, the NAR would decrease, as occurred in the fall and the winter planting, indicating a decrease in the velocity of accumulation of biomass by local leaf units, but negative values that would allow a mass gain did not occur, which in this case would be in roots as indicated by the TBMF increase.

These results indicate that the harvest point of the roots could be later immediately after flowering, as was reported by Oliveira and Nishimoto (2004) and Seminario, Valderrama, and Manrique (2003).

Another interesting fact that deserves mention is that in the mountainous region, the CGR maximum in the fall planting occurred later (180 days) compared to the spring planting (120 days). This behaviour indicated that the plants began the stage of the greater translocation of reserve substances to the tuberose roots at the same time of year, and this would allow cycle closure and the root harvest point to occur practically simultaneously. This result reinforces field observations in the region. This means that whether the planting in the fall or the winter, the harvest would occur at the same time.

This behaviour is also affected by temperature because it is known that the maximum CGR depends on the genotype and environmental factors (Conceição et al., 2005). Because the genetic material was constant in this study, the change in the maximum CGR occurred due to the environment, mainly because of the temperature.

Conclusion

Planting yacon in the fall in a mountainous or lowland region favours yacon growth and development.

Cultivation in a mountainous region independently of the planting time promotes greater mass accumulation in the tuberose roots, which reflects greater productivity.

Both fall and winter planting in the mountainous region allows root harvesting at the same time.

Acknowledgements

To Capes for my scholarship. To CNPq and FAPES for their the financial help. To FAPES for the Capixaba researcher second author scholarship.

References

Campos, D., Betalleluz-Pallardel, I., Chirinos, R., Aguilar-Galvez, A., Noratto, G., & Pedreschi, R. (2012). Prebiotic effects of yacon (*Smallanthus sonchifolius* Poepp. & Endl), a source of fructooligosaccharides and phenolic compounds with antioxidant activity. *Food Chemistry*, *135*(3), 1592–1599. doi: 10.1016/j.foodchem.2012.05.088

Cancellier, L. L., Adorian, G. C., Rodrigues, H. V. M., Siebeneichler, S. C., & Leal, T. C. A. B. (2010). Doses de potássio nas respostas morfofisiológicas de alface. *Revista Caatinga*, *23*(4), 21-27.

Conceição, M. K., Lopes, N. F., & Fortes, G. R. L. (2005). Análise de crescimento de plantas de batata-doce (*Ipomoea batatas* (l.) lam) cultivares abóbora e da costa. *Revista Brasileira de Agrociência*, *11*(3), 273-278.

Empresa Brasileira de Pesquisa Agropecuária [Embrapa]. (2014). *Sistema brasileiro de classificação de solos*. 4. ed. Rio de Janeiro, RJ: Centro Nacional de Pesquisa de Solos.

Ferreira, D. F. (2014). Sisvar: a Guide for its Bootstrap procedures in multiple comparisons. *Ciência e Agrotecnologia*, *38*(2), 109-112. doi: 10.1590/S1413-70542014000200001

Finger, F. L., Fontes, P. C. R., & Pauiatti, M. (2005). Dormência e tuberização. In P. C. R. Fontes (Ed.), *Olericultura teoria e prática*. Viçosa, MG: Suprema Gráfica e Editora.

Genta, S. B., Cabrera, W. M., Habib, N., Pons, J., Carillo, I. M., Grau, A., & Sánchez, S. S. (2009). Yacon syrup: Beneficial effects on obesity and insulin resistance in humans. *Clinical Nutrition*, *28*(2), 182–187. doi: 10.1016/j.clnu.2009.01.013

Habib, N. C., Honoré, S. M., Genta, S. B., & Sánchez, S. S. (2011). Hypolipidemic effect of *Smallanthus sonchifolius* (yacon) roots on diabetic rats: Biochemical approach. *Chemico-Biological Interactions*, *194*(1), 31-39. doi: 10.1016/j.cbi.2011.08.009

Hunt, R. (1982). *Plant growth curves: The functional approach to plant growth analysis*. London, UK: Edward Arnold.

Instituto Capixaba de Pesquisa, Assistência Técnica e Extensão Rural [INCAPER]. (2014). Retrivied on Dec. 20, 2014 from http://hidrometeorologia.incaper .es.gov.br/? pagina=bol

Kerbauy, G., (2013). *Fisiologia vegetal*. 2. ed. Rio de Janeiro, RJ: Guanabara Koogan.

Moura, N. A. de, Caetano, B. F. R., Sivieri, K., Urbano, L. H., Cabello, C., Rodrigues, M. A. M., & Barbisan, L. F. (2012). Protective effects of yacon (*Smallanthus sonchifolius*) intake on experimental colon carcinogenesis. *Food and Chemical Toxicology*, *50*(8), 2902-2910. doi: 10.1016/j.fct.2012.05.006

Oliveira, F. L., Araujo, A. P., & Guerra, J. G. M. (2011). Crescimento e acumulação de nutrientes em plantas de taro sob níveis de sombreamento artificial. *Horticultura Brasileira*, *29*(3), 291-298. doi: 10.1590/S0102-05362011000300006

Oliveira, G. O., Braga, C. P., & Fernandes, A. A. H. (2013). Improvement of biochemical parameters in type 1 diabetic rats after the roots aqueous extract of yacon (*Smallanthus sonchifolius*) treatment. *Food and Chemical Toxicology*, *59*, 256-260. doi: 10.1016/j.fct.2013.05.050

Oliveira, M. A., & Nishimoto, E. K. (2004). Avaliação do desenvolvimento de plantas de yacon (*Polymnia sonchifolia*) e caracterização dos carboidratos de reservas em HPLC. *Brazilian Journal of Food Technology*, 7(2), 215-220.

Pezzopane, J. E. M., Castro, F. S., Pezzopane, J. R. M., & Cecílio, R. A. (2012). *Agrometeorologia: aplicações para o Espírito Santo*. Alegre, ES: CAUFES.

Pohl, S., Lopes, N. F., Braga, E. J. B., Silva, C. P., Silva, F. S. P., & Peters, J. A. (2009). Características de crescimento de plantas de batata, cv. Baronesa, e seu genótipo transformado geneticamente para resistência ao PVY. *Revista Ceres*, *56*(6), 736-743.

Rodríguez-Falcón, M., Bou, J., & Prat, S. (2006). Seasonal control of tuberization in potato: conserved elements with the flowering response. *Annual Review of Plant Biology*, *57*, 151–180. doi: 10.1146/annurev.arplant.57.032905.105224

Santana, I., & Cardoso, M. H. (2008). Raiz tuberosa de yacon (*Smallanthus sonchifolius*): potencialidade de cultivo, aspectos tecnológicos e nutricionais. *Ciência Rural, 38*(3), 898-905. doi: 10.1590/S0103-84782008000300050

Seminario, J., Valderrama, M., & Manrique, I. (2003). *El yacon: fundamentos para el aprovechamiento de un recurso promisorio*. Lima, Peru: Centro Internacional de la Papa (CIP), Universidad Nacional de Cajamarca, Agencia Suiza para el Desarrollo y la Cooperación (COSUDE).

Vaz-Tostes, M. G., Viana, M. L., Grancieri, M., Luz, T. C. S., Paula, H., Pedrosa R. G., & Costa, N. M. B. (2014). Yacon effects in immune response and nutritional status of iron and zinc in preschool children. *Nutrition, 30*(6), 666-672. doi: 10.1016/j.nut.2013.10.016

Vilhena, S. M. C., Câmara, F. L. A., & Kakihara, S. T. (2000). O cultivo de yacon no Brasil. *Horticultura Brasileira, 18*(1), 5-8. doi: 10.1590/S0102-05362000000100002

Vitali, M. S., Sancho, G., & Katinas, L. (2015). A revision of *Smallanthus* (Asteraceae, Millerieae), the "yacón" genus. *Phytotaxa, 214*(1), 1-84. doi: 10.3989/ajbm.2392

PERMISSIONS

 Every chapter published in this book has been scrutinized by our experts. Their significance has been extensively debated. The topics covered herein carry significant findings which will fuel the growth of the discipline. They may even be implemented as practical applications or may be referred to as a beginning point for another development.

The contributors of this book come from diverse backgrounds, making this book a truly international effort. This book will bring forth new frontiers with its revolutionizing research information and detailed analysis of the nascent developments around the world.

We would like to thank all the contributing authors for lending their expertise to make the book truly unique. They have played a crucial role in the development of this book. Without their invaluable contributions this book wouldn't have been possible. They have made vital efforts to compile up to date information on the varied aspects of this subject to make this book a valuable addition to the collection of many professionals and students.

This book was conceptualized with the vision of imparting up-to-date information and advanced data in this field. To ensure the same, a matchless editorial board was set up. Every individual on the board went through rigorous rounds of assessment to prove their worth. After which they invested a large part of their time researching and compiling the most relevant data for our readers.

The editorial board has been involved in producing this book since its inception. They have spent rigorous hours researching and exploring the diverse topics which have resulted in the successful publishing of this book. They have passed on their knowledge of decades through this book. To expedite this challenging task, the publisher supported the team at every step. A small team of assistant editors was also appointed to further simplify the editing procedure and attain best results for the readers.

Apart from the editorial board, the designing team has also invested a significant amount of their time in understanding the subject and creating the most relevant covers. They scrutinized every image to scout for the most suitable representation of the subject and create an appropriate cover for the book.

The publishing team has been an ardent support to the editorial, designing and production team. Their endless efforts to recruit the best for this project, has resulted in the accomplishment of this book. They are a veteran in the field of academics and their pool of knowledge is as vast as their experience in printing. Their expertise and guidance has proved useful at every step. Their uncompromising quality standards have made this book an exceptional effort. Their encouragement from time to time has been an inspiration for everyone.

The publisher and the editorial board hope that this book will prove to be a valuable piece of knowledge for researchers, students, practitioners and scholars across the globe.

LIST OF CONTRIBUTORS

Saul de Andrade Júnior, Rodrigo Sobreira Alexandre, Edilson Romais Schmildt and Fábio Luiz Partelli
Programa de Pós-graduação em Agricultura Tropical, Departamento de Ciências Agrárias e Biológicas, Centro Universitário Norte do Espírito Santo, Universidade Federal do Espírito Santo, Rod. BR 101 Norte, Km 60, 29932-540, Bairro Litorâneo, São Mateus, Espírito Santo, Brazil

Maria Amélia Gava Ferrão
Empresa Brasileira de Pesquisa Agropecuária, Vitória, Espírito Santo, Brazil

Aldo Luiz Mauri
Instituto Capixaba de Pesquisa, Assistência Técnica e Extensão Rural, Fazenda Experimental de Marilândia, Marilândia, Espírito Santo, Brazil

Keila Regina Hossa
Programa de Pós-graduação em Agronomia, Universidade Estadual de Maringá, Av. Colombo, 5790, Bloco J-45, 87020-900, Maringá, Paraná, Brazil

José Walter Pedroza Carneiro and Alessandro Lucca Braccini
Departamento de Agronomia, Universidade Estadual de Maringá, Maringá, Paraná, Brazil

Terezinha Aparecida Guedes
Departamento de Estatística, Universidade Estadual de Maringá, Maringá, Paraná, Brazil

Rosangela Dallemole-Giaretta
Departamento de Ciências Agrárias, Universidade Tecnológica Federal do Paraná, Pato Branco, Paraná, Brazil

Leandro Grassi de Freitas and Silamar Ferraz
Departamento de Fitopatologia, Universidade Federal de Viçosa, Viçosa, Minas Gerais, Brazil

Everaldo Antônio Lopes
Instituto de Ciências Agrárias, Universidade Federal de Viçosa, Campus de Rio Paranaíba, Rod. MG-230, Km 7, 38810-000, Rio Paranaíba, Minas Gerais, Brazil

Marliane de Cássia Soares da Silva and Maria Catarina Megumi Kasuya
Departamento de Microbiologia, Universidade Federal de Viçosa, Viçosa, Minas Gerais, Brazil

Fernando Panison, Luís Sangoi, Daniel Fernando Kolling, Cileide Maria Medeiros Coelho and Murilo Miguel Durli
Centro de Ciências Agroveterinárias, Universidade do Estado de Santa Catarina, Avenida Luiz de Camões, 2090, 88520-000, Lages, Santa Catarina, Brazil

André Monzoli Covre, Fábio Luiz Partelli and Robson Bonomo
Universidade Federal do Espírito Santo, Rodovia BR-101, Km 60, 29932-540, São Mateus, Espírito Santo, Brazil

Heder Braun
Universidade Estadual do Maranhão, São Luiz, Maranhão, Brazil

Cláudio Pagotto Ronchi
Universidade Federal de Viçosa, Viçosa, Minas Gerais, Brazil

Paulo Roberto Coelho Lopes and Inez Vilar de Morais Oliveira
Empresa Brasileira de Pesquisa Agropecuária, Embrapa Semiárido, Petrolina, Pernambuco, Brazil

Raissa Rachel Salustriano da Silva
Programa de Pós-graduação em Solos e Nutrição de Plantas, Universidade Federal do Piauí, Bom Jesus, Piauí, Brazil

Ítalo Herbert Lucena Cavalcante
Departmento de Engenharias, Campus "Profa. Cinobelina Elvas", Universidade Federal do Piauí, Bom Jesus, Piauí, Brazil

João Paulo Nobre de Almeida and Cydianne Cavalcante Silva
Universidade Federal Rural do Semi-Árido, Mossoró, Rio Grande do Norte, Brazil

Bruno Franca da Trindade Lessa
Universidade Federal do Vale do São Francisco, Petrolina, Pernambuco, Brazil

Charles Lobo Pinheiro, Fernanda Melo Gomes and Sebastião Medeiros Filho
Universidade Federal do Ceará, Fortaleza, Ceara, Brazil

Bárbara Nogueira Souza Costa, Gabrielen de Maria Gomes Dias, Irton de Jesus Silva Costa, Franscinely Aparecida de Assis and Flávia Aparecida da Silveira
Programa de Pós-graduação, Departamento de Agricultura, Universidade Federal de Lavras, Av. Doutor Sylvio Menicucci, 1001, Kennedy, 37200-000, Lavras, Minas Gerais, Brazil

Moacir Pasqual
Departamento de Agricultura, Universidade Federal de Lavras, Lavras, Minas Gerais, Brazil

Irton de Jesus Silva Costa, Bárbara Nogueira Souza Costa, Franscinely Aparecida de Assis, Adalvan Daniel Martins, Leila Aparecida Salles Pio and Moacir Pasqual
Departamento de Agricultura, Universidade Federal de Lavras, Av. Doutor Sylvio Menicucci, 1001, Kennedy, 37200-000, Lavras, Minas Gerais, Brazil

Isabela Cristina Gomes Honóri and Filipe Pereira Giardini Bonfim
Faculdade de Ciências Agronômicas, Universidade Estadual Paulista "Júlio de Mesquita Filho", Rua José Barbosa de Barros, 1780, 18610-307, Botucatu, São Paulo, Brazil

Sebastián Giraldo Montoya and Vicente Wagner Dias Casali
Departamento de Fitotecnia, Universidade Federal de Viçosa, Viçosa, Minas Gerais, Brazil

João Paulo Viana Leite
Departamento de Biologia Vegetal, Universidade Federal de Viçosa, Viçosa, Minas Gerais, Brazil

Paulo Roberto Cecon
Departamento de Estatística, Universidade Federal de Viçosa, Viçosa, Minas Gerais, Brazil

Raquel Leszczynski
Programa de Pós-graduação em Genética e Melhoramento, Universidade Estadual de Maringá, Maringá, Paraná, Brazil

Alessandro de Lucca e Braccini and Carlos Alberto Scapim
Departamento de Agronomia, Universidade Estadual de Maringá, Av. Colombo, 5790, 87020-900, Maringá, Paraná, Brazil

Leandro Paiola Albrecht
Universidade Federal do Paraná, Palotina, Paraná, Brazil

Gleberson Guillen Piccinin and Lilian Gomes de Moraes Dan
Programa de Pós-graduação em Agronomia, Universidade Estadual de Maringá, Maringá, Paraná, Brazil

Camila Aparecida Lopes, Maria Laene Moreira de Carvalho, Michelle Conceição Vasconcelos, Ana Maria Pereira Ribeiro and Heloísa Oliveira dos Santos
Laboratório de Análise de Sementes, Departamento de Agricultura, Universidade Federal de Lavras, 37200-000, Lavras, Minas Gerais, Brazil

Lucas da Silva Domingues
Curso de Agronomia, Universidade Tecnológica Federal do Paraná, Dois Vizinhos, Paraná, Brazil

Nerinéia Dalfollo Ribeiro, Jerônimo Luiz Andriolo, Micheli Thaise Della Flora Possobom and Allan Emanoel Mezzomo Zemolin
Departamento de Fitotecnia, Centro de Ciências Rurais, Universidade Federal de Santa Maria, Avenida Roraima, 1000, 97105-900, Santa Maria, Rio Grande do Sul, Brazil

Soledad García Morales, Libia Iris Trejo-Téllez, Fernando Carlos Gómez Merino, Camila Caldana, David Espinosa-Victoria and Braulio Edgar Herrera Cabrera
Colegio de Postgraduados en Ciencias Agrícolas, Km 36.5, Carretera México-Texcoco, 56230, Montecillo, Texcoco, México

Andressa Strenske, Edmar Soares de Vasconcelos, Vanessa Aline Egewarth, Neusa Francisca Michelon Herzog and Marlene de Matos Malavasi
Centro de Ciências Agrárias, Universidade Estadual do Oeste do Paraná, Rua Pernambuco, 1777, 85960-000, Marechal Cândido Rondon, Paraná, Brazil

João Vitor Ferrari
Programa de Pós-graduação em Agronomia, Departamento de Fitotecnia, Tecnologia de Alimentos e Sócio Economia, Faculdade de Engenharia de Ilha Solteira, Universidade Estadual Paulista "Júlio de Mesquita Filho", Av. Brasil, 56, 15385-000, Centro, Ilha Solteira, São Paulo, Brazil

Enes Furlani Júnior and Ana Paula Portugal Gouvea Luques
Departamento de Fitotecnia, Tecnologia de Alimentos e Sócio Economia, Faculdade de Engenharia de Ilha Solteira, Universidade Estadual Paulista "Júlio de Mesquita Filho", Ilha Solteira, São Paulo, Brazil

Samuel Ferrari
Departamento de Agricultura, Universidade Estadual Paulista "Júlio de Mesquita Filho", Campus de Registro, Registro, São Paulo, Brazil

Thaís Cristina Silva, José Eduardo Brasil Pereira Pinto, Fernanda Ventorim Pacheco, Heitor Luiz Heiderich Roza, Diogo Mendes da Silva Lordêllo and Suzan Kelly Vilela Bertolucci
Universidade Federal de Lavras, Departamento de Agricultura, Lavras, Minas Gerais, Brazil

Edison Hitoshi Hirosse, José Eduardo Creste, Ceci Castilho Custódio and Nelson Barbosa Machado-Neto
Laboratório de Cultura de Tecidos, Faculdade de Ciências Agrárias, Universidade do Oeste Paulista, Rod. Raposo Tavares, Km 572, 19067-175, Presidente Prudente, São Paulo, Brazil

Maria da Conceição Santana Carvalho and Adriano Stephan Nascente
Empresa Brasileira de Pesquisa Agropecuária, Embrapa Arroz e Feijão, Km12, Zona Rural GO-462, 75375-000, Santo Antônio de Goiás, Goiás, Brazil

Emanoela Pereira de Paiva, Salvador Barros Torres, Tatianne Raianne Costa Alves, Francisco Vanies da Silva Sá, Moadir de Sousa Leite and Jeferson Luiz Dallabona Dombroski
Centro de Ciências Agrárias, Departamento de Ciências Agronômicas e Florestais, Universidade Federal Rural do Semi-Árido, Av. Francisco Mota, 572, Costa e Silva, 59625-900, Mossoró, Rio Grande do Norte, Brazil

Marcel Viana Pires, Alex-Alan Furtado de Almeida, Abel Louro de Figueiredo, Fábio Pinto Gomes and Margarete Magalhães Souza
Departamento de Ciências Biológicas, Universidade Estadual de Santa Cruz, Rod. Ilhéus-Itabuna, km 16, 45650-000, Ilhéus, Bahia, Brazil

Raphael Lemes Hamawaki
Coordenação de Aperfeiçoamento de Pessoal de Nível Superior, Ministério da Educação, Setor Bancário Norte, Quadra 2, Bloco L, Lote 06, Asa Norte, 70040-020, Brasília, Distrito Federal, Brazil
Department of Plant, Soil and Agricultural Systems, Southern Illinois University Carbondale, 1205 Lincoln Dr., Carbondale, Illinois, 62901, USA

Stella K Kantartzi
Department of Plant, Soil and Agricultural Systems, Southern Illinois University Carbondale, 1205 Lincoln Dr., Carbondale, Illinois, 62901, USA

Giovana Bomfim de Alcantara, Roberson Dibax, Ricardo Augusto de Oliveira, João Carlos Bespalhok Filho and Edelclaiton Daros
Laboratório de Micropropagação de Plantas, Setor de Ciências Agrárias, Departamento de Fitotecnia e Fitossanitarismo, Universidade Federal do Paraná, Rua dos Funcionários, 1540, Cabral, Curitiba, Paraná, Brazil

Diego Mathias Natal da Silva, Fábio Luiz de Oliveira, Paulo Cezar Cavatte, Mateus Augusto Lima Quaresma and Bruno Fardim Christo
Departamento de Agronomia, Centro de Ciências Agrárias e Engenharias, Universidade Federal do Espírito Santo, Alto Universitário, s/n° Guararema, 29500-000, Alegre, Espírito Santo, Brazil

Index

Q

R

S

T

www.ingramcontent.com/pod-product-compliance
Lightning Source LLC
LaVergne TN
LVHW080648210926
848485LV00003B/349